CLEVELAND
BOTANICAL
GARDEN®

D1283731

The National Audubon Society Almanac of the Environment

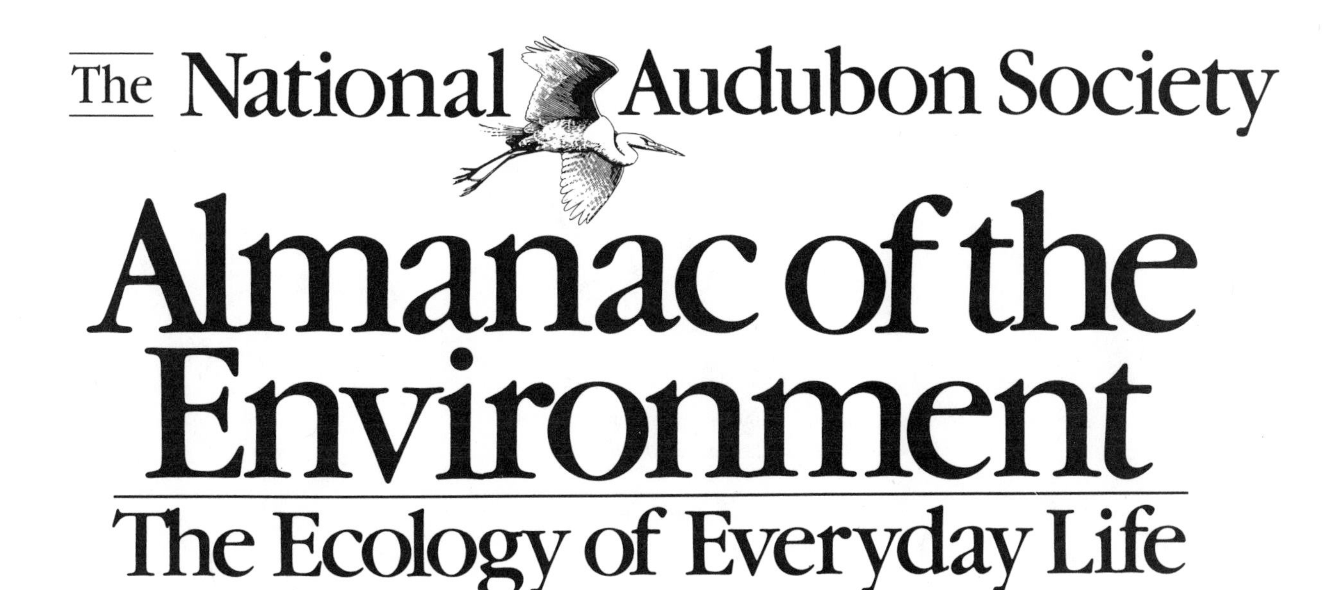

Written and compiled by

Valerie Harms

A Grosset/Putnam Book

Published by G. P. Putnam's Sons

New York

GE
195.7
.H37
1994

National Audubon Society and the egret
are trademarks of the National Audubon Society, Inc.

A Grosset/Putnam Book
Published by G. P. Putnam's Sons
Publishers since 1838
200 Madison Avenue
New York, NY 10016

Book design: H. Roberts Design
Illustrations: Howard R. Roberts
Cover design: One Plus One Studio
Cover photograph © by Carmona Photography/FPG International

Copyright © 1994 by National Audubon Society, Inc.
All rights reserved. This book, or parts thereof,
may not be reproduced in any form without permission.

Published simultaneously in Canada.

Library of Congress Cataloging-in-Publication Data
Harms, Valerie
The Audubon almanac of the environment / written and compiled by
Valerie Harms.
p. cm.
Includes bibliographical references and index.
ISBN 0-399-13942-7
1. Environmental responsibility. 2. Environmentalism. I. Title.
GE 195.7.H37 1994
363.7--dc20 93-6243CIP

Printed in the United States of America
1 2 3 4 5 6 7 8 9 10

This book is printed on recycled paper.

Contributors to the *Almanac*

Valerie Harms, Science Editor at NAS, author of eight books (overall editor and author of all chapters except the following):

Frank Graham, Jr., Field Writer for NAS, author of *Audubon Ark* and other books (chapters on Our Public Lands, Wildlife, Global Family)

Mercédès Lee, NAS Environmental Policy Analyst (chapter on Office Buildings)

Tensie Whelan, formerly Vice President of Conservation Information at NAS, now Executive Director of the New York League of Conservation Voters (chapter on Recreation and Tourism)

Dr. Jan Beyea, Chief Scientist at NAS, author of the Afterword, Eco Koans, and other contributions cited in the text

Work from the following writers is excerpted from Audubon publications:

Chapter 1
Paul Underhill, *Activist,* 5/91
Jon Luoma, "Soil Is Not a Factory," *Audubon* Magazine, 11/89
Ted Williams, "Frankenstein's Fish," *Audubon* Magazine, 9/87

Chapter 2
Paul Ehrlich, "The Sex Life of Birds," *American Birds,* Spring 1991
Jon Luoma, "Acid Murder," *Audubon* Magazine, 11/88
Jon Luoma, "The Human Cost of Acid Rain," *Audubon* Magazine, 7/88
Marguerite Holloway, "Perils of Chlorine," *Audubon* Magazine, 11/91
Christopher Hallowell, "Plants That Purify," *Audubon* Magazine, 1/92

Chapter 3
Jessica Maxwell, "Redressing Plastic Man," *Audubon* Magazine, 11/92
George Laycock, "Going for the Gold," *Audubon* Magazine, 7/89
Joel Vance, "Something to Lean On," *Audubon* Magazine, 11/86

Chapter 4
James Simmons, "The Most Useful Plant in the World," *Audubon* Magazine, 1/86
John Fleischman, "News from the Inland Island," *Audubon* Magazine, 3/86

Chapter 5
Peter Steinhart, "Whistle Up a Breeze," *Audubon* Magazine, 7/89
Suzanne Winckler, "The Platte Pretzel," *Audubon* Magazine, 5/89

Chapter 6
Peter Steinhart, "Pull of the Moon," *Audubon* Magazine, 9/89
Jon Luoma, "Saving Seeds," *Audubon* Magazine, 11/86
Glen Martin, "A Pond Is Born," *Audubon* Magazine, 6/90

Chapter 7
Peter Steinhart, "Waterway Watchdogs," *Audubon* Magazine, 11/90
Frank Graham, "Superfund Sanctuary," *Audubon* Magazine, 3/92

Chapter 8
Michael Harwood, "In Praise of Starlings?" *Audubon* Magazine, 3/88

Chapter 9
John DeCicco, James Cook, Dorene Bolze, Jan Beyea, *CO_2 Diet for a Greenhouse Planet,* Audubon Society Report, 1990
Rob Lester and J. P. Myers, "Global Warming, Climate Disruption, and Biological Diversity," *Audubon Wildlife Report,* 1989–1990

Chapter 10
Jolee Edmondson, "Golfing/Hazards of the Game," *Audubon* Magazine, 11/87

Chapter 12
Donald Jackson, "Every State Should Have a Leo Drey," *Audubon* Magazine, 7/88
Laura Riley, "A Man and His Refuge," *Audubon* Magazine, 9/86

Chapter 13
Ted Williams, "The Last Bluefin Hunt," *Audubon* Magazine, 7/92
Kenneth Brower, "State of the Reef," *Audubon* Magazine, 3/89

Chapter 14
Paul Ehrlich, "Birds in Jeopardy," *American Birds,* Summer 1992
Frank Gill, "Flocking/To Eat or Be Eaten," *American Birds,* Fall 1991
Malcolm Browne and Eric Fischer, "War and the Environment," *Audubon* Magazine, 9/91

Chapter 17
Walter Sullivan, "Aurora Borealis," *Audubon* Magazine, 1/87
Delta Willis, "Drawing on Nature," *Audubon* Magazine, 5/92
Frank Gill and Joseph Ewing, "Man and Birds," *American Birds,* Spring 1991
William Zinsser, "A Guide to Roger Tory Peterson," *Audubon* Magazine, 11/92

PERMISSIONS

Walter Sullivan, "Researchers Draw Back the Curtain on the Mysterious Aurora Borealis," 2/4/86. Copyright © 1986 by The New York Times Company. Reprinted by permission.

Frank Gill, "Flocking/To Eat or Be Eaten" and "Man and Birds," from *Ornithology*. Copyright © 1990 by W. H. Freeman and Co. Reprinted with permission.

Anne LaBastille, excerpt from *Beyond Black Bear Lake*. Copyright © 1987 by Anne LaBastille. Reprinted with permission.

Natalie Anger, "In Recycling Waste, the Noble Scarab is Peerless." Copyright © 1991 by The New York Times Company. Reprinted with permission.

Tensie Whelan, ed., excerpts from "Ecotourism and Its Role in Sustainable Development," from *Nature Tourism*. Copyright © 1991 by Island Press. Reprinted with permission.

Edward O. Wilson, excerpts from *The Diversity of Life*. The Belknap Press of Harvard University Press. Copyright © 1992 by Edward O. Wilson. Reprinted with permission.

Lynn Margulis, exerpt from "Power to the Protoctists," in *Earthwatch*. Copyright © 1992 by Lynn Margulis. Reprinted with permission.

J. P. Myers, excerpt from "Sex and Gluttony on Delaware Bay," in *Natural History*. Copyright © 1986 by J. P. Myers. Reprinted with permission.

Charles H. Callison, excerpt from *Overlooked in America*. Aperture. Copyright © 1991 by Charles Callison. Reprinted with permission.

Aldo Leopold, excerpts from *A Sand County Almanac: And Sketches Here and There*. Copyright © 1949, 1977 by Oxford University Press, Inc. Reprinted with permission.

Loren Eiseley, excerpt from *The Immense Journey*. Copyright © 1957 by Loren Eiseley. Reprinted by permission of Random House, Inc.

Terry Tempest Williams, excerpt from *Refuge: An Unnatural History of Family and Place*. Copyright © 1991 by Terry Tempest Williams. Reprinted by permission of Pantheon Books, a division of Random House, Inc.

Matthew Coon Come, excerpt from his speech quoted with permission.

Annie Dillard, excerpt from *Pilgrim at Tinker Creek*. Copyright © 1974 by Annie Dillard. Reprinted by permission of HarperCollins Publishers Inc.

Wendell Berry, excerpt form *Sex, Economy, Freedom and Community*. Copyright © 1993 by Wendell Berry. Reprinted by permission of Pantheon Books, a division of Random House, Inc.

Libby: The Sketches, Letters & Journal. Copyright © 1987 by Betty John. Originally published by Council Oak Publishing. Excerpt reprinted with permission.

Linda Hogan, excerpt from *Ariadne's Thread*. Copyright © 1982 by Lyn Lifshin. Reprinted with permission by HarperCollins Publishers Inc.

William Zinsser, excerpt from *"A Field Guide to Roger Tory Peterson."* Copyright © 1992 by William K. Zinsser. Reprinted by permission of the author.

Contents

ACKNOWLEDGMENTS

This *Almanac* is indebted to Jan Beyea for his multiple wits.

I am especially grateful to literary agent Jean Naggar for her praise of the proposal, which sustained the hard months of creation, and to publisher Jane Isay, whose "skillful means" cleared the path of obstacles. It was especially pleasing to have designer Howard Roberts' work grace these pages.

I wish to thank the following individuals at National Audubon for their invaluable suggestions and support:

> Carl Safina, Jim Cook, Katherine Santone, Carol Hyatt, Geoff LeBaron, Brock Evans, Tom Exton, Elaine O'Sullivan, Marilyn England, Karen Blumer, Pat Waak, Nikos Boutis, Maureen Hinkle, Wayne Mones, Gene Knoder, Jean Porter, Phil Schaeffer, Dusty Dunstan, Steve Kress, Jane Lyons, Jesse Grantham, Tom Bancroft, Carol Hyatt, Angie Farmer, Pueng Vongs, David Day-Kitazano, Barbara Linton, Harry LeGates, Mike Wyson, Vicki Shaw, Chris Indoe, Graham Cox. Also the resources of the *Audubon Activist, Audubon* Magazine, *American Birds*, Audubon Environmental Education Department, International Department, Sanctuaries, Regional Centers, and Chapters.

In addition, I thank John DeCicco at ACEEE for his review, Dr. Eric Fischer, at the National Academy of Sciences, and the numerous organizations that supplied useful information. Many people helped root out errors; any that are left are mine.

It is important to note that the opinions belong to those who express them and do not reflect the official policy of the National Audubon Society unless so stated.

Note to the Reader

Make a trusty companion out of this book. Not an encyclopedia of facts, this *Almanac* is designed to sensitize you to the fascinating web of ecology in interesting bits and pieces. You may browse through it in bed, in the bathroom, at the office, and of course in the great outdoors.

Have fun with it. It can be read from the start or the end or the middle! If you want to look up something in particular, use the index. Otherwise, each chapter—and its rich menu of features—places our activities in the larger context of ongoing nature cycles. You will find on the left-hand page of each chapter opening an explanation of the Ecosystem Pathways that the facing illustration depicts visually. With this guidance you will gain much of what you need to know to preserve our earthly habitat. Hopefully, your love for nature will be vigorously renewed as well.

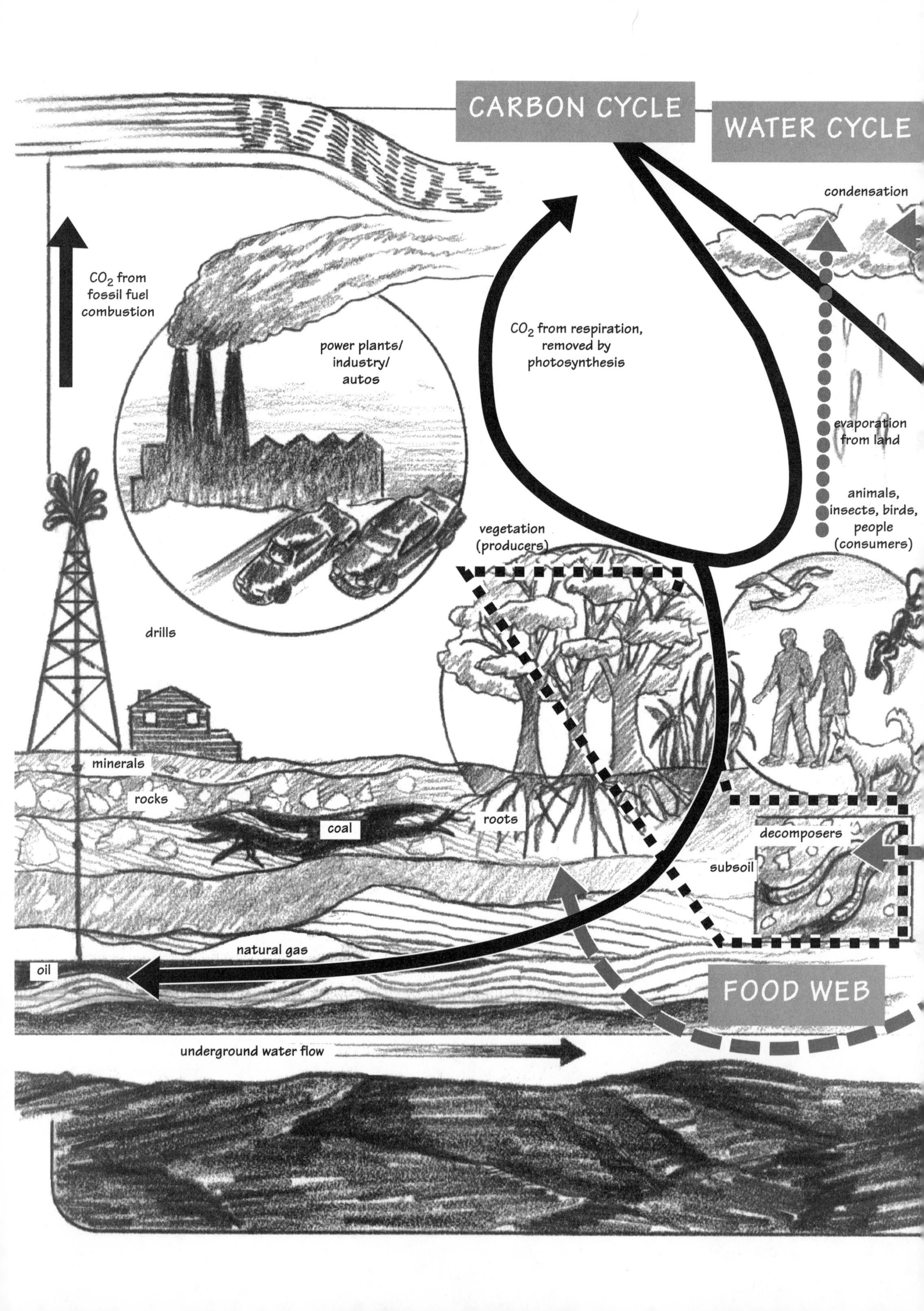
WINDS
CARBON CYCLE
WATER CYCLE
condensation
CO_2 from fossil fuel combustion
power plants/ industry/ autos
CO_2 from respiration, removed by photosynthesis
evaporation from land
animals, insects, birds, people (consumers)
vegetation (producers)
drills
minerals
rocks
coal
roots
decomposers
subsoil
natural gas
oil
FOOD WEB
underground water flow

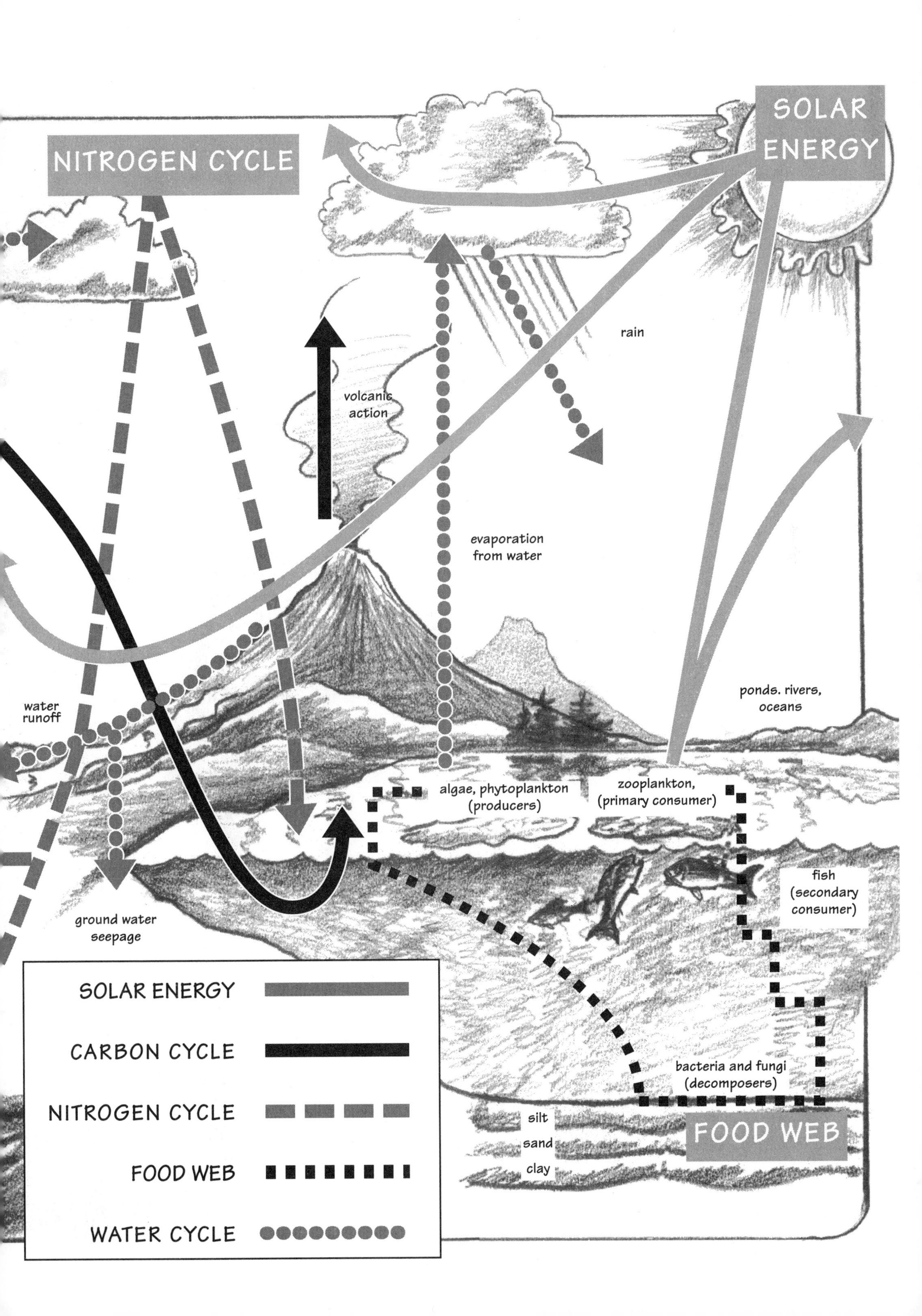
NITROGEN CYCLE
SOLAR ENERGY
rain
volcanic action
evaporation from water
water runoff
ponds. rivers, oceans
algae, phytoplankton (producers)
zooplankton, (primary consumer)
fish (secondary consumer)
ground water seepage
bacteria and fungi (decomposers)
silt
sand
clay
FOOD WEB
SOLAR ENERGY
CARBON CYCLE
NITROGEN CYCLE
FOOD WEB
WATER CYCLE

The mission of the NATIONAL AUDUBON SOCIETY is to conserve and restore natural ecosystems, focusing on birds and other wildlife, for the benefit of humanity and the Earth's biological diversity.

In the vanguard of the environmental movement, AUDUBON has more than 600,000 members, fourteen regional and state offices, and an extensive chapter network in the United States and Latin America, plus a professional staff of scientists, lobbyists, lawyers, policy analysts, and educators.

Through a nationwide sanctuary system it manages 150,000 acres of critical wildlife habitat and unique natural areas for birds, wild animals, and rare plant life.

Its award-winning *Audubon* Magazine, published six times a year and sent to all members, carries outstanding articles and color photography on wildlife and nature, and presents in-depth reports on critical environmental issues, as well as conservation news and commentary. It also publishes *Audubon Activist*, a monthly newspaper, and *American Birds*, an ornithological journal. *Audubon Adventures*, a bimonthly children's nature newsletter, reaches more than 500,000 elementary school students.

Its award-winning "World of Audubon" television shows, airing on TBS and public television, deal with a variety of environmental themes; companion books and educational computer software give viewers an opportunity for further in-depth study of the subjects covered by the television series. Audubon's travel program sponsors many exciting trips every year to exotic places like Alaska, Antarctica, Baja California, Galápagos, Indonesia, Japan, and Patagonia.

For information about how you can become a member, please write or call: NATIONAL AUDUBON SOCIETY, Membership Dept., 700 Broadway, New York, New York 10003. 212-979-3000.

PREFACE

by Paul R. Ehrlich and Anne H. Ehrlich

Center for Conservation Biology
Stanford University

A rich main course is best complemented by a spartan appetizer; we'll go to the extreme by starting your banquet with the bones. What follows is rich indeed, providing informative and entertaining detail on a wide variety of aspects of humanity's environmental crisis. But in all its complexity, the entire assemblage is based upon a handful of fundamental circumstances.

The most important is that the scale of the human enterprise has grown too large for the capacity of Earth's life-support systems to maintain it for long. There are too many people; in the rich nations the average person consumes much too much; and the technologies used to support that consumption are often needlessly inefficient and environmentally destructive. The human economy has grown too large physically; it must soon either shrink or it will destroy its natural foundations.

Whether you are concerned with the excess fat in your diet, the threat of AIDS and other novel virus diseases, overflowing landfills, pollution-spewing energy technologies, a nation ruled by automobiles, the abuse of Western public lands, global warming, or the disappearance of Eastern forest songbirds (or of the ozone shield), you should be aware that all are elements of the same general human predicament.

There is no way that these and other "environmental problems" can ever be solved without dealing with the fundamental problem of scale. That problem is embodied in the I = PAT equation: the scale of the human impact (I) on our environment is the product of population size (P), consumption per person, which is a measure of affluence (A), and a measure of the damage done by the technologies (T) selected to supply each unit of affluence. Since 1850 the human population has multiplied fivefold, from slightly over a billion people to five and a half billion. The use of energy per person (a rough surrogate for per capita impact, the product of A times T) has increased fourfold. As a result, the total impact of humanity has increased some twentyfold. Some foresee a five- to tenfold further increase in the coming century as they visualize population rocketing toward 12 billion, with booming consumption by people in both rich and poor countries.

Such an expansion clearly is not in the cards, since the capacity of Earth to sustain our economy is already showing signs of crumbling. It is now not possible even to support 5.5 billion people on income—it can be done only by exhausting a one-time bonanza of natural "capital" that *Homo sapiens* inherited from the planet. The most critical elements of that capital are fertile agricultural soils, fossil groundwater, and biodiversity. Soils are generated on a timescale of inches per millennium, and are being destroyed in many areas on a timescale of inches per decade. In many areas subterranean waters are being pumped from the ground at many times the rate of recharge. And we are now witnessing the beginning of an extinction episode caused by human destruction of habitats that may well exceed the calamity that exterminated the dinosaurs and many other life forms 65 million years ago. It took tens of millions of years for Earth's stock of biodiversity to be restored after that event.

Unhappily for us, there are no substitutes for good soils, fresh water, and the other organisms that are working parts of the ecosystems that supply indispensable services to society. Humanity cannot get along without natural ecosystems that, among other things, maintain the balance of gases in the atmosphere, generate and preserve soils and make them fertile, regulate the hydrologic cycle that brings us fresh water, and maintain a vast genetic library from which humanity has already extracted the very basis of civilization in the form of crops, domestic animals, medicines, and industrial products. Humanity also cannot get along without the dependable functioning of the agricultural systems that supply sustenance to burgeoning numbers of people, yet the scale of the human enterprise is undermining those systems as well.

There are only two basic choices facing humanity. One is to continue on today's trajectory until a combination of hunger and disease halts the growth of the human population with a catastrophic crash and puts an already shaky, growth-centered economic system into a terminal decline. That "death-rate" solution would be nature's way. Or human beings could use their brains and culture to halt population growth around 10 billion people and start a gradual decline by regulating their birthrates. Simultaneously, wasteful consumption would be curbed among the rich and more efficient technologies deployed so that the gap between rich and poor could be closed. Such a reduction in consumption could clearly be accomplished while increasing the quality of life among the rich, but it would take considerable social and political will to accomplish the required changes.

It would also take great effort to suppress racism, sexism, religious prejudice, chauvinism, and gross economic inequality. There is no way that humanity can see its way through to a sustainable society without cooperation that extends far beyond the ethnic fragments into which the human species increasingly seems to be regressing.

These global problems, which have barely been outlined here, are the context in which we all must live. But as we act locally, the global picture must always be kept in mind. No amount of recycling, careful husbandry of resources, or humane attention to the problems of society can possibly preserve civilization unless the basic problem of scale is solved—and solved soon. Everyone, but especially politicians and other decision makers, must perpetually keep this in mind. You should keep this in mind as you savor the richness of detail in the pages that follow. And it is the duty of all of us to remind our leaders constantly of the basic ecological facts of life. Whatever other efforts you make to preserve society, that campaign should be relentless.

Note: More details on the issues discussed here, and references to the literature, can be found in the Ehrlichs' books, *The Population Explosion* and *Healing the Planet.*

PART I

Body Ecology

ECOSYSTEM PATHWAYS

The following points explain the illustration on page 3, tracing the sources of our food.

• Everything starts with the sun, that great ball of hot gases (mostly hydrogen), about 864,000 miles in diameter and 93 million miles away from Earth. The sun gives off tremendous amounts of light and ultraviolet energy. The Earth receives only about one-billionth of the sun's total energy output. Much of it is either reflected back into space or absorbed by chemicals in the atmosphere, so that most of the harmful ultraviolet radiation does not reach our planet's surface. Its radiant light is experienced as heat on Earth, and as such it powers the major cycling of carbon, nitrogen, oxygen, water, and minerals.

• A tiny fraction of the sun's energy is captured by green plants and bacteria on land and by algae in water, and is used in the process of photosynthesis to make sugar, starches, and the organic compounds that many organisms must eat to live. From the sugar and starches animals produce other substances, such as proteins, oils, fats, and vitamins that we and other creatures need to thrive. All this happens on the relatively thin, twelve-mile zone that extends from the deepest ocean floor to the tops of the highest mountains. This layer, called the biosphere, caused the astronauts to marvel at its fluid and fragile beauty.

The term "ecosphere" emerged as an attempt to combine the concepts of biosphere and ecosystem. As the Earth orbits annually around the sun, creating seasonal cycles, it also spins completely around on its own axis, creating a cycle from light to dark.

• The Earth's atmosphere is the thin, gaseous envelope that surrounds the planet. Most of our air is found within eleven miles above sea level. Air has been cycling for billions of years and is a precious necessity for all of us. We participate in its circulation by breathing it in and out. Weather cycles are created by air movements (winds) that start when warm air from the Earth expands and becomes lighter than cold air. The pressure and pull of gravity on colder heavier air causes it to flow under the warm air, pushing it upward. Such winds keep air circulating from place to place. Since the Earth's axis is tilted at an angle of 23.5 degrees, as the planet rotates it is heated unevenly. Because many vertical and horizontal winds are thus set in motion, weather differs greatly from place to place. Variations in climate led to a wide range in the distribution of animal and plant communities across the biosphere.

• Large areas with similar vegetation, birds, animals, insects, and microorganisms are called biomes. Most familiar to us are tundra in arctic regions, coniferous forests with evergreen trees such as spruce and fir, deciduous forests with trees such as oak and maple where the leaves fall off annually, grasslands, deserts, chaparral, and tropical jungles. Each of these large communities has species that have adapted to its conditions of soil, water, and temperature. Polar bears thrive in the arctic. Cactus plants have thick "skins" for storing water. Birds' bills, feet, legs, tails, and wings are shaped to enable them to survive where they live. Biomes may be seen as large ecosystems, but ecosystems are also found in places as small as a puddle.

Ecosystems are a dynamic complex of plant, animal, and microorganism communities and their nonliving environment interacting as a functional, largely self-sustaining unit but also in conjunction with the larger cycles of nature. What comes in and goes out of ecosystems is important as well. This is true for forests as well as cities. Their future depends as much on the external life-support environment as on internal activities.

Ecosystems have to stay in balance or else they fail. No community has the capacity to carry more organisms than it has food, water, or shelter to accommodate. The balance of food and territory is often maintained by fire, disease, and the predator-prey ratio. Each creature has a niche or role to play.

• The food cycle is actually a complex web, but briefly it consists of (1) the sun as energy source, (2) producers—green plants and trees, (3) consumers—herbivores, carnivores, parasites, and scavengers, (4) decomposers—mostly bacteria and fungi that convert dead matter and gases such as carbon and nitrogen back into the air, soil, or water. Without decomposers, such as molds and worms, most of which are deemed horrible by humans, the world would be buried in litter. Decomposers are invaluable for recycling the nutrients that can be used by the producers.

Thus, when we spray pesticides around, we imperil the food chain. Also, our habit of packaging food in throwaway containers has generated more trash than our decomposers can handle. Since food provides the energy that all living things must have in order to carry out their purpose, we need to be cautious about the food chains, which are the orderly processes of nature by which essential energy is passed from one organism to another. These connections are as intricate as a spider's web.

• Carbon is the basic building block of the DNA, RNA, and proteins essential for life. Carbon is cycled by green plants during photosynthesis and into the food web. Animals release carbon dioxide back into the air or water as one of their waste products. Decomposers also release carbon dioxide into the atmosphere. If it were not for decomposers, all carbon would eventually become locked up in organic matter that could not decay. Carbon is tied up deep in the Earth in fossil fuels—coal, petroleum, natural gas—for long periods of time. When these are extracted and burned, carbon dioxide is released into the air. Volcanoes also release carbon dioxide. Carbon dioxide is soluble in water. Some of this gas is taken in by the oceans and returned to the atmosphere; some is taken to form calcium carbonate to build shells and rocks and skeletons of tiny protozoans and corals.

We humans have interfered with the carbon cycle by removing forests and other vegetation without replanting. For instance, the world has

(Continued on page 4)

1

ALL ABOUT WHAT WE EAT

Almanacs traditionally present seasonal information limited to monthly temperatures and phases of the moon. This Almanac takes a wider view, focusing on the ecological systems that affect our life on Earth. For we have reached a critical point in history wherein our survival depends on knowing ecology.

Food is the first essential need of all. It's common knowledge that our food is dependent on healthy soil, air, and water. It is less understood how it is dependent on the presence of food for all levels of biological life (biodiversity). In this book "biodiversity" means the entire range, known and unknown, of plant, mammal, insect, amphibian, and bird species. Because our food sources depend on widespread biodiversity, it's important that we get to know and protect its variety, habitat, and needs.

The message written on the wall of the future says that if we humans use the Earth for human consumption only, the entire human habitat will become so degraded that we are doomed. Putting people above all leads to scarcity for all. It's dangerously shortsighted to see ourselves as separate from all food sources. Our goal must be the protection of the ecological support systems.

We are what we eat. Simple sentence, with complex implications. The items that we buy in the store originate in the cycles of nature. The major ones will be described here and again throughout this Almanac, for we need to see our human activities in their full context.

Food—Picking Our Way Through the Perils

Like other animals, we are part of the food chain, obtaining our food by eating plants or the animals that feed on plants. Our food, as we have seen, comes out of the complex cycling of chemicals and nutrients. But how the

ECOSYSTEM PATHWAYS—cont'd

(Continued from page 2)

lost 40% of its tropical forests to meet our demand for coffee, chocolate, sugar, and hamburger. By 2000 there could be no such forests left. In this way we rob nature of its ability to absorb carbon dioxide (CO_2) and leave all of us vulnerable to the dangers of global warming.

- Nitrogen is an essential part of all the amino acids, proteins, and DNA in living cells. Nitrogen in the atmosphere is made available to plants by the nitrogen-fixing bacteria and algae of the soil. These microscopic plants use nitrogen and convert it to nitrogen-containing salts called nitrates. The nitrates are released into the soil and then may be taken up, dissolved in soil water, by the roots of plants. Once plants have used nitrogen in the manufacture of their proteins, it can circulate through the rest of the living world by way of food chains. The cycle also occurs when organisms excrete waste products or die, and the decomposers break them down and release nitrogen in the form of ammonia. The nitrifying bacteria use the ammonia and convert it into nitrites and then nitrates, which are available to plants through their roots. In water nitrogen is found mostly in the bottom sediments, which is why shallow estuaries are so rich in nutrients.

But when humans cause nitrogen overload, ecosystems are threatened. We disrupt the nitrogen cycle by emitting large quantities of nitric oxide into the atmosphere when wood or fuel is burned. Nitric oxide combines with oxygen gas to form nitrogen dioxide, which can react with water vapor to form nitric acid. This acid has been damaging trees and killing fish. Other disturbances occur when we use certain inorganic fertilizers, overharvest nitrogen-rich crops, mine mineral deposits, and allow industrial, agricultural, and urban runoff and untreated sewage into our waters. Plant communities saturated with nitrogen become too acidified. In lakes, rivers, and oceans excess nitrogen depletes oxygen, causing the algae blooms and red tides that destroy fish and other vital organisms.

- Minerals, such as potassium, phosphorus, calcium, sulfur, and magnesium, also circulate between organisms and their surroundings—from producers to consumers to decomposers to producers again. Most of these processes, which are aided by climate and winds, take a great deal of time.

Mining phosphates to produce fertilizers and detergents and permitting runoff of animal waste and fertilizers from farms into our waters have severely disturbed aquatic life. Excess sulfur dioxide gets into the atmosphere from our burning coal and oil for electricity, petroleum-refining, and the smelting of metallic minerals into copper, lead, and zinc.

- The water cycle is so universal that it too is as likely to be taken for granted as the aforementioned. Fluids, such as blood in animals and sap in plants, are carried mostly by water in which food and other materials needed by the cells are dissolved. The chemical reactions of life take place in water. Water also supplies the hydrogen and oxygen that make up living things. Water falls upon the Earth in the form of rain, hail, or snow. It either sinks into the soil or runs off the surface into marshes, swamps, rivers, lakes, and eventually the sea. Much of it is taken from the soil by the roots of plants to make carbohydrates. Most of the water, carrying dissolved nutrients such as nitrates and minerals, moves up through the stems to the leaves and is evaporated into the atmosphere through small pores. A mature tree may transpire as much as 1500 gallons of water a year. Water is also evaporated from the surface of the land and from bodies of water. It circulates in the atmosphere as water vapor until it condenses and falls again as rain, snow, or hail.

We interfere with the water cycle by withdrawing huge quantities of freshwater and depleting our supplies. Also, by clearing vegetation from land for the sake of roads, parking lots, etc., we reduce seepage of water into the ground for storing and thus increase the likelihood of flash-flooding and surface runoff, which causes soil erosion.

Cycles Matter—Matter Cycles

foods we find in our local grocery store are grown, processed, or shipped is generally unknown to most of us. We hear reports that chicken may contain salmonella, that milk, fruit, and even bottled water contain hazardous chemicals. Uncontaminated food is hard to find. Even if we seek fresh food, untreated with preservatives, we then run the risk of its growing carcinogenic molds if we leave it around too long. Imported foods, so many of which appeal to our gourmet taste, undergo even fewer safety inspections than homegrown.

Yet, inspection methods of the Environmental Protection Agency (EPA), the Food and Drug Administration (FDA), and other agencies are less than reassuring. Food processing plants in the United States are inspected for microbiological contaminants every five years or so. The FDA surveys grocery items four times a year, focusing on foods that have caused problems, unable to check out the rest. Inspection by sight, smell, or touch can be of little use in detecting chemical hazards. No agency is obliged to examine fish.

What can we do? We don't have to fall for the promotion of picture-perfect foods. If we allowed for even slight blemishes in skin quality, two thirds of pesticides could be eliminated. We don't have to indulge in unseasonal fruits and vegetables, such as strawberries and asparagus in winter. We can wash our vegetables and fruits and demand that regulating agencies resist the pressures of chemical companies to add more chemical processing, such as irradiation.

The process of irradiation involves placing food on a conveyor belt that runs through a sealed chamber. There, fuel rods of cobalt-60 or cesium-137 emit radioactive gamma rays that kill bugs, fungi, bacteria, and viruses. At times irradiation has its uses. Very high levels of exposure can sterilize food for hospital patients who need contamination-free environments. Lower levels can control or eliminate insect and bacterial infestations.

The theory is that irradiation, when applied to food, extends its shelf life because it stunts the ripening of fruits and vegetables and stops microbial action that leads to spoilage. For the ordinary consumer, irradiation in addition to pesticides used in growing food is risky. The FDA allows the use of food irradiation at levels up to 100,000 rads (considered a "low dose") on a number of different foods. Higher doses (up to 5,000,000 rads) are allowed for spices. Even the FDA agrees that vitamins are lost in the process. Irradiated food may look fresh, but as we know, looks can be deceiving.

We can expect food manufacturers to be ecologically sensitive. Just as public pressure induced tuna processors to sell "dolphin-safe" tuna, so we, along with environmental groups, business leaders, and scientists can work for responsible harvesting, respectful treatment of animals, and more useful labeling of foods. The term "natural" is a misnomer, for instance. Seedless fruits are not natural; they were developed by people to make them easier to eat. Since vegetables, fruits, and animals have been manipulated for centuries, it is very difficult to find any food in a "pure" state. Finally, we can demand that manufacturers use less wasteful packaging, and we can recycle and compost our garbage.

From Bean Sprouts to Broccoli: There's a World of Good in Greens

Paul Underhill

This year, the average American will eat 168 pounds of red meat and poultry—more than twice as much as, say, the average Brazilian. That comes to nearly half a pound a day—thanks to which, at least in part, we can expect an increased risk of heart disease, clogged arteries, strokes, and certain forms of cancer.

If that isn't sufficient to make you consider reducing the amount of meat in your diet, here's another reason: It might just be better for the planet.

Increasingly, the evidence indicates that by lowering our consumption of animal products, as well as by demanding more organically and locally grown products, we can reduce our impact on the environment—and make a significant contribution to a sustainable future.

With a little practice, you can make vegetables into a hearty and delicious main course, not the overdone, tasteless side dish they have often come to represent. And don't fret about protein. Nutritionists have found that only between 4 and 9 percent of our diet need be protein. Even a purely vegetarian diet provides a minimum of 10 percent. Pasta with vegetable sauce and cheese, peanut butter and jelly sandwiches, beans and rice with yogurt and salsa, and a good vegetable soup with bread are all quick, protein-rich meals for active people.

Eating lower on the food chain can also save your money. Reinvest your savings in organically and locally grown products. They taste better (really) and have less impact on the environment. The money you spend generally goes to small farmers, not to supermarket chains, food-processing companies, and other intermediaries.

One of the most nourishing foods since the beginning of time has been seeds—the pits of fruits, the grains, the nuts, and the beans. Other people in the world find good nutrition in these foods from the low end of the food chain. For instance, quinoa seeds, which are 16% protein, 60% carbohydrates, 8% oil; tarwi seeds, which are like soybeans but have 40% more protein; and other beans and fruits that someday may be more widely used by us as well. Even insects have been hailed as the last large unexploited food source. How about feasting on live honeypot ants, mealworm dip, wax worm fritters, cricket tempura, roasted Australian grubs, sautéed Thai water bugs, and insect sugar cookies?

Back to the Farm

Here is where food production begins. In our day, farms have become very highly managed ecosystems. Sometimes because of our ignorance concerning ecology and the ways of biodiversity, decisions are made that are tantamount to risky experiments.

The soil cycle is crucial. It works through a succession of layers. Topsoil forms dense food webs consisting of roots, decayed matter, leaves, twigs, bacteria, fungi, and numerous decomposers. Here gophers and moles make their home, help to loosen the soil. Humus is made from the partially decayed organic material, giving structure to the soil and retaining moisture and nutrients. Subsoil, the second layer, contains more minerals. The minerals of rocks and soil are dissolved in water, which plants then take up through their roots to use in foodmaking. The third layer down consists of solid rock or larger rocks that can be many feet thick; it is lighter in color than the above layers. Rocks are broken down by heat, cold, water, chemical action, air, and plant life. Cyclical freezing and thawing, wetting and drying, and rain blending with air to form a weak acid help break down hard rock. Winds blow and water carries surface particles away. When enough tiny particles of rock accumulate in one place, plants move in and start to grow. Bacteria, fungi, and lichens move in early and live on minerals in the rock particles, as well as on air and water. As they live and die, they build up new soil.

Soil Is Not a Factory

Jon Luoma

"There's much more interest in sustainable agriculture than even a year ago," says Youngberg, director of the nonprofit Institute for Alternative Agriculture. "There's no question that it's gaining momentum among both farmers and scientists."

Most share an uneasiness about "conventional" agriculture—the style of farming tied to the high use of fossil fuels, pesticides, and synthetic fertilizers that came to dominate American farming in the 1950s. There is an increasing concern that synthetic chemicals are polluting lakes, streams, groundwater aquifers, the air, and soils, and perhaps harming the health of farmers, their families, and consumers. For many, the greatest concern of all is that chemically intensive farming degrades soils, reducing their natural fertility by destroying microbes and reducing organic content.

According to [farmer] Kirschenmann, it means recognizing that our metaphors are out of whack: "The conventional approach treats the land as if it's a factory," he says. "But soil is not a factory, it's an organism." Many farmers are now paying attention to sustainable farming out of sheer economic pragmatism: The national farm economy has suffered in recent years, and chemicals are expensive.

To make the kinds of improvements they insist need to be made, the sustainable farmers are bringing into play a host of techniques, some of them ancient, some modern. Since pre-Columbian times, for example, Central American Indians have interplanted corn, squash, and beans—the corn serving as a natural trellis for the beans, the legumes fixing nitrogen to enrich the soil, and the squash plants covering the soil, abating erosion and suppressing weeds.

Now, farmers and researchers are experimenting with modern variations on the technique. In one recent controlled experiment in Nebraska, rows of corn were planted between rows of sugar beets, offering the beets shelter from high winds and the corn more access than usual to sunlight. The result: Beet yields rose 11 percent, and corn yields by about 150 percent. In experiments elsewhere, rye, a grain that releases a natural herbicide into soils, is being interplanted with nitrogen-fixing soybeans.

More commonly, sustainable farmers are paying close attention to crop rotations to build richer soils, including idling some fields during "soil building" years. Despite the lost year of productivity, many claim they quickly make up the costs in improved fertility without the high cost of synthetic fertilizers. Other farmers have turned to new equipment, including such devices as ridge-tillers and rotary hoes that control weeds mechanically, and even a new machine designed to vacuum insects from plants....

Most experts agree, however, that it's not always that simple on every farm, especially at first. Kirschenmann has suggested that, to maintain yields, farmers be prepared to use some chemicals, particularly herbicides, in the early years of soil building. Others, notably fruit and vegetable growers, have found it difficult to produce marketable products without using some pesticides. . . . The point for many farmers is to find a sound compromise....

Reports are mixed on whether sustainable farming is a better short-term economic deal for farmers or not. Rex Spray says his methods require more labor, but that he comes out ahead by saving the cost of chemicals and receiving premium prices for organically grown crops. Kirschenmann says in average years he does at least as well economically as his neighbors. But he is convinced his balance sheet comes out far better in troubled years. During the disastrous drought year of 1988, he harvested a partial crop of wheat even though "a lot of my neighbors never had reason to pull a combine out of the shed." He attributes much of that to the ability of his spongier, higher-quality soils to conserve water.

More important, he says, he didn't waste tens of thousands of dollars on chemicals that would not have helped in a drought. "There's no question in my mind that over the long term your economic survivability improves," he says.

Christine Eliason, who farms two hundred acres in Nebraska, suggests there are even more important reasons: "Children who haven't walked yet have a right to the productivity of my land."

Soil is affected by the kind of plants grown on it. Moreover, it is thin on mountain slopes, thick in deep river bottoms. It can be red, yellow, brown, black, or gray, according to climate cycles. Desert soil is dry, reddish-brown, with closely packed pebbles; grassland soil is more alkaline and rich in humus; tropical rain forest soil is more acidic, with a subsoil of iron and aluminum compounds mixed with clay; deciduous forests have thick leaf litter and a subsoil of thick clay; a coniferous forest has more acid litter and humus.

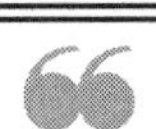

Every day in the U.S. four square miles of our nation's prime farmlands are shifted to uses other than agriculture. The thief is urban sprawl.

—National Agricultural Lands Study, 1979

Problem: Disappearing Prime Farmland

Today soil is eroding faster than it is being formed. Erosion occurs on slopes where water does not soak in. The topsoil is carried off into streams and rivers. Land needs to be managed so that the soil is protected and holds water, as in contour farming. Erosion is caused by agricultural practices that remove plant cover, and by overgrazing, poor logging, and off-road vehicles. Like the federal deficit, it keeps getting worse with the passage of time; yet at some point the consequences have to be faced. The desertification of lands spreads like a plague once it gets started.

Problem: Failure of Fertilizers and Pesticides

> Farmers didn't set out to put on more and more chemicals. It's U.S. farm programs that forced farmers to increase yields per acre per year. . . . Increased yields of these commodities translated to accumulation of excessive stocks of corn and wheat way above domestic and export use and emergency needs. . . . When this occurs, the resulting surpluses depress crop prices as well as net farm income, and raise commodity program expenditures to outrageous heights. . . . This deliberate creation of glut and surpluses creates huge environmental problems. . . . The dark side of our food supply is the probability that 60 percent of herbicide usage today is carcinogenic or potentially carcinogenic.
>
> *—Maureen Hinkle, Audubon's Director of Agricultural Policy*

Plants normally get nitrates from nitrogen-fixing bacteria that live within nodules of the roots of the legume family. When farmers tried to speed up the process by dumping artificial fertilizers on the soil to increase crop production, they learned that because fertilizers are noncyclical (do not degrade or dissipate well) the regeneration process was not sustaining and ultimately causes erosion. Rather than through fertilizers, nitrogen can be put back in the soil with manure or crop rotation.

Dangerous farm products are not adequately regulated by the EPA. Alachlor, the most widely used herbicide in the country, marketed under the name Lasso, a product of Monsanto Company, is an example. Here are the reasons we should lobby for the banning of alachlor:

- Over 90% of Lasso is used on corn and soybeans, crops that are in such excess that American taxpayers paid $13 billion on corn alone last year to reduce production.

• Alachlor accounts for one tenth of all pesticides applied in the United States—an estimated 100 to 130 million pounds per year.
• Alachlor causes cancer of the stomach, thyroid, and nose in laboratory animals. Even short-term exposures to alachlor produces cancer in rats. It also breaks down into two mutagenic compounds.
• Residues of alachlor are regularly found in ground and surface water, drinking water supplies, and rain.
• Thousands of farmers face excessively high levels of risk. Farming generally is considered a healthy occupation, but recent studies have shown that farmers have higher than normal rates of various kinds of cancer. Cancer mortality has contributed to the recent elevation of farming by some experts as the most dangerous occupation, surpassing even mining.
• The public at large cannot avoid alachlor in normal diets or in drinking water supplies since removal technologies are deficient.

The risk of continued use of alachlor is greater than for any pesticide now in use, and the benefits cited by EPA are exaggerated. The EPA not only misunderstands agricultural economics, but also misreads public sentiment. In Iowa, where corn is king, the *Des Moines Register* conducted a poll in November 1986. Of those polled, 78% favored limiting farm chemicals even if it means lower production. Moreover, 52% regarded farm chemicals as the biggest threat to water quality.

An example of the way pesticide use can backfire is in the attempt to wipe out the imported fire ant with the pesticides heptachlor and dieldrin. According to Hinkle, the eradication program not only failed, it actually contributed to the spread of the fire ant, and enabled its dominance over native ants. The fire ants "became territorial defenders, eliminating all competitor ants in their territory. Those areas became environmental sinks—populated only by imported fire ants."

It was during the 1950s and 1960s that the unrestricted use of chemical pest controls surged. The chlorinated hydrocarbons had helped win World War II by killing lice and thus preventing typhoid fever. The potential to end the plague of pests and to feed a needy world appealed to both patriotic and humanitarian goals. The publication of *Silent Spring* by Rachel Carson in 1962, however, precipitated a decade of congressional hearings about the nontarget effects of pesticides. Evidence mounted about the capacity of insects not only to develop, but also to pass on to their progeny, resistance to pesticides as they were deployed. As a result, Integrated Pest Management grew more popular as a method of controlling pests by plan rather than blanket attack with poisons. The idea is for scientists to study a problem and devise a program of predators (other creatures that prey on insects), rotation of crops, variation of planting times, companion planting, and limited, targeted spraying.

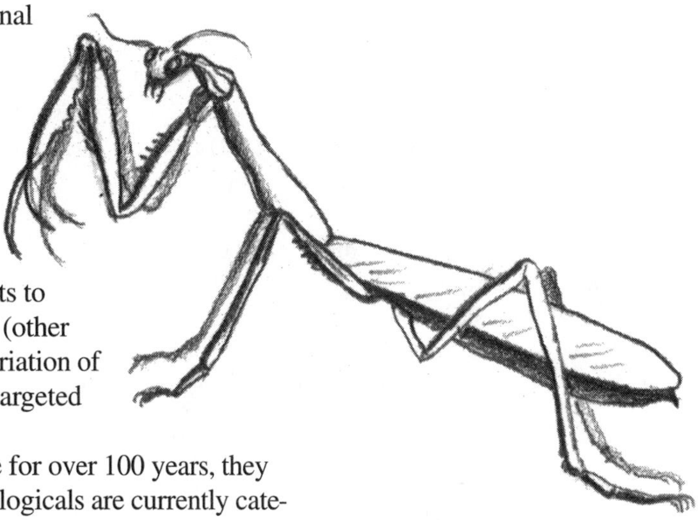

While biological controls have been available for over 100 years, they never have gotten enough attention or funding. Biologicals are currently categorized as follows: (1) parasite and predator species; (2) naturally occurring biochemicals that may attract, retard, or destroy pests—e.g., pheromones,

THINKING LIKE AN ECOSYSTEM

In "Thinking Like a Mountain" Aldo Leopold coined the phrase that would sum up the awareness of the balance required in maintaining food supplies. Here is an excerpt:

A deep chesty bawl echoes from rimrock to rimrock, rolls down the mountain, and fades into the far blackness of the night. It is an outburst of wild defiant sorrow, and of contempt for all the adversities of the world.

Every living thing (and perhaps many a dead one as well) pays heed to that call. To the deer it is a reminder of the way of all flesh, to the pine a forecast of midnight scuffles and of blood upon the snow, to the coyote a promise of gleanings to come, to the cowman a threat of red ink at the bank, to the hunter a challenge of fang against bullet. Yet behind these obvious and immediate hopes and fears there lies a deeper meaning, known only to the mountain itself. Only the mountain has lived long enough to listen objectively to the howl of a wolf. . . .

My own conviction on this score dates from the day I saw a wolf die. We were eating lunch on a high rimrock, at the foot of which a turbulent river elbowed its way. We saw what we thought was a doe fording the torrent, her breast awash in white water. When she climbed the bank toward us and shook out her tail, we realized our error: it was a wolf. A half-dozen others, evidently grown pups, sprang from the willows and all joined in a welcoming melee of wagging tails and playful maulings. What was literally a pile of wolves writhed and tumbled in the center of an open flat at the foot of our rimrock.

In those days we had never heard of passing up a chance to kill a wolf. In a second we were pumping lead into the pack. . . .

We reached the old wolf in time to watch a fierce green fire dying in her eyes. I realized then, and have known ever since, that there was something new to me in those eyes—something known only to her and to the mountain. I was young then, and full of trigger-itch; I thought that because fewer wolves meant more deer, that no wolves would mean hunters' paradise. But after seeing the green fire die, I sensed that neither the wolf nor the mountain agreed with such a view.

Since then I have lived to see state after state extirpate its wolves. I have watched the face of many a newly wolfless mountain, and seen the south-facing slopes wrinkle with a maze of new deer trails. I have seen every edible bush and seedling browsed, first to anaemic desuetude, and then to death. I have seen every edible tree defoliated to the height of a saddlehorn. Such a mountain looks as if someone had given God a new pruning shears, and forbidden Him all other exercise. In the end the starved bones of the hoped-for deer herd, dead of its own too-much, bleach with the bones of the dead sage, or molder under the high-lined junipers.

I now suspect that just as a deer herd lives in mortal fear of its wolves, so does a mountain live in mortal fear of its deer. And perhaps with better cause, for while a buck pulled down by wolves can be replaced in two or three years, a range pulled down by too many deer may fail of replacement in as many decades.

So also with cows. The cowman who cleans his range of wolves does not realize that he is taking over the wolf's job of trimming the herd to fit the range. He has not learned to think like a mountain. Hence we have dustbowls, and rivers washing the future into the sea.

In this Almanac the feature "Thinking like an Ecosystem" will be used to point out the principles that we need to think about in preserving balance in our ecosystems.

hormones, sterilizers, and plant-growth regulators; (3) microorganisms such as viruses, bacteria, protozoa, and fungi.

Example: A biological program of 283 species of parasitic insects successfully controlled three species of weed and thirty-eight species of insect pests of sugarcane in Hawaii.

Example: Neem seed oil. A tree called neem that grows widely in Africa and India and has been revered for centuries has generated enormous excitement among Western scientists. "Chemicals derived from the seeds ward off more than 200 species of insects, including some of the world's most tenacious pests: locusts, Gypsy moths, and cockroaches. But toxicological testing to date shows that the neem products are much safer for other species than are synthetic insecticides, many of which work by poisoning nerve cell functions."

Rachel Carson described the way nature controls its own in this section on how insects serve as predators on other insects:

> Some are quick and with the speed of swallows snatch their prey from the air. Others plod methodically along a stem, plucking off and devouring sedentary insects like the aphids. The yellowjackets capture soft-bodied insects and feed the juices to their young. Muddauber wasps build columned nests of mud under the eaves of houses and stock them with insects on which their young will feed. The horse guard wasp hovers above herds of grazing cattle, destroying the blood-sucking flies that torment them. The loudly buzzing syrphid fly, often mistaken for a bee, lays its eggs on leaves of aphid-infested plants; the hatching larvae then consume immense numbers of aphids. Ladybugs or lady beetles are among the most effective destroyers of aphids, scale insects, and other plant-eating insects. Literally hundreds of aphids are consumed by a single ladybug to stoke the little fires of energy which she requires to produce even a single batch of eggs.
>
> Even more extraordinary in their habits are the parasitic insects. These do not kill their hosts outright. Instead, by a variety of adaptations they utilize their victims for the nurture of their own young. They may deposit their eggs within the larvae or eggs of their prey, so that their own developing young may find food by consuming the host. Some attach their eggs to a caterpillar by means of a sticky solution; on hatching, the larval parasite bores through the skin of the host. Others, led by an instinct that simulates foresight, merely lay their eggs on a leaf so that a browsing caterpillar will eat them inadvertently.

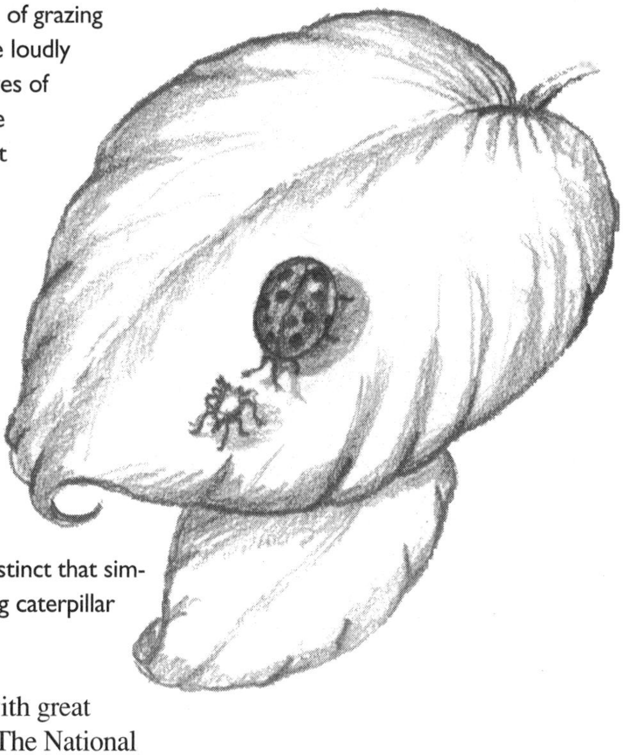

Community composting is a relatively new development with great potential to increase the fertility of agricultural and forest lands. The National Audubon Society has found in pilot communities that 70% of household trash can be recovered by combining composting and recycling. These compostable materials, separated at the source, derived from yard and food wastes—wet and soiled paper, diapers, pet wastes, and paper packaging— can be returned to the soil for its enrichment.

Interest in biologicals and low-impact techniques, however, may be threatened in the future by the hasty commercialization of genetically engineered products. These products furthermore may be used on an intensive wide-scale basis, posing a danger to nontargeted organisms. We have to watch out that in turning against so-called pests, we do not also turn against the Earth itself.

Biotech: Will It Bring Miracles or Disaster? Either Way, Wildlife Will Be Affected

In the bold claims being made for the prospects of a brave new biotechnical world, we hear echoes from technology optimists of the past. Once upon a time, atomic power was going to solve the world's energy problems with power "too cheap to meter." Chemical pesticides were going to make possible an agriculture that would end world hunger. The revolutions in these and other technologies have immeasurably improved the standard of living throughout the industrialized world, but environmentalists have learned that with new technologies sometimes come unforeseen and even catastrophic consequences. What environmental problems might a biotechnical agriculture pose?

If the new industry lives up to its billing, food and forage crops will be developed that can thrive in formerly inhospitable climates and soils. For example, a strain of orange trees tolerant of freezing temperatures and requiring very little water could lead to the conversion of prairie grasslands or even the Great Basin desert into vast orange groves. A cold- and saline-resistant tomato plant might lead to the cultivation of remaining coastal salt-marshes, transforming them into rows of tomato plants. On the other hand, if nitrogen-fixing grasses are developed, allowing them to flourish in poor soil, the felling of the last of the tropical forests might be forestalled by making already cleared forest land permanently productive for crops or grazing—but even this heartening possibility is fraught with potential abuses that could change forever the ecological succession of these priceless ecosystems.

If we too narrowly confine agriculture or industry to engineered products, we are in danger of losing the gene pool. In the last 600 million years, the Earth has experienced five big extinctions of life, usually linked to climatic change, in which 35% to 95% of all species then on the planet disappeared. Scientists know from fossil records that in each case it took 10 million years or more for the Earth to regain its biological diversity. Many biologists fear that humans may now be precipitating an extinction of comparable scale. We need to preserve as much genetic multiplicity as possible in order to keep alive possibilities for not only life, but also medicines and foods that may be needed in the future.

Millions of dollars are being poured into agriculture and engineered food. But crops designed to fend off insects could become useless in a decade when the insects become immune. New foods are considered a panacea for the exploding populations all over the world, but the chemical impacts of these products are as yet unknown. Furthermore, when the genetic makeup of

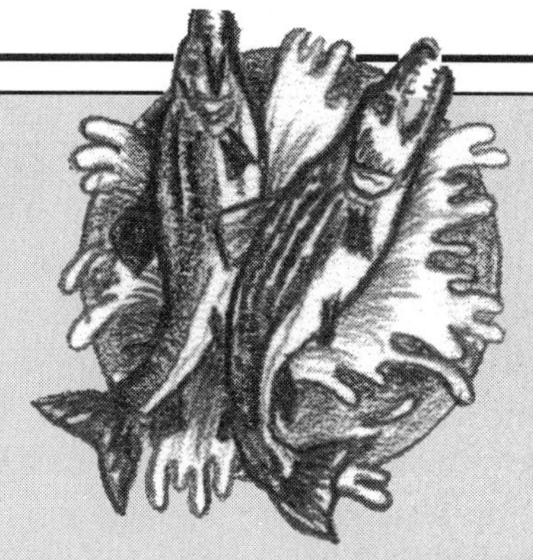

Frankenstein's Fish

Ted Williams

It strikes Robert H. Smith and me that those who pursue and manage fish in these United States are forgetting things. Like what fish are and why they are important. Take, for example, our favorite fish—trout. God made them last, someone told me when I was a child, after He had practiced on all the others. But now fish managers are making them over, and everyone save me and Robert H. Smith thinks it's swell.

The other day he was moaning and groaning about how all the trout streams in the West have been fouled with hatchery genes. You hike ten miles up a canyon only to find that the pure strain of cutthroats you thought persisted in the headwaters has been hybridized to extinction by domestic rainbows whose distant wild ancestors hailed from the Pacific Coast.

These days I catch hatchery trout only by accident—genetic wrecks that penetrate, like coliform bacteria, to the forgotten rills I find by first locating the state stocking route and then proceeding upstream and in the opposite direction. Recently, I cut open one of these fish to see what it had been feeding on; it had been feeding on cigarette butts.

Hatchery trout are everything real trout are not. They are selected not to consume living fauna but processed pellets, not to thrive amid clean gravel, rocks, and brush, but in filthy, crowded, cement troughs that wear away scales and fins. The flaglike dorsal, because it is on top, is the one fin that cannot be worn away. But hatchery trout attend to this themselves, nipping any dorsal in sight so that virtually all become fleshy, withered stumps. Walk along a wild trout stream and fish will streak for cover. Walk along a hatchery raceway and fish will streak to your side in the hope of getting fed. After years of inbreeding, hatchery trout tend to be deformed. Gill covers don't fit, jaws are bent, tails pinched. They are wriggling sausages flung to the public like gurry flung to gulls, and consumed as raucously. In my state, Massachusetts, some stocked trout actually bear tags which may be exchanged for prizes.

Most denizens of this planet have in each cell two sets of chromosomes. Sperm and eggs must therefore reduce their chromosomes to a single set to prevent the embryo from having four. But fish managers have learned how to foil this process by shocking eggs with heat, hydrostatic pressure, or chemicals just as they are about to shed one chromosome set (which in the egg happens after fertilization). Shocking results in hatchlings with three sets of chromosomes.

Why would anyone want to do this? Well, you have to think like a manager. First, "triploid fish," as freaks are called, are sterile; their odd number of chromosomes cannot synapse in natural gametogenesis. Second, the gonads are structurally deformed, and nutrients that normally would go into their development are rechanneled to body bulk. Bigness, in and for itself. That is what the angling public demands, and that is what the managers serve up.

And now researchers have perfected a technique to destroy genetic material with ultraviolet light. They can strip fish sperm of its single set of chromosomes, use it to fertilize eggs, shock the eggs to prevent the expulsion of the second set of female chromosomes, and produce fertile fish with two sets of the maternal chromosomes—clones.

That is not the worst of it. Female salmon are being shot full of testosterone and transformed into males which, because they retain the genetic makeup of females, sire females only.

Diploid salmon of grotesque and unnatural proportions are being contrived by dosing the fry with bioengineered chicken and cow growth hormones. Next step, now in progress: Equip salmon with the genes to make the hormones themselves. . . .

Finally, America is addressing the Divine error of fish bones. One of the grandest things about fried or broiled trout is the way one can fillet them with a fork, peeling away the sweet, firm meat and leaving a perfect, bare skeleton of the sort trafficked in by cartoon cats. There is, in this simple act, something eminently natural and satisfying, like popping meat from a lobster tail. Even those who never have eaten whole fish can, if possessed of fingers and functional intelligence, be taught in seconds to debone a cooked trout. When people say they do not like trout bones, it means they have mutilated the trout they have eaten.

To deal with such incompetence, the Clear Springs Trout Company in Buhl, Idaho, is producing a fish to match—the boneless trout. Already culturists have done away with the pin bones in the abdomen. Inevitably, the boneless trout—the ultimate symbol of our current taste in "natural" objects—will find its way into state stocking programs. What, I wonder, will Smith say when he hauls one of these amorphous blobs of protoplasm onto the bank and it collapses like a jellyfish?

These megacompanies are using genes just as earlier corporate powers used land, minerals, or oil. In many ways, DNA is the ideal corporate resource: It can be patented and wielded as property. . . . DNA can be used to produce tremendous quantities of rare and expensive products for pennies.

—Jack Doyle, Friends of the Earth

an organism is altered, surely a chain reaction follows in related organisms. Many fear that the more money invested in engineered products, the less money and effort will go into finding more sustainable solutions.

Should we allow genes to be bought and sold by multinational corporations? If so, organic farmers will have a still more difficult time surviving, and people will not have much choice in foods available to them. Regulations could control the situation, but huge companies have hired staffs to figure out ways around them.

PROBLEM: GROUNDWATER POLLUTION

Flowing through cracks, wormholes and fissures, nitrates and pesticides have been detected ten to fifteen feet below the surface just two days after application. In 1988, forty-six pesticides were detected and confirmed in the groundwater of twenty-six states due to agricultural use. Our food and water could get safer—not totally safe, for that is impossible—but improved by 80% by eliminating residues from a few high-risk compounds.

Sediment is the most abundant of all U.S. surface water pollutants, resulting in pollution damages estimated to be as high as $1 billion annually. It fills up reservoirs, increases the frequency and seriousness of floods, clogs navigation facilities and canals, interferes with industrial hydraulic equipment, increases the cost of treating drinking water supplies, destroys aquatic wildlife, encourages unpleasant algae growth, and diminishes recreational potential of downstream waters. Taxpayers end up paying again for costs of erosion control on these lands.

If we don't protect our freshwater, as supplies become scarce, countries are likely to fight over them.

Integrated Pest Management can help clean up our water enormously at its many sources. We can also require farmers to keep records of their pest controls, which would go a long way to evaluating environmental impacts.

Above all, we can support the needs of ecosystems. For, as Paul Mankiewicz, Director of the Gaia Institute, says, "Nature's cycles still far surpass the hard-won, yet inefficient human technologies that turn resource into waste. By turning to the study of natural systems, we may learn to ground our industries on the profound yet humble principles of interchange and reciprocity, and so generate a lasting wealth based on returning what we receive."

PROBLEM: HUNGER AND THE LOSS OF BIOLOGICAL DIVERSITY

Donella Meadows, environmental science teacher at Dartmouth College and author of *The Limits to Growth* and *Beyond the Limits*, says the saddest chart she has plotted is the one that shows food production in different parts of the world.

> In Africa it has doubled in just twenty years; in Asia it has tripled. But after twenty years of Green Revolution in the Third World, the amount of

> food *per person* [her italics] has barely changed—except in Africa, where it has steadily declined.
>
> More food is raised now, but it does not feed hungry people. And the cost to the environment has been high. Food output has been increased by clearing forests, damming rivers, irrigating fields, and pouring chemicals onto the land. Between 1970 and 1990 the total use of fertilizers in the world rose from 70 million to 145 million metric tons per year. The use of pesticides rose from 1.3 to 2.9 million metric tons per year.
>
> How many more doublings will it take of fertilizers, pesticides, and food production to feed the world's growing population? How many more doublings are possible?

We feed upon food that is eked out of ecosystems. As we alter or obliterate them, more and more species are crowded out, sickened, or destroyed. We humans must use not only our brainpower but also our deep affection for nature to sustain the habitat that nourishes us all.

Wes Jackson, co-director with his wife of The Land Institute, acknowledged how he learned at mealtime that land was the source of sustenance and health:

> In our farm kitchen scarcely a meal began without a prayer of thanksgiving for food and other blessings or ended until the plate had been wiped clean with bread. During dishes and clean-up following the meal even so little amount of food as half of a fried egg would be returned to the refrigerator; it would appear on the table at another meal, usually in a hash consisting of left-over potatoes, gravy, some onion and a few meat scraps. This wasn't poverty, just frugality well-managed by an imaginative mother. I have become increasingly aware that more values about land and its relationship to people are taught in dining rooms than anywhere else in America.

E c o Q u i z

Do you know...?

What is in the water you drink?

That the world loses 7% of its topsoil per decade?

That it can take 5000 years to produce five inches of topsoil?

That billions of microorganisms are present in a single teaspoon of topsoil?

The creatures we need to love not hate for the job they do in making precious soil? Invertebrates, such as centipedes, millipedes, dung beetles, earthworms, snails, slugs. Fungi and molds should also be on our gratitude list.

Where the food you buy is grown? You can tell your grocer that you want locally grown food or you can deal directly with a local farmer. You can influence the way food is produced by getting to know the people who produce it.

How nature cycles persist in cities? We see them when grass thrusts through cracks in the sidewalk, when asphalt is broken up by weathering during the alternation of heat and frost, when algae forms in puddles, when dirt and debris pile up and decompose, when rain starts channels running in the streets, when trees transpire air along congested streets. With a better sense of ecology, people could plan cities to allow for food, shelter, and work in better conjunction with the flow of nature cycles.

In this picture we can see the many pathways of connection between ourselves and nature cycles. Because we are so intrinsically bonded, the health of one depends on the health of the other.

- We are part of the carbon cycle as we breathe in oxygen and exhale carbon dioxide.
- Our body wastes are at some point dumped into the ground or water, thereby entering the food and water cycles.
- The food we eat is taken from the land, which in turn is affected by soil, water, nitrogen, phosphorus, and mineral cycles. The way our food is grown, processed, and preserved greatly affects our health. The presence of pesticides, radiation, and carcinogenic molds causes disease. The degradation of soil and water leads to loss of the biodiverse plants, animals, and microorganisms, an impoverishment that in some parts of the world has already led to famine.

The mass extinctions which the Earth is currently facing is a threat to civilization second only to the threat of thermonuclear war.
—National Academy of Sciences

- Our reproductive rate has led to such a rise in population that the cycles and ecosystems are thrown out of balance. The sheer numbers of us make it difficult for ecosystem "services" to control the proportion of gases in the air, circulate water, pollinate, control pests, generate soil, and recycle waste.
- Our patterns of consumption have led to personal and industrial pollution of air, soil, and water, which in turn makes us and other creatures sick. Air pollution, for instance, eats away at the atmosphere's protective ozone layer, permitting ultraviolet rays to form cancer cells in our skin. Furthermore, our lifestyle is a disastrous model for the lesser-developed countries, for if all nations consumed at the rate of the United States, we would destroy ourselves.
- Our longer lifespan contributes to the higher numbers of people who must coexist on Earth at one time.
- Our bodies can be considered ecosystems, with populations and niches and cycling of gases, food, and liquids. In order to understand this concept, let us briefly review the microscopic components of our wondrous bodies.

Atoms are the basic structure of our bodies as well as all matter; atoms have electrons, protons, and neutrons in constant motion around their nucleus. On the atomic level a great deal of interaction among elements is taking place. Atoms can lose or gain electrons to form ions (atoms with one or more positive or negative electrical charges). Combinations of atoms became known as "compounds."

The term "organism" generally means any form of life—plant, bird, insect, animal, fish, human, and the tiny microorganisms. (Only a small percentage of organisms have ever been identified.) Molecules, made up of atoms, form the protoplasm from which the cells of organisms are made. Cells are the basis of our body's tissues, organs, and systems—digestive, circulatory, nervous, musculature, etc. Cells contain our DNA.

Our bodies contain microorganisms that move in and out of us as we eat, breathe, excrete, copulate, and die. For instance, the saliva in a healthy mouth is said to have 10 million to a billion bacteria per cubic centimeter. Harmless mites live exclusively at the base of our eyelashes. Some mites can carry viruses. Dust mites can cause dust allergies. "E. coli" is the name of bacteria that live in our large intestine, benignly helping themselves to food.

Sometimes an imbalance of organisms creates ill-health. If we eat food that has been sitting around too long, certain parasites will attach to and live on our intestinal walls, causing stomachache and diarrhea. Candida is a yeast that thrives on mucous membranes of our body. If it gets too numerous, it causes infection. Usually lactic acid bacteria keep it in check.

The food web created by producers and consumers exists in our bodies as well. An important cycle of our body cells involves the flow of matter and energy. In a slow burning process oxygen is used to release the energy stored in the chemical bonds of carbohydrates (such as glucose) and other compounds. Without this energy-yielding process none of our body's functions could occur, but it can be disrupted by poisonous chemicals or radiation from the outside.

Where does the Earth end and I begin? you may well wonder. Once you know about cycles, you will find it hard to pinpoint the place. For there is no barrier; the fact is that we are made from the Earth. If we comprehend this truth in all its amazing ramifications, as this Almanac shows, we will be much better off in caring for ourselves and our planet/nest, the ecosphere.

2

SEX, BIRTH, DEATH...AND THE HEALTH CONNECTION

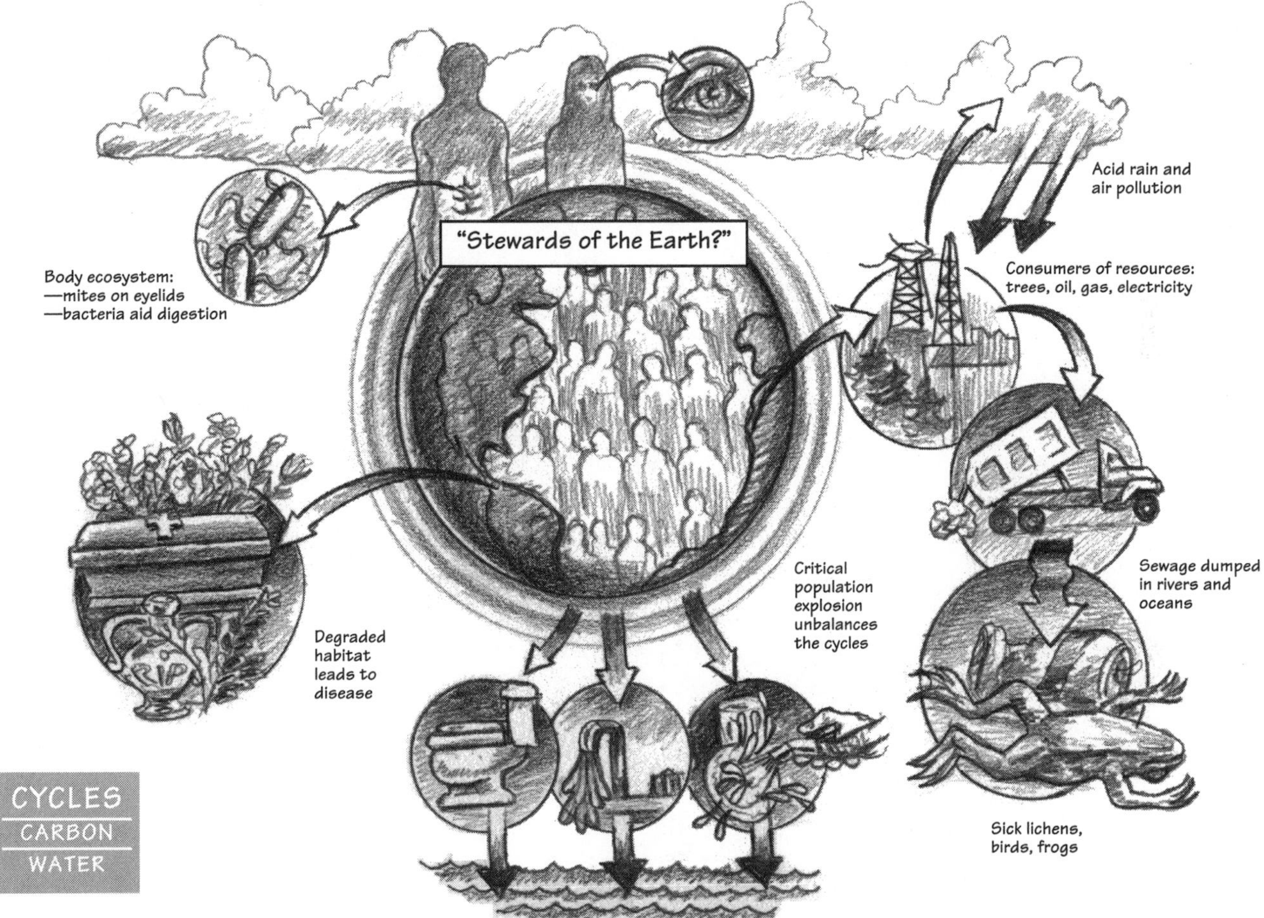

CYCLES
CARBON
WATER

HEALTH DEPENDS ON SYMBIOTIC RELATIONSHIPS

The term symbiosis has come to include the many ways in which nature cooperates, whether in parasitism (one destroys the other for its benefit), commensalism (both partners coexist without injury or aid), or mutualism (both partners benefit). We live in a paradoxical world in which attack and defense become a system of support and reciprocity. Parasites, for instance, bore into our skin (attack), but in living out their lives can produce starches and sugars that we need in order to thrive. Likewise with competition and cooperation. For a long time we've thought the world evolved only through competition, and even made it the basis of our economy. But nature teaches us that survival also depends on cooperative behavior. The kinship map for survival is tightly constructed, featuring a balance between predator and prey, death and life, that we have yet to accept.

THINKING LIKE AN

ECOSYSTEM

Bacteria

Microbiologist Lynn Margulis describes how bacteria are not only the building blocks of life but also indispensable occupants of every living being.

In the first two billion years of life on Earth, bacteria—the only inhabitants—continuously transformed the Earth's surface and atmosphere and invented all life's essential, miniaturized chemical systems. Their ancient biotechnology led to fermentation, photosynthesis, oxygen breathing, and the fixing of atmospheric nitrogen into proteins. It also led to worldwide crises of bacterial population expansion, starvation, and pollution long before the dawn of larger forms of life.

Bacteria survived these crises because of special abilities that eukaryotes lack and that add whole new dimensions to the dynamics of evolution. First, bacteria can routinely transfer their genes to bacteria very different from themselves. . . . Bacteria can exchange genes quickly and reversibly, in part because they live in densely populated communities. Consequently, unlike other life, all the world's bacteria have access to a single gene pool and hence to the adaptive mechanisms of the entire bacterial kingdom. . . . The result is a planet made fertile and inhabitable for larger life forms by a worldwide system of communicating, gene-exchanging bacteria.

This symbiogenesis, the merging of organisms into new collectives, is a major source of evolutionary change on Earth. The results of these first mergers were protoctists, our most recent, most important—and most ignored—microbial ancestors. Protoctists invented our kind of digestion, movement, visual, and other sensory systems. . . . These complex microscopic beings and their descendants even developed the first male and female genders, and our kind of cell-fusing sexuality involving penetration of an egg by a sperm.

The lichen is an exceptional example of symbiosis, in which two separate organisms—a species of fungus and a species of algae, coevolved into a single growth, dependent yet separate. The fungus extracts nutrients from harsh, lean environments; the algae take those nutrients and put chlorophyll to work in the food factory of photosynthesis, manufacturing carbohydrates, which in turn nourishes the fungus. The lichen is extremely sensitive to sulfur and hydrocarbons. Because it is one of the first to suffer from air pollution, its destruction signals a threat to the rest of us.

The health and survival of other species, such as birds and frogs, are also good indicators to people about when habitat is being sickened. We are so interconnected that when these creatures decline, we can be sure we do too.

Many symbiotic relationships between insect and plant further pollination, the bee and flowers being the most familiar example. Some fish feed on small organisms that dwell on other fishes' bodies; in this way the big fish get cleaned. Large animals, such as rhinos, are always seen with birds that eat insects off their bodies. Some birds help animals find food.

As in sports, a team can get ahead—until a new strategy is mounted against it. Cowbirds have learned to lay their eggs in other birds' nests in order to get the hosts to raise them. In this parasitic way they multiply their progeny and reduce that of others. So far they've succeeded quite well.

Sometimes when a non-native species ("exotic") is introduced to an ecosystem, and there is no natural predator or decomposer, the exotic species will grow wildly rampant. Such are zebra mussels, which voraciously gobble algae and starve out other mussels and clams, and gypsy moths. Only a few gypsy moths were brought into the United States from Asia; but they quickly multiplied and became a major problem. Exotics in this way can create huge imbalances.

Sex Is a Special Form of Symbiosis

The sex stories of plants, insects, birds, and animals are full of fascinating moments of symbiosis. Cells dock, exchange nuclei, and generate DNA. Insect queens may be serviced by many males. Males may inseminate many females. Sexual habits evolved to best assure the safety of the next generation. The "best" habits for reproduction vary with the species' ecology. For instance, most species are polygamous, but most birds are monogamous because raising baby birds is such hard work that the full-time efforts of both parents are usually required.

In ponds, water fleas reproduce asexually (by cell division) throughout the summer, giving birth to females only. As the pond becomes crowded, males are produced. These males mate with their sisters, which then lay fertilized, drought-resistant eggs capable of withstanding a journey to a new pond stuck to the leg or bill of a passing bird, or of overwintering in the mud and hatching the next spring to begin a new sexual generation. Their sexual pattern is determined by their plan for survival.

Sex will find a way. Female Bonellias, marine creatures, live wedged into clefts between rocks. They have a large trunk to gather foods and males. The male, much smaller, travels down the trunk and into the female's gullet, boring into her genital tract, where he fertilizes her eggs. When the larvae are born, they have no sex at first. If they settle on a rock, they develop as females; if they settle on a female trunk, they become males.

Snails, which are hermaphrodites, meet and press their feet together, caressing and secreting slime. Each at some point of the dance will dart its penis into the other.

Birds use songs to attract mates. Fireflies flash for sex. Males flash only for sex, females when they either want sex or are hungry. If a female firefly wants a male, she'll flash a subtle "come and get me" code. The length of the pauses between flashes signal different messages. A female will lure a male of another species by imitating his flash pattern. When she gets him, she'll eat him in order to absorb some of his defensive chemicals into her. These chemicals will help protect her against predators.

Female choice appears to influence the physical characteristics of males. One theory is that among insects the need to be sure that they have copulated with the right species has stimulated genital changes, such as little bundles of hair, hooks, sacs, and knobs, in males' appearances. Among other kingdoms,

The Sex Life of Birds

Paul Ehrlich

While the vast majority of bird species are monogamous, some are polygynous, and a few are promiscuous, forming no pair-bonds at all. A few others are polyandrous, with one female bonding with more than one male. Among the latter, typified by phalaropes, the usual color schemes are reversed, and it is the females that have the bright nuptial plumage. All of these mating systems are interesting. . . . but I want to focus on monogamy, for it has recently been shown that avian monogamy is not quite what it was once thought.

The degree to which monogamous males participate in raising their young varies enormously among bird species. In many ground-nesting birds such as shorebirds, gulls, and geese, males are full partners and even put their lives in jeopardy when predators threaten their offspring. Many songbird males bring food to their young or to the nesting female. In a few monogamous species, however, the father's contribution is minimal. Male Willow Ptarmigans help the female only by serving as lookouts against danger. And male Eastern Bluebirds only supply a place to raise the young by defending a territory containing a suitable hole in which to nest. . . .

It turns out, though, that the formation of a monogamous pair bond in birds does not necessarily mean that all sexual activity takes place within the social pair. . . .

One of the fascinating questions of avian behavior is why infidelity has evolved. . . . The immediate reasons for cheating may differ for males and females, but the basic answer is presumably the same for both. It evolved when unfaithful individuals had, on average, more offspring reproducing in the next generation than faithful mates. For example, in situations where a female can successfully raise the young with little or no help from her mate, the male has everything to gain from sneaking around trying to copulate with others. If he is successful in fertilizing the eggs of other females, his contributions to the next generation will increase. If he is not successful, he will, in most cases, have lost little. . . .

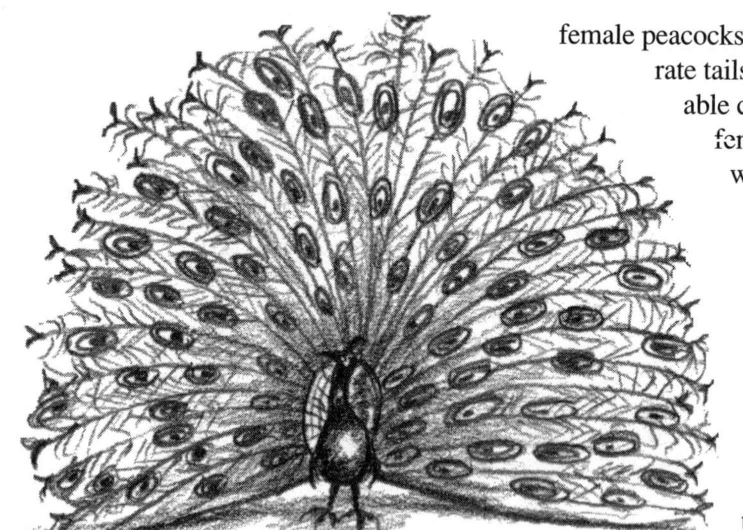

female peacocks prefer males with elaborate tails; grouse prefer expandable colorful throat sacs; female moose like bulls with large antler racks; and women like men with deep voices.

Both sexes are deeply motivated to perpetuate their species. Females are especially concerned with selecting the best father and with raising the young. Many males help care for the brood, but in some species they don't. Female crocodiles receive no help from the male. This seemingly ferocious creature becomes the model of maternal devotion. She will lay her eggs in a hole and cover them with sand, guarding them for several months without leaving even to feed. When the young are hatched, she digs them out, even helping them, with her huge mouth, to rupture their shells, and carrying them tenderly in her jaws.

Many sexual secrets have yet to be revealed, but we do know that successful breeding depends on climate cycles and protected habitat. Consider this story about the mating habits of horseshoe crabs and of migratory birds.

Sex and Gluttony on Delaware Bay

J. P. Myers

Late in May every year, sandy beaches along Delaware Bay in southern New Jersey and Delaware host unparalleled scenes of chaos, sex, and gluttony. . . .

Moving upward from the water, horseshoe crabs . . . resemble armor-clad tanks; their heavy outer carapaces protect them and their soon-to-be-laid eggs from numerous predators. At least eleven species of migratory shorebirds, three species of gulls, song sparrows, cowbirds, common grackles, mourning doves, house finches, and even starlings, house sparrows, and pigeons feed upon accessible eggs. But the list is not limited to birds. Nearby waters roil with small fish picking off eggs washed into the bay, while mollusks scavenge on the nearshore bottom. Many egg clusters are parasitized by a nematode worm, probably transferred from mother to egg at the time of laying. Mammals, too, join in the feast. Raccoon and fox tracks can be found along the beaches at dawn, and even mole mounds can be seen marking a trail from the dunes to the upper reaches of the intertidal zone where the eggs lie buried. . . .

Male crabs appear first as the tide starts to recede. They emerge from the bay to line the water's edge in a band several crabs wide, a crab or two high, and continuous for miles along beaches of suitable habitat. More than 10,000 can pack into just a half a mile, all waiting for the females to emerge and crawl up the beach toward a suitable egg-laying site. Each male vies to be the one to fertilize a female's eggs.

Females begin rising out of the bay within a hour of the turning high tide. As the females appear, the solid line of males breaks down into clumps of ten to fifteen or more males, each individual attempting to clasp a female's carapace with its pedipalps. Males scramble into heaps, male atop male atop female, washed to and fro by the surf. Sticking vertically out of the water, the tails wave back and forth like dancers in a chorus line. This is the moment when some crabs are flipped.

When a male succeeds in attaching to a female they inch up the beach together, the male holding on from behind. Both male and female must leave the water to mate. The female digs a hole, and then, as she lays her eggs, the male fertilizes them. Together they cover the egg mass with sand. Tides permitting, the pair then return to the water.

A single female may lay up to 80,000 eggs. Each egg is round, approximately a few hundredths of an inch in diameter, and may range from olive green to brown. The female divides her clutch into a number of clumps, each a tightly bundled mass from one-half inch to four inches across. Perhaps to confuse birds probing for eggs with their bills, the female mixes pebbles and bits of debris within each clump. After they are buried, the clumps of eggs lie clustered in a nest some two to eight inches beneath the surface. Laying activity is so dense that by season's peak, some beaches will have as many as fifty nests in each square yard of sand near the high-tide line.

Now the shorebirds begin to arrive. From the coasts of Brazil, Patagonia, and Tierra del Fuego, from desert beaches of Chile and Peru, and from mud flats in Suriname, Venezuela, and the Guyanas, shorebirds have flown days on end to reach these egg-strewn beaches on Delaware Bay. Some birds arrive in early May.

By the end of the first week of June, few birds remain.

Why the rush? Breeding grounds beckon in the high Arctic of Canada. There the migrants face stiff competition for breeding territories and only a few weeks in which to mate, nest, raise young, and prepare for southward departure. . . .

Enough is known to place the total at between 500,000 and 1,500,000 shorebirds within the bay in the latter half of May. During the brief stay in the bay, the shorebirds gorge ravenously on eggs. Some birds literally double their weight from fat accumulated. . . .

Delaware Bay underpins the entire migration system of New World shorebirds. Without these sites, the stupendous migrations undertaken by North America's shorebirds would be impossible.

Staging sites—separated by vast expanses devoid of suitable foraging sites—work as links in a chain connecting wintering areas with breeding grounds. Studies of food and habitat availability indicate that shorebirds have no alternative but to gather in these limited areas. . . .

The health of entire animal populations depends upon the continuing ability of Delaware Bay and other key staging sites to support migratory birds.

Delaware Bay from May 20 to June 2 evokes prehistoric epochs—horseshoe crabs from the depths, bird hordes from afar. How it has persisted through these centuries of human population growth defies reason.

Birth

Copulation inevitably leads to progeny and hence to populations. But population size depends on death rates as well as birthrates. In our country the death rate has gone down faster than the birthrate and therefore we have a growing population. Other parts of the world have higher death rates as well as higher birthrates. Unfortunately, as people have increased, we have forced other species out.

In 1950 the human world was one-half the size (2.5 billion) it is now. In thirty-five years the population doubled. Now it is over 5 billion and expected to be 10.2 billion by 2100. In everyday life imagine how this translates into hordes of people needing homes, driving cars, congesting roads, crowding in subways and stores, applying for jobs, waiting to get through to someone on the phone.

Q: How many people can the world support?
A: That depends on what you think about the quality of life. If you mean standing up and having one meal a day, the number is enormous. If you're talking about the way we live now—which means misery for half the people in the world, a so-so life for a few others, and an attractive life for a few hundred million—then I think we've already exceeded the maximum.

—Dr. Carl Djerassi, Stanford University

Just what the carrying capacity (environmental tolerance) of Earth is, and at what standard of living, are questions being borne by us now. Already one out of three persons does not have enough fuel to keep warm or cook food; one half of the world does not have sanitary toilets. Some people hope that human ingenuity will find the solutions we need in order to survive. This attitude reflects the notion that humans are at the center of the universe and lord of it, which has so often proven to be a vain misperception. Past experience in ecological history shows us that when a species reproduces at a rate that is beyond the support of its food, water, and territorial needs, its population crashes.

The health of females is imperative for the survivability of all kinds of ecosystems. It is hoped that human males will lend their support to women who want to reduce birthing and excessive consumption of resources.

Pat Waak, director of Audubon's Population Program, has written:

> Women are critical to this solution, not just as the bearers of children, but as the primary managers of the household and land economy. As infant mortality drops, women tend to have fewer children because they know the ones they have will survive. An increase in women's education, status and access to income has a direct correlation with reduced fertility.
>
> In the year 2000, four hundred million more women around the world will be of childbearing age and in need of contraceptive services. Promoting universal access to safe, affordable voluntary family planning will enable millions of women to make essential contributions to healthy families and to their local political and economic systems, as well as lead to stabilization of world population.

While women do the childbearing, men have often been making the decisions about how it is to be done. According to Stephanie Mills, author of *Whatever Happened to Ecology?*, things might be different if women were more fully compensated participants in the economic community and men had more responsibility for their offspring. "As it stands now, sexism is such a pervasive force in the world that a preference for male babies even drives Chinese couples into violations of their country's stringent one-child-per-family policy. And in most of Asia, North Africa and Latin America the death rate for young women is higher than that of men as a consequence of neglect."

Population's Insidious Impact on the Health of the Planet

Without a stable population, people become as endangered as other species. Achieving stability should be our goal.

The term "demographic transition" refers to the way birthrates and death rates have historically proceeded from high to low levels. Such studies of population trends can be useful but are not necessarily predictive. Transition appears to occur in three phases. At first, both birthrates and death rates are high or in balance, so little, if any, population growth occurs. In the second phase, death rates fall because of improved living conditions—better health care, increased food production, expanded social services—but birthrates remain high. In this stage, populations grow rapidly. Finally, in the third phase, economic and social gains combined with lower infant mortality rates reduce the desire for large families and birthrates decline. As the gap between birthrates and death rates narrows, population growth slows. When a population has reached the point of equilibrium between birthrates and death rates, and is able to maintain a state of zero population growth (i.e., there are no net gains in the number of people in a given population), then the population is said to have stabilized. Some countries have even advanced to a fourth stage in which death rates are higher than birthrates and population has declined.

The problem in the developing world is that most of these countries get stuck or "trapped" in the second phase of the demographic transition. Modern medical technology and improved diets reach these populations and decrease death rates before a modern economy has been able to develop and encourage lower birthrates. Populations continue to grow at a rapid pace, outstripping the ability of these countries to meet the needs of such a large number of people. This places undue stress on the land and natural resources as well as negating the modest economic gains of an incipient modern economy. Per capita income declines. If this trend is accompanied by declines in per capita food production (as in much of sub-Saharan Africa over the last decade), rising food imports will increase external debt, putting further stress on the economy. People lose all hope of attaining a better life for themselves, and political instability results. The population then has grown beyond the capacity of the immediate environment to peacefully sustain it.

Facts: The average American uses fifty-four times more resources than the average citizen in a developing country. Citizens of the developed world spend 14% to 30% of their income on food, while those in the developing countries spend 50% to 70%.

Overconsumption of resources as in the United States is just as hard on the land's ability to support life as is overpopulation in India. Degradation is caused by what we consume and produce multiplied by how many people we are, mitigated by the carrying capacity of the land.

Poverty, environmental degradation, and overconsumption of resources cause health threats on many fronts. Foremost is the systematic emission of greenhouse gases into the atmosphere from increased deforestation and burning of fossil fuels. The depletion of the ozone layer means not only more skin cancers in humans but destruction of plants, fish, and plankton without protective immune systems. Air and water pollution intensify. Diseases, such as AIDS, spread.

Health Threats from Radiation and Pollution

> I happen to believe that being a naturalist solves all problems and soothes all heartaches. It relieves insomnia, stops headaches and heals sundry discomforts, cures boredom, and offers an open-ended, continually expanding, fascinating exploration into distant deserts of the mind and hillsides of the heart
>
> —Ann Zwinger

A newly recognized illness has been variously dubbed multiple chemical sensitivities (M.C.S.), environmental illness, total allergy syndrome or, more dramatically and ominously, 20th-century illness.

—Robert Reinhold, *When Life Is Toxic*

How horrible it is to many nature lovers to find that their cherished landscapes, lakes, and rivers can make them sick.

Fact: U.S. industries pump at least 2.4 billion pounds of chemicals into the air every year. —EPA

Cancer is a major environmental hazard. The carcinogenic action of viruses, radiation, chemicals, and asbestos (the term for a number of fibrous silicate minerals) alters cells. Children, if they are exposed to a carcinogen, are more likely to develop cancer later in life than adults who are exposed to the same substance for several reasons: (1) cancers take a long time to develop, usually several decades, (2) children have more years than adults over which to be exposed more than once to carcinogens, (3) children are growing, during which time cells are dividing and synthesizing DNA, which can be subject to mutations and the propagating of mutations to many cells. Old people are also more susceptible because of their weaker immune systems.

Acid rain's impact on forests and water ecosystems is evident around the world; it also adversely affects human health. Coal-burning power plants are responsible for most of the sulfur dioxide. Automobiles are the chief source of nitrogen oxides. Natural rainfall has a pH level between 5.0 and 5.6. Below 5.0 rain becomes "acid." A pH of 4.7 or lower kills the tiny plants and animals on which fish feed, causing the fish to starve and die off. A key cause of death to forests is the spread of acid mosses and soils. Science writer Jon Luoma writes:

> The mosses produce organic acids which appear to gang up with inorganic acids in polluted rain to mobilize aluminum naturally present but harmlessly bound up in most soils. Additional aluminum appears to be falling out of the atmosphere bound to dust particles. The mobilized aluminum is toxic to the fine feeder roots of most trees. In fact, mats of killer mosses inevitably overlay networks of dead feeder roots. Furthermore, sphagnum acts as a sponge, saturating the soil just beneath the moss and creating an anaerobic, or oxygen-starved, soil environment, which also helps kill roots. The mosses occur naturally in forests and may even be part of an extremely slow plant-succession process that, over centuries or millennia, turns old forests to bogs. But the present moss invasion appears to be promoted and greatly speeded up by acidic rainfall.

The pollutants that cause acid rain are linked to cancers of the colon and breast as well as to many respiratory problems:

> Take a deep breath. Some ten million times each year you ventilate about a pint of atmosphere deep into trachea, bronchi, bronchial tubes, bronchioles, and moist alveolar membranes for purposes of gas exchange—oxygen diffusing into blood, carbon dioxide diffusing out.
>
> The majority of us don't pay much attention to the mechanics of it, vital function though it be. But a significant minority must. According to the American Lung Association, breathing problems afflict about one out of every five of us noticeably and many more of us subtly, insidiously, and eventually. The obvious problems can range from an intermittent condition doctors call "twitchy lungs" to the chest tightness and wheezing of asthma, to chronic bronchitis, cystic fibrosis, and emphysema. The less obvious problems include diminished lung function and a slow alteration in lung architecture that can go unfelt and undiagnosed for years but can nevertheless lead to miseries ranging from moderate discomfort to difficulty in exercising or doing heavy work outdoors, to lost workdays, to considerable pain, to early death.
>
> Some leading health experts warn that the same pollutants that cause acid rain can instigate or exacerbate these health problems. They also suggest that respiratory disease would almost certainly be diminished as a side benefit of any congressional action that attempts to protect lakes, streams, forests, fish, and wildlife by sharply reducing the pollutants responsible for acid rain.
>
> Each year in the United States, largely while burning or refining fossil fuels for energy, we dump some 25 million tons of sulfur dioxide, a pollutant gas, into the air. The furnaces of coal-burning utilities and nonferrous metal smelters are the major sources of this pollution. Sulfur dioxide, at sufficiently high levels, has long been considered a threat to public health—irritating the lungs, stressing the heart, and lowering the body's resistance to respiratory infections.

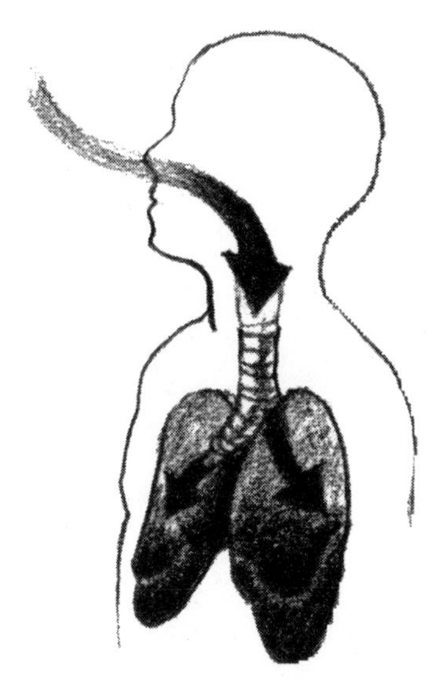

Audubon's Citizens Acid Rain Network helped to document the fall of acid rain around the world. Citizens with measuring kits sent in data from all over the United States and points in Russia, South America, and Canada. Audubon would also like to do an epidemiological study of coal plant emissions, using computer models correlating meteorological information with the transmitting of pollutants. These models were successfully used in tracking radiation and studying the epidemiology from the Three Mile Island nuclear facility at which releases occurred from a valve that was stuck open in March of 1979.

While no convincing evidence turned up that showed radiation releases at TMI influenced cancer risks, the National Audubon Society's current position on nuclear power is that it is still not a safe enough technology to invest in large scale in order to avoid carbon dioxide emissions. However, the option should still be open to some research, only as a possible "insurance policy" in case solar power unexpectedly fails us in the next century.

Because of the threat of releases into the air and water, many people are rightfully afraid of radioactive wastes dumped near where they live. In the *Audubon Activist* Chris Wille reported:

> During its four decades of operation, the American weapons-making industry has labored in secrecy, a closed, almost cultish society shielded by national security laws and bureaucratic laxity. The recent revelations have been stunning.
>
> At the uranium-processing plant near Fernald, Ohio, runoff from waste pits has carried tons of radioactive waste into drinking water wells and the Great Miami River. Radioactive particles have been released into the air. As early as 1958, the plant's operator told the government that tanks containing thousands of pounds of radium and other radioactive wastes had developed cracks and were leaking. The tanks are still in service, but the plant has been temporarily closed.
>
> At the Hanford plutonium processing plant in Washington, 200 billion gallons of low-level wastes have been dumped into pits, and huge quantities of radioactive iodine were released into the air. The Hanford reactors have been shut down.
>
> The Savannah River facility in South Carolina has contaminated the underground aquifer and once very nearly suffered a meltdown. The three 34-year-old reactors also have been shut down.
>
> In an unusual display of self-examination and candor, the Department of Energy made public a long list of accidents, equipment failures, procedural errors, contamination incidents, and disputes between engineers and management. As the list grew, so did the public's feeling of betrayal and mistrust.
>
> The questions now are what to do about the millions of gallons of radioactive waste stored in bomb-building plants, how to restore public confidence in the system, how to pay for the clean-up that is necessary, and how many other $100-billion tabs are about to come due from years of environmental neglect.

In the following excerpt from Terry Tempest Williams' book *Refuge*, she describes the horror of her awakening to cancer connections in her family:

> **I belong to a Clan of One-Breasted Women. My mother, my grandmothers, and six aunts have all had mastectomies. Seven are dead. The two who survive have just completed rounds of chemotherapy and radiation.**
>
> **I've had my own problems: two biopsies for breast cancer and a small tumor between my ribs diagnosed as a "borderline malignancy."**

This is my family history.

Most statistics tell us breast cancer is genetic, hereditary, with rising percentages attached to fatty diets, childlessness, or becoming pregnant after thirty. What they don't say is living in Utah may be the greatest hazard of all.

We are a Mormon family with roots in Utah since 1847. The "word of wisdom" in my family aligned us with good foods—no coffee, no tea, tobacco, or alcohol. For the most part, our women were finished having their babies by the time they were thirty. And only one faced breast cancer prior to 1960. Traditionally, as a group of people, Mormons have a low rate of cancer.

A little over a year after Mother's death, Dad and I were having dinner together. . . . Over dessert, I shared a recurring dream of mine. I told my father that for years, as long as I could remember, I saw this flash of light in the night in the desert—that this image had so permeated my being that I could not venture south without seeing it again, on the horizon, illuminating buttes and mesas.

"You did see it," he said.

"Saw what?"

"The bomb. The cloud. We were driving home from Riverside, California. You were sitting on your mother's lap. She was pregnant. In fact, I remember the day, September 7, 1957. We had just gotten out of the Service. We were driving north, past Las Vegas. It was an hour or so before dawn, when this explosion went off. We not only heard it, but felt it. I thought the oil tanker in front of us had blown up. We pulled over and suddenly, rising from the desert floor, we saw it, clearly, this golden-stemmed cloud, the mushroom. The sky seemed to vibrate with an eerie pink glow. Within a few minutes, a light ash was raining on the car."

I stared at my father.

"I thought you knew that," he said. "It was a common occurrence in the fifties."

It was at this moment that I realized the deceit I had been living under. Children growing up in the American Southwest, drinking contaminated milk from contaminated cows, even from the contaminated breasts of their mothers, my mother—members, years later, of the Clan of One-Breasted Women.

Aside from radiation and chemicals, oil spills are severely damaging to ecosystems. Since the *Exxon Valdez* spill, at least 10,000 spills have occurred in the United States, dumping 15 to 20 million gallons of oil. The 1991 spill in the Persian Gulf was the largest in history. Victims are tens of thousands of birds, mollusks, aquatic vegetation, as well as salt marshes, swamps, and coastal ecosystems. Oil spills also occur on land when pipelines leak or break. Oil is difficult to remove from soil, and it leaches into nearby creeks and lakes. Oil-contaminated water is unfit for use.

Water is also contaminated by sediments laden with toxic chemicals, such as DDT, PCBs, and dioxins. These chemicals, which settle on the bottoms of lakes, rivers, and estuaries, are released back into the water by dredging, wave action, and burrowing organisms. Once released into the water, they can cause skin lesions and birth defects in exposed wildlife and accumulate in the tissues of fish, shellfish, and birds. People can be exposed to these pollutants when they eat contaminated fish or shellfish, drink the water, or swim in it. Nature's wetlands, however, are natural filters, and protecting wetlands is one of the most effective ways of controlling runoff pollution, for wetlands filter out pollutants, especially nutrients that are choking the waters. Barnacles help filter water with their legs, snails and slugs rasp at algae and bacteria, fish collect food from the crevices of both rock and peat.

Each year 10 million deaths are directly attributable to waterborne intestinal diseases. One-third of humanity labors in a perpetual state of illness or debility as a result of impure water; another third is threatened by the release into water of chemical substances whose long-term effects are unknown.

—Philip W. Quigg,
Water: The Essential Resource

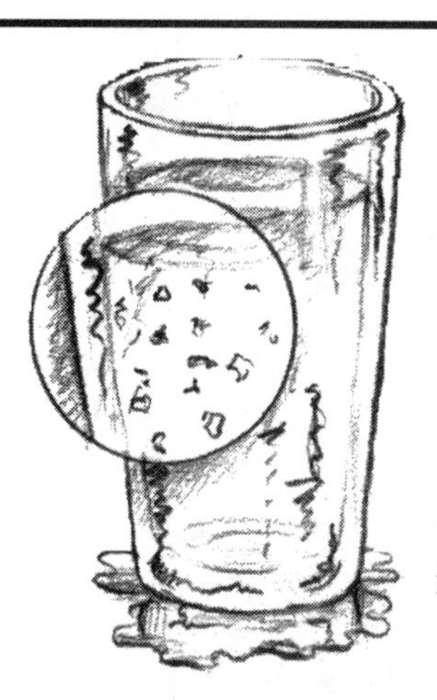

Part of protecting our water involves conserving it. In the past we have tried to increase water supplies through expansive water projects, such as dams, but the costs for these projects are exceeding their benefits. "New" water can be provided by using water for multiple purposes (for instance, water from baths and sinks could be used in toilet tanks or for washing cars), incentive efficiency programs for farmers, efficient irrigation techniques, and efficient showers and toilets. How water is treated is an important health issue. Consider chlorine:

THE PERILS OF CHLORINE

Marguerite Holloway

Sometimes a stream can be just too clean.

Chlorine has been used in the United States as a disinfectant for wastewater since early in the century. The lethal green gas or liquid is routinely mixed with treated sewage shortly before the concoction is discharged into streams, rivers, lakes, and coastal waters. Because it is such a potent poison, municipalities use chlorine to kill microorganisms that cause typhoid, cholera, and hepatitis, thereby protecting drinking water. The disinfectant is also used to protect waterways where people swim or harvest shellfish that could carry disease.

But chlorine can kill more than deadly microorganisms. In the 1970s nationwide reports of fish kills, of streams seemingly pristine but void of life, and of harmful compounds formed as by-products of chlorine prompted many scientists and the Environmental Protection Agency to investigate. In 1976 the EPA reported that chlorine poisoned fish and that it could lead to the formation of carcinogenic compounds, such as chloroform. It urged sewage treatment plants to reevaluate their use of chlorine and to use the disinfectant only when absolutely necessary.

If food and drink don't get you, things will.

Alternatively, a viable option is a natural sewage system. About 300 exist today. *(See article on page 29)*

We take in toxic metals through the food we eat, air we breathe, and water we drink. They tend to accumulate in the brain, liver, and kidneys and do their damage unnoticed.

Lead poisoning is an especially serious threat, particularly to the poor and working class communities in urban areas. Adults can suffer from hypertension and heart attack; fetuses from prenatal birth defects. Children can receive irreversible learning disabilities, retardation, and even death from eating chips of leaded paint. Lead emissions come from leaded gasoline, coal burning, mining, waste incineration, and manufacturing processes. Millions of people are poisoned when lead leaches into water from solder in the piping. It is also tracked onto carpets from outdoors and may be found in ceramics. Burning color comics or Christmas paper can also release lead.

Plants That Purify: Nature's Way to Treat Sewage

Christopher Hallowell

[The town of] Benton's rock-reed filtration works on the same principles as any natural system, and it goes like this: About 200,000 gallons of raw sewage a day are pumped into a 10-acre oxidation pond. A floating, wind-driven aeration device anchored in the pond's center churns and oxygenates the resulting soup, allowing aerobic bacteria to slowly break down the solids, which then settle to the bottom as sludge. In this airless gunk, anaerobic bacteria continue the decomposition process.

The time that bacteria are given to consume nutrients is one of the main differences between a natural and a mechanical sewage-treatment system. Benton's sewage remains in its settling pond for three months, ample time for bacteria to chew on and break down pollutants. An average conventional plant flushes sewage through the system within 48 hours or less, not enough time for bacteria to thoroughly reduce pollutants in the solids.

Benton's partially treated wastewater then is drawn by gravity from the pond into the rock-reed filter, a 1,200-foot-long gravel bed that wraps around two sides of the pond. Rows of nutrient-absorbing plants such as African calla lilies, water irises, arrowheads, and miniature and giant bulrushes grow in the waterlogged gravel. A small group of plants with large leaf surfaces, such as the African calla lily, can suck up water at a prodigious rate—about 1,000 gallons per day, depending on sunlight—and release it into the air through evapotranspiration. The wastewater spends a month trickling through the rocks and roots of the filter, where bacteria and microbes attached to plant roots further break down pollutants.

Visitors to these natural treatment plants will discover a few surprises: First of all, there is no unpleasant odor. In fact the systems emit no particular smell at all. Second, some of the facilities have become impromptu wildlife refuges; others double as recreational areas.

There are limitations to natural sewage treatment, however, and one is the large expanses of land required. This may be one reason the South has been the principal laboratory for natural treatment technology. Also, the systems depend on the continuous biological activity of plants and microbes, which slows greatly during the winter months in colder climates.

Yet in some instances small-scale natural sewage-treatment facilities—which principally consist of simple ponds and marshes—avoid certain problems endemic to conventional plants. They do not require the tons of concrete, the miles of iron pipe, the mechanical equipment, and the heavy doses of chemicals. Nor do they suffer the inevitable breakdowns.

They also help reduce the bane of all conventional plants: the quantities of sludge, often including heavy metals, that settle out of the liquid and must be incinerated or hauled off to overflowing landfills—both controversial practices. New York City produces enough of this goo every day, for example, to cover a football field to a depth of 10 feet. . . .

Another benefit of natural waste-treatment systems: They can be pockets of floral and faunal richness, of no small importance as wetlands vanish at the astounding rate of nearly 300,000 acres per year. . . .

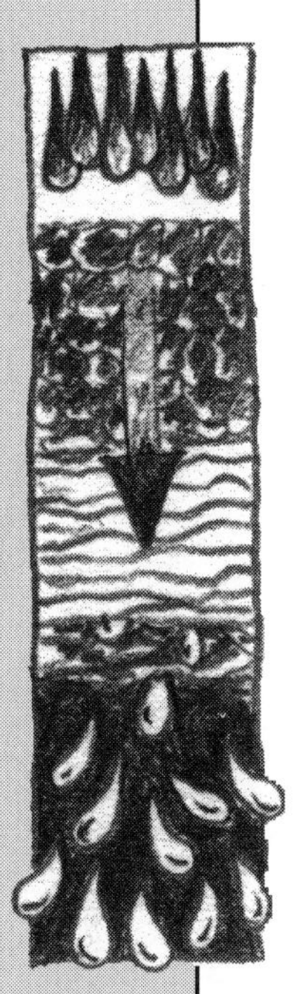

At least one person believes that large northern cities will someday be able to use natural waste-treatment systems, and he's constructed one as an experiment.

In a 30- by 120-foot greenhouse nestled in a grimy industrial wasteland a few miles from downtown Providence, Rhode Island, 16,000 gallons of raw sewage daily—the approximate output of 120 households—are rendered into water that is theoretically potable, but has not been approved for drinking by the state. Inside the greenhouse, rows of huge translucent cylinders contain water hyacinths, watercress, bald cypress seedlings, ginger, and philodendrons, as well as snails and tilapia (a fish species), which break down the waste as it flows from tank to tank.

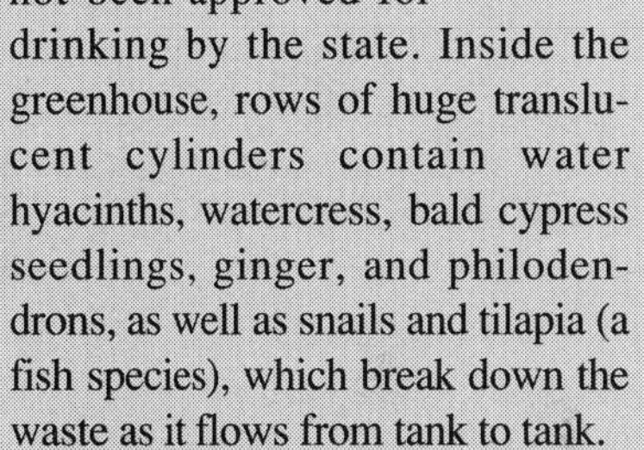

This unique system is the brainchild of John Todd, Ph.D., a visionary and former scientist at the Woods Hole Oceanographic Institution. Such small facilities, though viewed dubiously by state regulatory boards, have the potential to create permanent oases in urban neighborhoods, to treat sewage cheaply and effectively, and to recycle water.

The obstacles faced by any emerging technology result from human caution as often as ecological unknowns. But detractors serve an important purpose: They force the proponents of natural systems to make sure the alternative solutions to the nation's sewage problems are as safe and effective as possible.

NATURE'S CURES

Animals have adopted mechanisms to protect themselves against diseases and overpopulation. One way is dormancy. Bears, because their food source is scarce in winter, hibernate. While they sleep, their bones leak calcium, which is used to make new bone, and their urine is recycled to make essential proteins.

Sometimes a poison to humans can be a cure for others. Poison ivy and poison sumac, for instance, while immensely irritating to some of us, are food for quail, chickadees, flickers, mockingbirds, bees, mice, deer, and cattle.

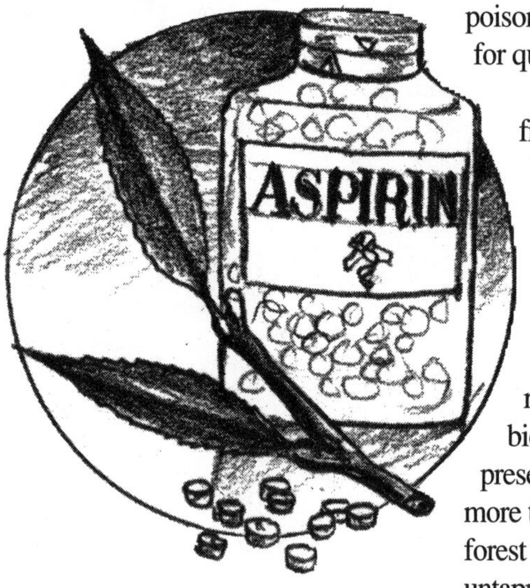

Many products in forests have yet to be discovered. Taxol—derived from the bark, foliage, and berries of the Pacific yew tree—has recently been discovered to have cancer-fighting qualities. As priceless rain forests and cultures are decimated and the loss of biodiversity becomes a crisis, ethnobotanists are scurrying with greater urgency than ever to study the numerous plants in South America and other parts of the world for their medicinal qualities.

Tropical rain forests occupy only 2% of the Earth's surface, in a narrow band encircling the equator, but they are home to over half of all the biodiversity found on the planet. To consider the potential for medicines, at present some 121 useful prescription medicines are obtained from plants, and more than one third of these come from rain forests. Yet fewer than 1% of rain forest plants have ever been analyzed for their possibly useful properties. The untapped potential for new medicines is staggering!

GIVING DEATH ITS DUE

When someone in our culture dies, his or her remains are usually subjected to expensive, and needless, funeral practices. Even cremation, with the ashes spread over hillside or seashore, means paying an undertaker and obeying certain laws. Many people are persuaded that no funeral is complete without embalming chemicals. On the other hand, Orthodox Jews hold that rich and poor alike should be buried in simple white shrouds and plain pine boxes. This practice affirms that all are equal in death and returned to the Earth from which they sprung.

Audubon scientist Carl Safina wrote about how nature maintains a balance between birth and death:

> Nature is very beautiful, but it is also, frequently, indifferent to life in ways we would call "inhumane" if humans were involved. For example, most animals have far more young each breeding season than would be necessary to produce in an entire lifetime in order to keep population levels constant. Some fish lay millions of eggs each year. In many species, including mammals, most of the young die before adulthood. This condition is so universal that Charles Darwin called it "the struggle for existence," and wrote extensively about it. The ecologist Paul Errington, who studied muskrats, spoke of the "doomed surplus." There are a number of unusual twists on the general theme of disposable individuals in nature. For example, the famed suicidal lemmings are actually individuals made homeless by population explosions. They are not suicidal, they are on a vain and desperate search for greener pastures.

In some birds, there is a system called brood reduction, the evolutionarily planned starvation of most chicks. Some birds that can find food enough for only one chick lay two or three eggs. It seems that one or two chicks are usually just "backups" in case the first-hatched dies. If the first-hatched lives, the other two often starve. In some eagles, the first-hatched chick always kills its younger sibling. This is not pretty by human standards, but it is a part of nature in a system that is vastly more stable than our own culture.

Nature teaches that death for species, including humans, is essential for the biological recycling of matter. Unless interfered with, when plants, animals, or humans die, the nutrients are released into the soil, water, or air to begin another journey through other living things. Ecosystems are dependent on this cycling between death and life. We come full circle. Death is the necessary stage for birth, or the renewal of life.

Thus, in nature "life after death" is a reality. The atoms of all matter have been recycling—living, dying, being reborn—for millions of years. Many a creature's atoms may be locked up for long periods in rocks, or miles deep in the ocean, or high in the atmosphere. Atoms from a dead opossum may become part of a berry bush. The processing of life after death is so hidden to us that we rarely think about it, and sometimes we don't want to.

Perhaps if death were understood ecologically, humans could learn from it better. In the seasons of nature, winter or the dying phase is essential for the setting of the seed for rebirth in the spring. We all like life to be in full bloom. But the dignity of death lies in its gift to life. Death makes more life possible.

EcoQuiz

Q: How can we be assured that our children will inherit a better life?

A: By adapting the Indians' rule that decisions be guided by consideration for the "seventh generation" hence.

Q: Suppose you own a pond on which a water lily is growing. The lily plant doubles in size each day. If the lily were allowed to grow unchecked, it would completely cover the pond in thirty days, choking off the other forms of life in the water. For a long time, the lily plant seems small, and so you decide not to worry about cutting it back until it covers half the pond. On what day will that be?

A: On the twenty-ninth day. You have one day to save the pond. (This example, from *Limits to Growth*, illustrates how exponential growth works.)

Q: Do you believe that we are responsible for other countries' problems?

A: Our dependency on other countries for food, minerals, metals, fabrics, medicines, fuels, etc., makes it in our interest to ensure a decent standard of living in other parts of the world. The need for water, air, and healthy ecosystems does not respect man-made boundaries.

Q: Why is poverty a health issue?

A: Where people are poor, there is rapid deterioration of renewable resources and ecosystems, followed by starvation, disease, and death.

E C O S Y S T E M   P A T H W A Y S

The clothes, cosmetics, and jewelry that we wear interfere with and alter natural cycles in numerous ways, as the picture suggests.

- Fibers grown for clothing constitute the single most polluting crop in the world. These fibers account for the heaviest use of pesticides and salinization of the soil, which in turn contaminates water and air.

- Since ancient times, cotton, wool, linen, and silk have been "natural" sources for clothing, but today clothing made purely from these materials is very rare. Not only do these materials undergo chemical treatment in their growth stages, but also in their manufacturing. The manufacturing of clothing, articles made from precious metals, and jewelry disturbs and destroys ecosystems. It also spews noxious chemicals into the air and into water cycles, which in turn affects climate cycles.

Furthermore, an article of clothing might contain fiber from Europe, decorations mined in Canada, and petrochemicals from Alaska or Kuwait. Very little manufacturing is done locally; it is based on an elaborate transportation and information network that requires vast quantities of fuel and other resources to operate.

- Demand for exotic materials, such as alligator skin, ivory, feathers, and furs leads to mass murder, theft, and possible extinction of wildlife, thereby throwing habitats and food cycles out of balance.

- The making of synthetic fabrics, such as polyester and acrylic, and cosmetics involves extensive extraction and processing of non-renewable fossil fuels, such as oil, natural gas, and coal. Once fossil fuels are burned, they are gone forever. Synthetics and cosmetics also use limited mineral resources, such as iron, copper, and aluminum, and nonmetallic minerals such as salt, gypsum, clay, and sand. In bold letters logic tells us that we must utilize renewable sources to run our industries. Many opportunities exist for recycling and reusing materials.

- Mining minerals and extraction of fossil fuels by definition intrudes on the rock cycle. Mining violently disturbs land, causes erosion and air and water pollution, and generates heavy solid wastes, some of which may be radioactive.

3

WHAT WE WEAR

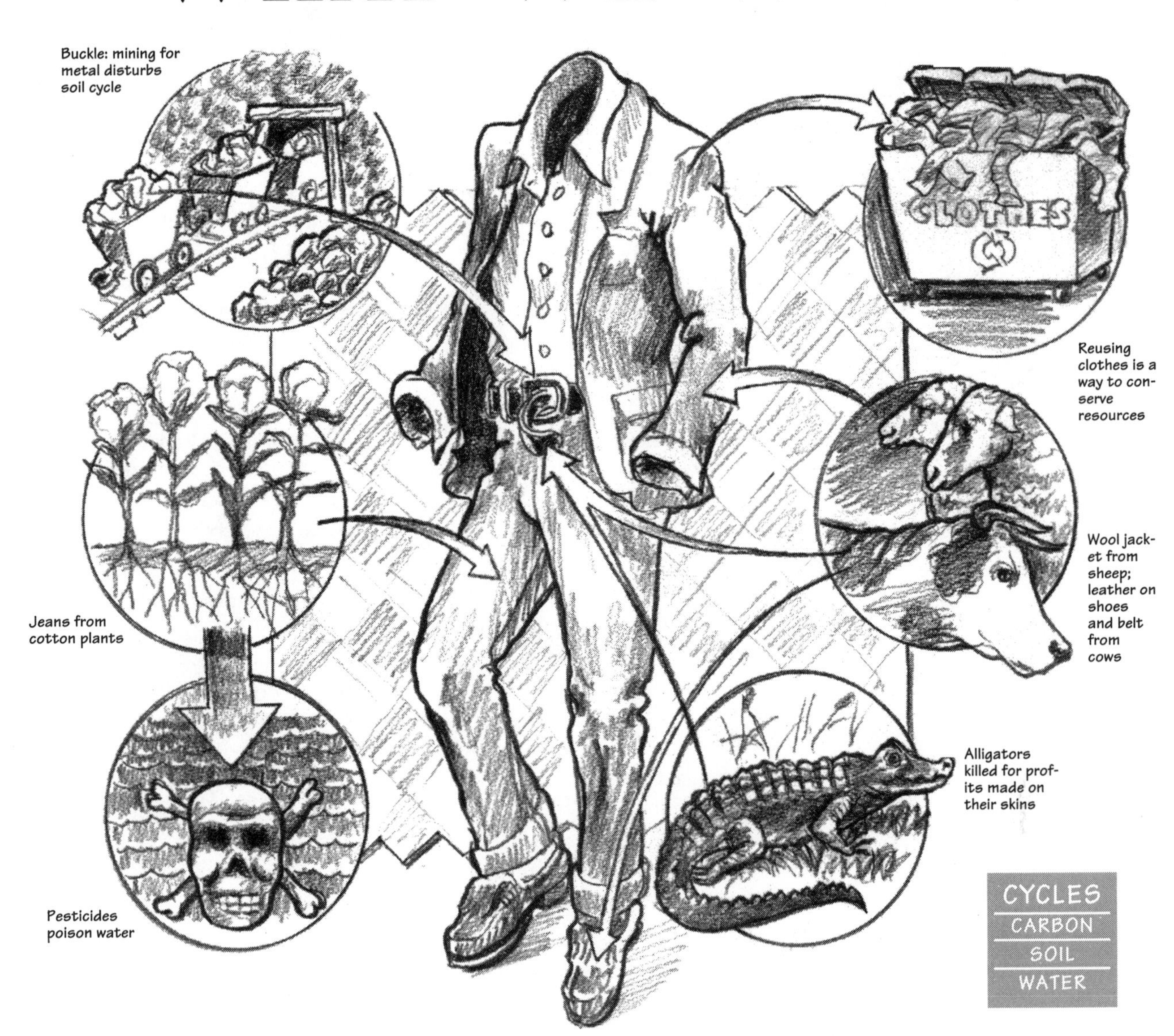

A GREEN REVOLUTION FOR CLOTHES?

Fashion trends are created seasonally but often with little awareness of the Earth's resources. However, signs of a new ecological consciousness are emerging.

Jeans are one of our most ubiquitous items of clothing. A pair of denim jeans consists of a heavy twill fabric made from tightly twisted cotton yarn, usually woven with blue threads mixed with white or with unbleached filling threads. If they have been stone-washed or pre-faded, they were subjected to a process that required strip mining of volcanic rock (pumice), which endangered the habitats of plants and animals. However, more recently some of the major brand-name companies have experimented with making jeans out of naturally colored cotton, unbleached and undyed.

New fibers are sure to come from large investments in biotechnology. New pest controls are being tested as well. For instance, a microbe is being created that will feed only on a single plant parasite and then self-destruct. Nonetheless, there are potential dangers in introducing life forms that are outside the known ecological web.

Lichens, which are invaluable in the weathering of rocks to form soil, have long been a source of dyes. A few are also important in the perfume trade, being used to make the many ingredients found in perfumes evaporate together, with pleasant effects. We must use them judiciously.

Color dyes and formaldehyde finishes to make fabrics colorfast, wrinkleproof, or shrink-resistant (as in "time-saving" polyester/cotton blends) are hazardous to people and to the environment. Most commercial dyes are made from nonrenewable petrochemicals and often involve heavy metals. Formaldehyde has been related to cancer, birth defects, headaches, respiratory problems, and insomnia. Yet, the technology exists for preshrinking cotton without chemicals and bleaching with hydrogen peroxide. To protect yourself you can wash synthetic fabrics before using them. Better yet, look for natural, untreated fibers.

Some of our most familiar items of clothing undergo toxic treatments to make them what they are. In the making of leather, for example, to remove hairs from the hides of cattle, buffalo, sheep, goats, pigs, horses, seals, or walruses, the hides are suspended in calcium hydroxide and sodium sulfide. Leather is then "tanned" in order to make it water-resistant, flexible, and permeable. Tanning, until the nineteenth century, was done with astringent organic substances derived from bark, seed pods, and roots of trees. Then the tanning action of chromium salts was discovered. Since the 1970s the most widespread method is a chromium sulfate solution, which at certain levels can be carcinogenic. In addition, the sulfur dioxide used eats away at the protective layer of ozone in the atmosphere.

To make gloves white, a synthetic tanning process is used that is based on treatment with hot sulfuric acid—of certain coal-tar derivatives such as phenol, cresol, and naphthalene, and their subsequent condensation with formaldehyde.

Nylon production is responsible for up to 10% of the nitrous oxide in the atmosphere, which is 200 times more effective than carbon dioxide in trapping heat (i.e., contributing to global warming). Think of all those nylon stockings produced over the last fifty years!

Dry-cleaning clothes presents a problem because the solvents used—PERC (perchlorethylene), ethylene dichloride, naphtha, and benzene—are hazardous to your health. They can be absorbed through your skin, get into your bloodstream, and can disturb your central nervous system. When you bring dry-cleaned clothes home, first air them in a garage or hallway to vent the fumes.

Perhaps animals know something that we don't, since many of them clean their fur, feathers, or fins with their mouths without hazard. Animal "clothes" may also be brilliantly and beautifully colored, as many birds and fish attest. Sometimes their outer coverings are protection against predators, such as the quills of porcupines or the camouflage colors of grouse. Polar bears have a unique system of insulation to protect them against the cold. Their hair traps ultraviolet light, converting 95% of it into heat, which their black skin efficiently absorbs.

Evolution produced an intricate protective wrap in a bird's feathers. Close to a bird's body soft downy feathers serve as "underwear." Then it has an overcoat of feathers that trap heat in cold weather and let it escape in warm weather. Constructed to help a bird fly, the lightweight feathers twist and grip the air, pulling a bird forward. Tail feathers can be used to steer or to slow it down.

The fashion industry thrives on the consumer's desire for something new every season: something that is costly, colorfully decorated, and easy to care for. But consumers can be smarter than these producers and save their money and the Earth's resources. Clothes don't have to be thrown onto the garbage heap every season. We can look for fabrics that last. We can buy pieces that can be interchanged to make new outfits. We can recycle clothes by giving them to the Salvation Army, passing them on to other people (in the United States or other countries), and frequenting thrift shops.

Catalog houses such as Seventh Generation and Smith & Hawken are experimenting with taking used wool sweaters, breaking down the fabric, and then reconstructing new sweaters out of the old wool. Some companies support indigenous economies, for example, by using tagua nuts from Ecuador for buttons. Others recycle tires, particularly truck inner tubes, to make bags, belts, and boots. Nike has devised a way to grind up old athletic shoes and use 20% filler (rubber) material in new shoes. Recycled fluff (fiber) can be used to stuff futons or pad briefcases.

Some synthetic fibers, such as acetate and rayon, are derived from the cellulose in trees and plants; they are examples of environmentally sustainable synthetics because they are renewable.

There is plenty of room for a green revolution in clothes, just as there has been in food. Just as consumers learned that not demanding picture-perfect fruits and vegetables meant that less pesticides would have to be applied to them, we can learn to appreciate clothes that are not perfectly white, so they don't need to be chemically treated. We can wear "organically grown" fabrics. If dyes are used, they can be derived from vegetables. Creative inspiration can come from the beauty of nature's colors. Clothes can be less ornate and more durable—less toxic to ourselves and the Earth. We can hope that as Promostyl, a fashion-trend forecaster, declared, "Ecology will become the single strongest inspiration to textile and interior industries as they move toward the end of the century."

Cotton—Truth Is Stranger Than Fiction

Cotton has been a longtime favorite of ours. The plant has been subjected to much head-scratching experimentation as cultivators tried to keep it from being eaten in the field by the boll weevil.

Let's look at the story of cotton from the point of view of the boll weevil, for a change. Prior to the invention of the cotton gin, which separated the seed from the lint, cotton was a backyard crop for British colonists of the New World. Seeking to step up production, farmers introduced cotton in the Texas Brazos River floodplain in 1822; cultivation quickly spread because of the crop's value to the European textile market. Originating in Mexico, a small gray insect with a long snout moved out of the semi-tropical Rio Grande Valley into the cotton acreages. It had the ability to fly—people said fifty miles or so—and thrived on the cotton fruit, which was now growing abundantly all the way to the Atlantic. Unbeknownst to the boll weevil, an entomologist named Frederick Mally was brought to Texas to figure out how to get rid of it, but farmers paid little attention to his ideas at this time (1902). Eventually other entomologists would be amazed throughout their careers at the invincible march of the boll weevil. Funds spent on saving the South's promising cotton economy were without parallel.

Facts: Cotton is produced in eighteen states from California to Virginia, with major concentrations in Mississippi, Arkansas, Louisiana, and Texas. Acreage increased from 8 million acres before the Civil War to 44 million in the 1920s. To the mid-1960s acreage decreased but yield increased. Since then acreage has remained the same but production has decreased. The average is around 12 million bales.

—a bale = 500 pounds

—Japan is the largest importer of cotton; the United States and the former Soviet Union are the largest exporters.

The weevil found the cotton plant eminently suitable to its purposes. After emerging from winter hibernation in brush and forest, it moved into the fruiting cotton, puncturing "squares" where it would feed and lay its eggs. Fresh weevils would emerge daily through the growing season. Beginning in July, 100 weevils could spawn 100,000 by September. Growers grew alarmed when weevils numbered 1000 per acre, and would fight to keep weevils under that level. The weevil had no particular predators to beware of, but it experienced a setback when farmers developed a strain of cotton that had such a short fruiting season that the weevil couldn't hatch its eggs. Only the weevils born in late summer or fall could survive a winter, so when farmers also began to shred cotton stalks, they too were threatened.

Those measures were not enough for the farmers. They wanted to use a cheap, easy-to-apply insecticide, and calcium arsenate was their big hope. After getting enough applications of this stuff on it, the boll weevil suffered, but the growth of aphids, as well as bollworms and other creatures, was encouraged in huge numbers, causing farmers to tear their hair out even more. Then came DDT. Its developer, Paul Mueller, was awarded a Nobel Prize.

DDT was applied systematically, even if the boll weevil population was low enough to not need it, even at the risk of damaging the soil and other crops. Farmers were denied agricultural loans unless they participated in the DDT treatment schedule. By the 1950s the boll weevil had retrenched to give its genetic structure a chance to recoup and develop a counterresistance. And, much to everyone's consternation, bollworms and tobacco budworms became impervious to DDT and could chomp away at will. The assault with DDT turned out to be like the U.S. military found the situation in the Vietnam war: the enemy was elusive and kept popping up elsewhere. Despite government pressure on farmers to increase yields, this has not happened since 1961; in fact, they have slightly decreased.

While a slew of industries and researchers geared up to eradicate the boll weevil, despite past evidence of the insect's ability to resist pesticides, others practiced Integrated Pest Management. Farmers never considered discontinuing pesticides, only using them in reduced quantities. Some had observed that the boll weevil could not multiply if it did not have a chance to overwinter, and they developed short-season schemes. But what worked in Texas might not work in California. Even though there are known to be forty-two insect and mite species that parasitize the boll weevil, these biological predators have generally been ignored.

Over a 200-year war the boll weevil proved it could not be defeated. It is a living symbol of nature's ingenuity under stress.

Certified Organic Cotton

Recently some farmers have started to grow cotton with the same guidelines that are applied to organic food. They use compost instead of petroleum-based fertilizers, natural bacteria and insects instead of toxic insecticides. Sometimes these crops of cotton end up being mixed in mills with ordinary cotton, but they are occasionally processed in an environmentally sensitive way. This means that the ginning, or seed removal, is done without toxic solvents and the cotton is cleaned without caustic chemicals (orange peels have worked fine) and not bleached or dyed. For cotton to be labeled "certified" means it has to be grown on soil that has been treated organically for three years and processed in this way as well.

Sally Fox is an inspiring woman who has a brand of cotton products known as Foxfibre and an organic line called Natural Cotton Colors, which is being used by many manufacturers. She began as an entomologist to grow brown cotton in California because these plants were more resistant to pests than commercially grown white cotton. For almost a decade she struggled to improve the quality of this brown cotton, and in experimenting with selective breeding found that her brown cotton was hiding genes for green, red, and pink fibers. Because she was aware that ancient farmers may once have cultivated a wide range of shades, she began working to unmask recessive genes for other colors. She is developing a yellow-green, a blue-green, and has plans for orange, yellow, copper, blues, and matte and glossy finishes.

She believes that the easiest way to tell if cotton clothing is organically grown is by the integrity of the colors. An undyed colored cotton will darken upon washing. She is afraid that some clothing manufacturers will dupe the public by dying their cotton brown and calling it organic.

In 1991 one tenth of Sally Fox's crop was organically grown; in 1992, one fifth. Other farmers have started to make a commitment to organically grown cotton; in Arizona, for instance, the acreage has increased tenfold.

If we buy cotton that has been labeled certified and organic, more money can be reinvested in the processes of farming and manufacturing. Knowing what you are buying, and shopping accordingly, can go a long way to changing the clothing industry. But don't be fooled by the word "natural"—some people call plastics "natural" because they are made from petrochemicals that come from the Earth!

Petroleum-Based Synthetics

Oil is the source of many "miracle" fabrics. The article on page 39 is about the lengths to which humans will go in simulating protection against the cold.

The oil industry would like to exploit all possible opportunities for oil until presumably it is all used up. Since oil is a nonrenewable resource, sooner or later we must develop alternatives. Does it make sense to continue to destroy ecosystems until the last drop is found? Audubon believes that preserving ecosystems and their inhabitants is more important for the future of the ecosphere, and that efforts should be made immediately to develop renewable resources instead. That is why it goes to battle for such areas as the Arctic National Wildlife Refuge.

Cosmetics

Animals and insects contain pheromones, which are chemicals secreted in their urine, and communicate sexual messages to one another. Is our application of fragrances very different? Creatures also allure with color, and so do we. The trouble is that our charm-enhancing products often contain a bewilderingly long list of chemicals, coloring agents, and petroleum products.

The cosmetics industry is changing, though. One option is for them to make the manufacturing process more efficient so that fewer chemical by-products pollute the atmosphere. And "green" companies are producing cosmetics that deplete fewer natural resources, did not involve animal testing, contain fewer animal-derived ingredients, and use less-wasteful packaging techniques.

Many natural-cosmetic companies use unscented or naturally scented herbs and other botanicals that they declare are grown without pesticides. In an effort to avoid using nonrenewable petrochemicals in their products, such companies do not include such petroleum by-products as mineral oil, petrolatum, isopropyl alcohol, and microcrystalline wax and paraffin. They do not excessively package their products, shunning boxes for their bottles and offering pumps instead of sprays. Some companies offer large economy sizes, allowing the consumer to refill smaller bottles and jars. More changes will follow if we support these companies.

Cosmetic labels are long and perplexing, but a few warnings are in order. Avoid using ingredients such as toxic formaldehyde, the irritant talcum powder, artificial colors, artificial fragrances, fluoride, and mineral oil as a lubricant. The most common

Redressing Plastic Man

Jessica Maxwell

Plastic shaped our lives; it was only a matter of time till we began to wear it. . . .

The invention of plastic outerwear was predictable. Hairless Chihuahuas that we are, humans have forever sought asylum from the elements under second skins, often ones that originally belonged to someone else: buffalo, for instance, or deer. Then about 10,000 years ago, our first Neolithic farms launched the age of agriculture, and soon we were braving the worst winter storms in our homespun woolens and beeswax-impregnated linen and cotton.

With all due respect to W. L. "Gore-Tex" Gore, we survived quite nicely. But visions of the perfect second skin danced in our modernized heads. The allure of infinitely durable, waterproof-but-waterwicking, windproof, heat-trapping, machine-washable, non-shrinking petro-fabrics proved far too tempting in the age of oil.

The first serious attempt occurred in the early '60s. . . . A normal cotton canvas coat was constructed with 40 percent nylon thread in hopes of creating a happy hybrid of water-repellent breathability. It was a benign enough beginning.

Within a decade nearly 70 percent of all fiber used by U.S. mills was synthetic. There followed the litany of outerwear fabrics and finishes we have today: polyester, polypropylene, Capilene, Ripstop, Teflon, Synchilla, Polar Fleece, Drys-on-ya Fleece, Lycra, Spandex, Sympatex, Ultrex, Entrant, urethane coating, Durepel, Thermax, and of course, the heavily patented great-granddad of synthoskins, Gore-Tex. . . .

What we have, then, are herds of nature-loving, outdoorsy men and women hauling their synthetic sleeping bags/tents/backpacks all over the last wild places on Earth, dressed in more-breathable-than-thou petrochemical clothing whose very existence threatens—through pollution and through utter nonbiodegradability—the enduring existence of nature itself. Which leads to scenarios of undeniable contradiction, like the group of ripe and bearded kayakers, fresh off a 10-day trip down the Colville River on Alaska's North Slope, who stood tall in their Lycra pants and neon petro-fleece alongside their polyethylene kayaks, and openly denounced oil drilling in the Arctic National Wildlife Refuge. . . .

I fear that a deeper reason for our petro-wear infatuation springs from identifying with technology, from wanting to become the machines we worship. The packs of touring cyclists that blast down my country road in the summer remind me always of many-wheeled machines, Lycra-thigh pistons pumping, helmeted heads like NASA nose cones. And then I think of Ireland, of old Irish men dressed in those beautiful woolen tweeds woven by local farmers in the winter, pedaling slowly on ancient bicycles, greeting everyone, noticing everything. . . .

But the spell, if not broken, has been cracked by Patagonia founder Yvon Chouinard, who commissioned an essay called "Reality Check" for the company's spring 1991 catalogue. "Everything we make pollutes. . . . Period," it begins, thus announcing Patagonia's decision to launch what it calls a major "environmental audit."

"It's a really big problem," explains Megan Montgomery, director of corporate affairs at Patagonia. "We've looked at wool and found that sheep are very destructive to the land when they're not raised in their native habitats. Then wool is treated with formaldehyde and many chemicals in the dye-fixative process, and it has to be dry cleaned. . . .

We made it a long, long way with wool, down, linen, cotton, and with silk, which is "perhaps the warmest per weight in natural fabric," reports Gail Weisman, sales head at R.E.I.'s Seattle store. She also recommends natural Duofold—cotton against the skin, wool on the outside. "It's lightweight, warm . . . for natural underwear you can't beat it." . . .

Let me tell you the story of Bonnie Duncan's sweater. Bonnie was kayaking among Washington's San Juan Islands when a strong current flipped her over. Having been taught that a capsized kayaker is a double drowning waiting to happen, her partner kept her distance while Bonnie hung on to her kayak with one arm.

"Aren't you hypothermic yet?" her partner hollered several times.

"No," Bonnie replied. "I'm just fine."

Finally convinced that Bonnie was in control, her partner paddled the both of them to shore.

"It was early March," Bonnie recalls, "and I never was the least bit cold."

She was wearing a big Irish fisherman's sweater, knitted from undyed Irish wool, and not a stitch of plastic.

preservatives in cosmetics are parabens: methyl, propyl, and butyl. These are widely accepted. Purists prefer plant-based preservatives, such as propionic acid from fruits, leaves, and wood pulp, but products with these ingredients are rare. If a product—food, for example—contains no preservatives, it is subject to dangerous bacteria, yeasts, and fungi. It would have to be refrigerated and used quickly. Because consumers are not likely to do that, cosmetics will continue to include preservatives.

JEWELRY

Our need to adorn ourselves is a personal pleasure; the trouble is that fashionable demands are made by an ever-expanding population that is ravaging the planet. Think of the ads we read daily for items such as gold chains, watches, and bracelets.

The mining of gold, diamonds, and other gemstones generally results in deforestation, soil erosion, accumulation of silt in rivers, and high levels of pollution. The most noticeable environmental problem with mining is scarring of the land surface. In open-pit mining, machines dig holes and remove ore deposits; strip mining, mostly used for coal, consists of opening the Earth's surface in strips. Ocean bottoms are also mined. In the United States nonfuel mining produces at least six times more solid waste material than the total amount of garbage produced by all our towns and cities.

THINKING LIKE AN

ECOSYSTEM

Rock Cycles

Sedimentary rock forms from erosion, the remains of organisms, and the impact of wind, water, and ice. Gravel, sand, silt, and clay are examples. Some sedimentary rocks, such as limestone, are precipitated from solution. Bituminous coal is derived from plant remains.

Igneous rock forms when molten rock (magma) wells up from beneath the Earth's crust, as during a volcanic eruption, and hardens into rock. Granite is an example. Rapid cooling produces a fine-grained or glossy texture, such as in basalt and pumice. The most popular gemstones—rubies, diamonds, sapphires—are found within igneous rocks.

Metamorphic rock is produced when rock is subject to high temperature, or pressure, or active fluids. Marble, anthracite, slate, and talc are examples.

As rocks are exposed to varied temperatures and conditions, they can change from one type to another. The ones near the surface are gradually crumbled by the weathering action of heat, cold, rain, snow, and ice. The extremely slow cycling of rocks is responsible for concentrating the mineral resources we have used. Because they take so long to be made, they are considered nonrenewable.

Going for the Gold

George Laycock

The fact is that most of America's easily recovered gold was claimed long ago. Up to 1985, according to the U.S. Bureau of Mines, this country produced some 325 million ounces of the metal. The states yielding the most gold were, in order, California, Colorado, South Dakota, Nevada, Alaska, Utah, and Montana. Since then Nevada has become the top producer. By and large, today's goldfields have been known for a long time and have been periodically worked and abandoned. Now the hills are again crawling with prospectors, often highly educated geologists who have returned to old mines hoping that new technology might render low-grade ores profitable.

Because of newly opened mines, many of them huge, gold production in this country grew by nearly three hundred percent between 1979 and 1986. . . .

This gold rush was predictable. The governments of most advanced nations agreed in 1968 to stop buying newly mined gold and the ban against American citizens' owning gold was lifted in 1975. After that the freemarket rush was on.

Small-time creekbed miners have a heritage of broken dreams, squalid poverty, and ghost towns. But they are out by the thousands, their shovels pecking away at the rocky hillsides and their dredges sucking sand and gravel from the bottoms of mountain streams. "People are going to dig up every creekbed in central and western Montana," says Dave Alt (Professor at the U of Montana's Dept. of Geology). He calls gold mining 'a disastrous industry. . . .'

But as Congressman Nick J. Rahall, chairman of the House Interior Subcommittee on Mining and Natural Resources, recently learned, the selling of federal lands at giveaway prices still goes on. Rahall asked that the General Accounting Office study recent sales and report its findings. He found the situation "outrageous."

For as little as $2.50 an acre buyers could often acquire "mineral" lands near resorts and casinos. Twenty such sales since 1970 handed over federal lands valued at an absolute minimum of $13.8 million for less than $4,500. In 1983 the government issued title to 310 acres in the Las Vegas resort area. For this land the federal treasury received $775. The Bureau of Land Management recently appraised the property at $1.2 million. In another case detailed by the GAO, the government sold 160 acres near the Keystone, Colorado, ski resort in 1983 for $400, or $2.50 an acre. Forty-four acres of this land was offered for sale in 1988 for $484,000, or $11,000 an acre. Frequently the interest in mining evaporates as soon as these public lands become privately owned.

Representative Rahall labels this practice "profiteering" at public expense and says it is easy to understand why, according to the most recent figures, 265 applications were pending for some 80,000 acres of public land. The GAO recommends that Congress repeal the antiquated law that makes this brand of profiteering possible.

In addition to squatters, thousands of recreational miners are flocking to the streams to seek their fortunes. In recent years California has sold its $10 mining permit to more than 8,000 recreational miners annually. Some belong to mining clubs that are run as businesses and stake as much public land and water as possible for the exclusive use of members. . . .

Some forest rangers work closely with these clubs to minimize their impact on public property, but others see them as a misuse of the mining law. "Those claims are not filed for gold mining but for recreation," I was told by a federal employee in California. "They produce no gold commercially. The person who starts the club is running a recreation business on public land. It's no different from starting a commercial campground on public land. If you get enough of these clubs sewing up public property for private use, the person who doesn't belong to a gold-mining club has no place to go."

Trout fishermen claim that recreational gold miners create sediment that smothers trout eggs and fills up hiding places essential to mayflies and other trout food species.

In many western states environmentalists are disturbed by a recently developed, high-tech, large-scale gold-mining technique known as heap leaching. It depends heavily on the use of cyanide solutions to separate infinitesimal amounts of gold from many tons of rock. It usually involves creating huge open pits or tearing down mountains by strip-mining. With heap leaching, a mining corporation can make a good profit on ore that yields a few hundredths of an ounce of gold per ton. Methodical test drilling tells the company the extent and value of the ore deposit before production begins, giving a sense of security that earlier generations of gold seekers were denied. It is generally a rather short-lived operation: The average heap-leach mine closes after six or seven years. . . .

(Continued on page 42)

(Continued from page 41)

Typically, the process calls for piling up crushed ore, then sprinkling it with what a Bureau of Mines publication identifies as a "dilute alkaline-cyanide solution." The liquid goes through a series of steps calling for various combinations of activated charcoal, heat, chemicals, and electrolysis to remove the gold. The cyanide solution is then reconstituted and used until the ore is completely leached. Beneath the heap is a thick plastic pad to catch the solution plus any gold it carries. And beneath this liner there is normally a layer of clay. The system usually includes a built-in leak-detection arrangement.

If all goes according to plan, the pond of cyanide solution is contained by the liners. But some liners leak, and this worries both miners and environmentalists. Careless workmanship in installing the pads can almost guarantee leaks. Massive weight and shifting earth will, in time, cause bits of rock, sticks, or other foreign materials left in the clay to wear through the toughest liners. . . .

Environmentalists are also troubled about wild animals killed at the ponds. If migrating waterfowl and shorebirds, stressed by long hours in the air, suddenly spot a shimmering open pond in the desert, they may settle in, drink a swallow or two of the cyanide-laced water and die. Nobody knows how many birds are dying because of these ponds. In Nevada alone in recent years, more than 4,500 have been found dead around them. But these are figures volunteered to the Nevada Department of Wildlife by the miners themselves, and many see the reported numbers as a small fraction of the actual kill. "I think I'm as honest as they come," one biologist pointed out, "but why would I volunteer information that my gold mine is killing birds and risk getting into trouble? After all, it's a federal offense to kill migratory birds." Officials worry most about small, poorly financed companies and admit they don't always have a handle on what these outfits are doing. . . .

It is common for a heap-leach operation to recover gold for around two hundred dollars an ounce and either sell it at current prices, which late in 1988 hovered around four hundred dollars an ounce, or hold it until prices rise. Several environmentalists assured me that they are not against gold mining, which employs a lot of people; but they added, "The miners' profits are huge, and they can afford to put something back.". . .

Part of the threat from gold mining stems from the eagerness of distressed communities to embrace new industry at any cost. . . .

In the meantime, abandoned underground mines continue to leak acid waste into mountain streams. Montana environmentalists point with sadness to Beartrap Creek. In 1975 the flooding creek breached the dam holding the tailings of the old Mike Horse Mine and flushed the accumulated minerals downstream. The dam was rebuilt, but waters from the underground mine continue to escape into the creek, feeding it more toxic minerals, turning its water the color of rust, and even destroying vegetation along the banks. Beartrap's mineral brew empties into the Blackfoot River, and biologists suspect that it has been a major factor in the decline of fishing there in recent years. . . .

Meanwhile, the market for gold continues to be strong. The government projects an annual 2.4 percent increase in demand through the rest of this century. As long as the price remains high, gold seekers will work the hills and streams.

When mines collapse, roads buckle, houses tilt, and groundwater systems are disrupted. Mines and quarries are often abandoned. In the smelting of the metal from the ore, air pollution is often uncontrolled. Hydrogen sulfide kills vegetation, and there are totally gray landscapes for miles around smelting complexes. The large amount of cyanide used in processing gold can cause widespread water pollution and is toxic to fish. The burning off of sulfides in the gold produces sulfur dioxide, causing acid rain. Moreover, the supplies of our metals cannot last at the rate we are extracting them.

Wildlife Trade

Joel Vance wrote:

> **There is a great deal of pleasure in designing your own staff. I make mine of eastern redcedar heartwood. It's hard as flint, and brittle. But the glowing red wood needs no stain to beautify it, and a desiccated sapling, spared by circumstance from rotting before I find it, seems to have been created to be a staff. Its knobs provide wonderful handholds as well as individuality, and there is satisfaction in whittling out the sapwood, using a pocketknife and, especially, lathe turning tools which are heavy and keen and often shaped just right for gouging the valleys between the knobby ridges. I use a Stanley Shur-Form rasp for much of the work. Finally I sand it and apply a finish of tung oil. . . .**
>
> **A staff really is more than a handy tool, not merely the equivalent of, say, a hammer. It is more like an amiable old dog who plods along uncomplainingly and listens to everything you have to say.**

We like to take materials from our environment to suit our purposes. But, in our infinite folly we threaten the existence of certain wildlife just to obtain showy accessories and fetishes, such as coral, plume feathers, and horns. Because of illegal poaching for the aphrodisiac properties of their horns there are fewer than 4000 rhinos left. Soon we may see the last herd of wild elephants to roam Africa. We don't need ivory jewelry. Scrimshaw makes a good substitute. Many crocodilian species are sacrificed to supply a black market; their skins end up being sold as alligator belts and bags. Only the most knowing consumer would be able to recognize the illegal skins.

The National Audubon Society was founded in response to the large-scale slaughtering of birds during the nesting season to decorate women's hats. In 1886 Mrs. Augustus Hemenway scanned her Boston Blue Book and contacted those women most likely to wear feathers, plumes, and even entire birds on their heads. Barely one month later, a group of concerned socialites, sportsmen, and ornithologists met to organize as the Massachusetts Audubon Society. By 1889 similar alliances had sprung up in sixteen other states from Maine to California.

At that time virtually no laws, ethical or political, protected wildlife from unrestrained exploitation. Egg collectors were voracious robbers. Once they located a nest, it would be robbed repeatedly until the female gave up. George Grinnell became the first big-game hunter to plead for bird protection through the pages of his *Forest and Stream* magazine, and the public, once educated, responded in great numbers to conservation legislation. Theodore Roosevelt was a champion of the cause, establishing the nation's first wildlife refuge in Florida in recognition of the needs of migrating and nesting birds. As the years advanced, protection of birds expanded to include the protection of habitat, including air, water, biodiversity, and energy resources. (For a full history of Audubon, see Frank Graham's *Audubon Ark*.)

A woman who was able to accompany her husband on a seal-hunting expedition in the Pribilof of Islands off the coast of Alaska wrote in her diary about the effect that slaying had on the men who did it:

> **The men have begun the final week of killing. . . . The work is at its heaviest in order to get it all done soon, for the pelts get too gamey if not taken early in the season. Even this is a bit late, except that the weather has remained cool enough to prolong the work. Though I am spared the actual sight of the slaughter, the noise is so terrible that I am constantly aware of what is going on. It is an awful time to live through. The whole idea upsets me, and I am in a state of nerves that has me worried.**
>
> **While I watch John grow more morose and disgusted with his job, I have sensed a rising tension among the men of the company and among the natives. Perhaps the men have become so restive because of the very bestiality of the job they have to do—selecting the animals, directing the drives to the killing grounds, directing the flensing operations, making sure the skin is taken the moment the animal is killed, being right on the field in the thick of the blood and offal, coming in spattered with blood and dung, and reeking of decaying carcasses with almost nothing but the putrefaction of death in their eyes and minds and nostrils.**
>
> **Each night, by the time we arrive for dinner at the Lodge, they are gruff, less careful of their language, and less carefully attired. Though I have seen none of them take a drink, I have smelled alcohol on some breaths, and I have noted many bloodshot eyes.**
>
> **—Libby Beaman**

These men, it seems, denied the revulsion they felt for what they were doing. We have the power to act differently on behalf of our kindred—seals, who cannot speak for themselves. While thousands of them are killed annually for subsistence by native peoples, we can nonetheless safeguard those remaining from excess commercial "harvesting."

EcoQuiz

Q: How can I best plan my wardrobe?

A: Remember, progress, not perfection, is the goal. Look for fabrics that are "certified organic"; that means at least that they have been grown and manufactured without polluting chemicals. Avoid colored dyes as much as possible. Neutral colors are very attractive. That way you need fewer clothes. Wear recycled clothes.

Q: When oil and mining industries claim that the impacts of development will be minimal, what is the conservationist's response?

A: Impacts are never minimal. Even when environmental-impact studies are conducted, they are likely to be inadequate because we just don't know enough about the complexity of ecosystems. However, there is one area that cannot withstand any compromise, and that is wilderness. As author Anne LaBastille says, "The integrity of true wilderness can be maintained only by preserving it untouched by human structures, vehicles, and machines. Being wilderness is like being pregnant. There is no such thing as a little bit wild, or a little bit pregnant."

PART II

Home Ecology

E C O S Y S T E M   P A T H W A Y S

Home is where our heart is. "Hearth" is a word that combines "heart" and "earth." The word "ecology" comes from the Greek, meaning "study of homes." We want our private rooms to be safe and secure, warm and comfortable. When we also regard the Earth as our home, we feel responsible for keeping it clean, safe, and comfortable for all its inhabitants. Our personal home is set in the matrix of our larger home, the ecosphere. Let us count the ways:

- Electricity for lighting and appliances is powered by utilities that mostly draw upon coal, oil, or uranium. Electric transmission lines emit low levels of electromagnetic radiation.

- Oil, coal, or gas-fired heating/cooling systems depend on extracting resources from the Earth, which disrupts coastal habitats and pristine wildernesses. The transport of oil, coal, and gas is inefficient and burns additional energy. In our homes, lack of wall and window insulation as well as inefficient boilers and radiators also consume excess energy.

When fossil fuels are burned, only part of the energy is utilized; much is returned to the atmosphere as heat, along with combustion byproducts, such as carbon dioxide, carbon monoxide, sulfur oxides, hydrocarbons, nitrogen oxides, and solid particles. In this way the carbon cycle is disturbed, and the dispersion of sulfur and nitrogen oxides contributes to acid rain. The combined pollutants produce acidified clouds. When it rains, soil can be acidified to a depth of three feet, thus killing roots. The foliage on tree leaves dies because the normal process of photosynthesis is slowed and essential nutrients washed out. In this way multiple forests have perished.

- Unused solar energy—huge roof areas in towns and cities could collect solar instead of just absorbing it, reflecting it, and re-radiating it as heat into the sky.

- Our garbage is hauled off to landfills or incinerators, where its toxins may leach into the ground or, when burned, contaminate the air. Some sewage and wastewater enter the ground. Septic tanks, into which home toxics have been dumped, also leach underground into the soil and water, eventually contaminating food chains.

- The construction materials of our homes—floor- and wallboards, as well as furniture—come from faraway forests and factories. The glues, adhesives, and solvents waft toxic gases into the air. Asbestos insulation and lead paint are health hazards.

- Synthetic carpets are made with polyester fibers derived from oil that took millions of years to form in the Earth's crust. Our demand for such products exceeds the rate at which oil is naturally renewed.

- Pollen, spores, and bacteria are carried in our dust. Mites, mice, and moths compete with us for food and for the fibers of our clothing. Birds, such as sparrows, nighthawks, and barn owls, like to build nests around our homes. Spiders trap insects in webs built into corners of our rooms.

- Tap water comes to us from reservoirs fed by rivers and streams into which contaminants have been released. Further chemicals are added in the purification process. Wastewater, containing many kinds of detergents, goes out to septic tanks, leaches into the ground, and wends its way to streams and rivers. If the water is conducted to municipal sewage plants, sludge and phosphates are often dumped directly into seas and wetlands, polluting marine life.

It is possible to have healthier homes, with more environmentally benign interiors and less destructive impacts on our extended home. The home and its local ecosystem can be treated as a whole unit in planning the recycling of air, water, energy, and materials.

4 OUR HOMES

Do Forests Have to Pay the Price for Your Furniture?

Where does furniture come from? Often from tropical forests, where precious mahogany, teak, rosewood, and other woods have been logged so irresponsibly that tropical deforestation has provoked a worldwide outcry. Environmental groups are urging consumers to ask questions about the wood they buy, to use only recycled wood, and even to try plastic wood. In Europe, consumer pressure has already led to tropical wood labeling so that potential customers can be alerted to what they are buying.

The Rainforest Alliance has begun giving "Smart Wood" certificates to tropical wood from timber that has been cut responsibly. The designation is based on whether the harvester has maintained watershed stability and erosion control, has practiced sustainable-yield forestry, and has had a positive impact on the local community. The initial list includes plantation-grown teak, mahogany, rosewood, and pine harvested by the State Forestry Corporation in Java.

Alternatives to harvesting precious woods exist, and more can be found. The Yanesha Forestry Cooperative in Peru, with the aid of the World Wildlife Fund, uses a more environmentally sensitive technique, a shelter-belt system, for harvesting timber. A California company currently buys the wood for making musical instruments. The wood for teak furniture can come from plantations, rather than virgin forests.

Although there are no hard-and-fast rules for buying wood milled from domestic timber, here are some tips:

- Avoid buying cedar shakes for siding or roofing and use cedar boards instead. Cedar shakes are almost certainly from old-growth cedar trees.
- Use laminated boards rather than solid wood. Laminated beams, such as structural supports for floors and roofs, on the average contain about 80% second growth and 20% old growth, while solid beams are from old-growth trees, often from the ancient forests of the Northwest.
- Softwood doors and window frames are just as serviceable as those from tropical hardwoods. Softwoods include cedar, fir, hemlock, larch, pine, spruce, and yew. They may be preserved with natural resin-oil stains, varnishes, or paints.
- Sustainably grown hardwoods include: alder, apple, ash, aspen, beech, birch, blue gum eucalyptus, elm, hickory, lime, maple, oak, pear, poplar, sycamore, and walnut.
- Encourage home improvement stores to buy only sustainably grown and harvested woods.

Durable attractive furniture can also be made from rattan, reeds, cork, and bamboo. *(See article on page 51)*

Unfortunately we have grown accustomed to living in rooms with a variety of materials that contain hidden health hazards; they are also destructive to the environment. Once we know what they are, we can make better choices for our personal well-being. Do we really want to have synthetic carpets, foam-filled bedding, chemical air fresheners, synthetic paints, and vinyl wallpapers? For one thing, polyurethane foam in furniture is a fire hazard. No-iron sheets most likely contain formaldehyde, and petrochemical-based paints and vinyl wallpapers add to the chemical load. No, better to use natural fabrics, if untreated and unbleached. Latex foam, wool, and down make better stuffings. Ditto, paints made from plant oils, such as linseed, and natural ingredients, such as chalk, India rubber, and pure aromatic oils. A product called "lazur" is a porous shellac that emits no synthetic fumes and is biodegradable. Turpentine substitutes exist for thinning paints.

We've gotten into the habit of using many harmful household cleansers. With a little effort we can change our ways. (Remember, you don't have to do everything at once; a step at a time is fine for long-lasting results.) As a rule of thumb, use phosphate-free, biodegradable laundry powders and dishwashing liquids. Avoid chlorine-based scouring powders. For furniture polish buy pure beeswax polishes. As an alternative to metal polishes that contain highly irritating ammonia and sulfuric acid, here is a suggestion from *The Natural House Book.* Lay cutlery on aluminum foil in pan; cover with 2–3 inches of water, add 1 tsp. salt and 1 tsp. baking soda. Boil 2–3 minutes, rinse, and dry. Brass and copper can be cleaned with a paste of lemon juice and salt.

THE MOST USEFUL PLANT IN THE WORLD

James Simmons

Consider, if you will, a species of bamboo called Phyllostachys bambusoides. This plant blooms once every 120 years and then dies. We know from ancient Chinese records that P. bambusoides flowered in 919, and that it has flowered at roughly the prescribed intervals ever since. In between it reproduces asexually by sending up shoots from underground rhizomes. The fact that it reproduces faithfully according to a 120-year-old clock is wonder enough. But even more curious is that all specimens of P. bambusoides bloom together, controlled by the same clock, no matter where in the world they grow or when they sprouted. Thus, in the late sixties, the last time the plants of this species flowered, they did so simultaneously in China, Japan, England, Russia, and the United States.

Bamboo is the most useful plant known to man. The Japanese alone have discovered more than 1,500 ways to use the bamboo that grows profusely in their country. The asthmatic's labored breathing can be calmed with a bamboo potion, and a bamboo salve will soothe irritated skin. Bamboo is delicate enough to be shaved into phonograph needles yet strong enough to form cables and dams. When Thomas Edison was looking for a proper filament for his first electric light bulb in 1880, he tested more than 6,000 materials before settling on charred fibers from the common Japanese madake bamboo. (Remarkably, the first bulb still burns today, in the Smithsonian Institution.) . . .

In Central America a major threat to bamboo comes from cattle grazing. After a stand has bloomed and died, the young shoots of the next generation are completely unprotected. Cattle then eat everything in the area. And you lose your bamboo.

One animal species that could be wiped out by the loss of bamboo in its habitat is the giant panda of China. The panda's chief threat continues to be man himself and his tightening encirclement of the animal's final refuges. For the moment, however, the threat is not of man's making but of nature's. The giant panda eats between forty and ninety pounds of bamboo a day. Approximately one-quarter of the 1,000 wild pandas in China are threatened with starvation because their favorite food, the arrow bamboo, has begun a once-in-decades flowering cycle in which the edible adult plants wither and die. It will be several years before the new plants that are spawned mature.

In our homes we also bring in and take out a lot of trash, in addition to generating wastes in the kitchen and bathroom. Nature has an efficient waste-management system, in that decomposers (nature's garbagemen) break down human and animal biological waste into carbon dioxide, water, soil, sulfide compounds, and methane, among other substances. However, we have not devised technological decomposers to work on man-made products such as old refrigerators, plastic packaging, plastic-coated papers, metal cans, and glass bottles. Ideally we would cart our wastes back to solar-powered factories for recycling, in vehicles powered by renewable forms of energy, such as alcohol. At present recycling plants use energy sources for electricity such as gas, oil, and coal, or nuclear or hydropower, which stress the environment.

Audubon's ultimate goal is to put a sustainable system in place. The first step toward that goal is for individuals to learn how to recycle, reduce, and manage their "waste stream." People who have tried the following method have been amazed to learn not only how many different kinds of materials they have been accustomed to toss out thoughtlessly, but also how many materials can be redistributed.

Audubon's Seven-Day Garbage Test

The following test shows how large a fraction of our trash could be recycled or composted if appropriate programs were implemented. It emphasizes the correct disposal of household hazardous waste, which should not be going into the waste stream, and mixed materials that manufacturers should not be using until recycling technology can handle them.

Here is how the test can be done at home. It takes only one week of separating and saving trash in specially marked bags or boxes. Treat it as a game and it can be fun! The first step is to start with cleaned-out trash cans and wastebaskets and a supply of large thin garbage bags. A scale should be used—a postage scale that will weigh up to 10 pounds will do. Or, a heavy object can be put on a bathroom scale before weighing the garbage. The weight of the object should be subtracted from the total. Sort trash as follows:

BAG 1: *Returnables.* All materials that can be returned or recycled in your area.

Note that Bags 2–7 contain waste that could be recycled or composted in town programs.

BAG 2: *Organic Waste.* Food and fluids without any packaging whatsoever go into this bag. This waste may be stored in the refrigerator in a special bag in order to keep odors down. All of this material could be composted in backyard or municipal facilities and turned into rich humus.

BAG 3: *Paper.* Dry paper, including magazines and newspapers. This is all recyclable. Even glossy magazines can now be recycled into newspapers.

BAG 4: *Plastic.* Plastic, including bottles and wrappers. All of this material is capable of being recycled.

BAG 5: *Glass.* Bottles and other glass waste. All of this material could be recycled. Much of it could be rewashed and used without being melted down.

BAG 6: *Metal cans.* All cans that are not currently recyclable in one's community. This material could be recycled. Include aluminum foil here.

BAG 7: *Wood Waste.* This includes old furniture and other types of discarded wood. Wood furniture currently comprises 5% of the waste stream. Wood waste can be processed into chips.

BAGS 8–12 contain waste that is not recyclable with today's technology. (This waste should be reduced and reused wherever possible.)

BAG 8: *Paper/Food Waste.* This is for paper that has held food. If heavy metals can be removed from the inks used in paper packaging, this material could be composted along with the material in Bag 2.

BAG 9: *Plastic/Food Waste.* This is for plastic and any other substance but paper that has held food. The plastic in this bag can be recycled, although the fact that it may have to be thoroughly washed makes the process more expensive and water-intensive. If possible, food-contaminated plastic should be washed and put into Bag 4.

BAG 10: *Mixed Materials.* Products that contain more than one type of substance, i.e., milk cartons (paper with a plastic lining), and envelopes with glassine window. These materials are not currently recyclable, but regulations could minimize their use. They should be used less so that there is less to throw away. Manufacturers should be pressured to eliminate these products until recycling technology can handle them. Takers of this test should be encouraged to look at the types of things they are throwing into this bag. Can they cut down on anything? Are there nondisposable alternatives?

BAG 11: *Hazardous Household Waste.* Any containers with hazardous household waste, such as insecticides, drain cleaner, and batteries, should be placed in this bag. Store it in a safe place.

Bag 11 should go to the nearest hazardous-household-waste program. A better hazardous-waste-labeling system should be encouraged.

BAG 12: *Everything Else.* This material cannot be recycled using current technology. The things being thrown into this bag should be examined for ways to cut down. See if there are nondisposable alternatives. Include pet litter in Bag 12 (see below).

There are several other categories of waste that should be included if appropriate:

YARD WASTE: Estimate how many bags are generated per year and divide by 52 to get the weekly average. Alternatively, the national average figure of 0.7 pounds per day per person can be used. Yard waste should be deposited with a municipal composting program or composted on your land. Again, if no community program exists, it should be established.

CAT AND OTHER PET LITTER: Collect in a separate bag. Reduce volume by raking feces from the litter and flushing them down the toilet. This will extend the life of the litter.

After separating the trash, the next step is to weigh it. This can be done at the end of each day or week. Make a scorecard, listing each bag, the number of pounds for the day or week, and the percentage of the total volume. Divide the total waste by the number of persons taking the test and then by the number of days over which the test was taken. This gives the pounds per day. Multiply waste per day by 365 to get total waste per year per person. Multiply yearly waste by 70 to get lifetime totals per person.

If the test results show that the percentage being recycled is small, the community needs to get to work immediately on recycling items, most likely in Bags 8 to 12. The average American family of four can expect to collect between 21 and 28 pounds per person each week (84–112 pounds altogether). A wise and reasonable goal is to recycle or compost 90% of our waste.

HOME AS HABITAT

We share our ecological habitats—homes—with other living organisms, animals, and plants. Consider those African violets, ivies, jade plants, aloes, wandering Jews, spider plants, and ferns. These plants help clean our air. We may also have one or more cats, dogs, gerbils, parakeets, goldfish, or other pets. Less desirable as pets perhaps but nevertheless important are the flies, ants, mice, spiders, termites, bats, and nesting birds. A little anecdote about fleas: As fleas evolved to find their niche in feathers and fur, they lost their wings, but since they still had to get from animal to animal, they developed the ability to jump. They can jump 7 inches high and 12 inches long across a room.

In our kitchens we witness the natural processes of mold and decay as bread becomes stale and fruits soften. That yogurt was converted from milk by the action of bacteria in a temperature-controlled fermenting process. Cork from trees stoppers bottles. Brooms are made from grasses. Sponges soak up water through their pores. The graphite in pencils is technically an inorganic mineral that was once living. Gale Lawrence says in *The Indoor Naturalist*, "It is derived from plant matter that has been metamorphosed by heat and pressure into almost pure carbon. . . . Coal has been subjected to less heat and pressure, diamonds more."

We all cope with dust, from the dirt we track in to that generated by our activities, such as brushing our hair. In cities dust contains many more particles from traffic and industry. There is also cosmic dust, but we don't know much about that yet.

ANIMAL HOMES

Other animals build homes for the same reason we do—to be safe from enemies, to store food, to escape heat and cold, and to raise their young. But they are able to build sometimes amazing structures on instinct, with no need for compass or ruler or mechanical devices. Maybe they know some things we don't! Here are some findings taken from *Animal Architects*.

Some build mounds. Beavers are capable of gnawing down trees and damming streams with branches. Their work may destroy trees and flood the land, but their way does not have the negative impact that human-built dams do. The beaver-made pond creates a home for fish, waterbirds, and other animals that have coevolved. It prevents soil from washing downstream. The rich soil that builds up benefits a variety of plants. Years after the beavers have left, it has been found that a healthy new forest has been supported.

The mallee fowl of Australia dig a pit about three feet deep. In winter, the male and female pile leaves and twigs in it. After each rain they cover the vegetation with sand, sealing in the moisture. As the moist vegetation decays, the mound warms up. The couple spend eleven months adjusting the temperature. After the female lays her eggs, the male will keep the temperature exact by adding or subtracting layers of sand.

Some creatures weave homes and construct platforms. A weaverbird begins by holding one end of a blade of grass to a twig with its feet. With its bill it winds, weaves, and knots the other end. It builds a ring to stand on, then adds a chamber. Eventually the nest hangs from a branch like a piece of fruit. These birds also indulge in group housing. In Africa straw homes may stretch fifteen feet across and accommodate 125 nesting pairs.

Stickleback fish build a nest of algae and small waterplants in the sand. The male cements this structure with a sticky glue from his kidneys. At the entrance to his nest, he'll try to attract a female by displaying his colors. If she approves, she'll swim into his nest, lay her eggs, and then leave. The male will fertilize them, guard them, and watch over the young.

Ospreys and monkeys build homes at high altitudes and have commanding views. Some insects, such as caterpillars, make tents. (Our nylon tents are remarkably similar.) Rabbits, prairie dogs, and badgers have elaborate underground homes with tunnels and hidden entrances to keep them safe, warm, dry, and able to nurse their young. On sunny days they will carry wet bedding out to dry. Polar bears perhaps taught the Eskimos a thing or two about snow dens.

Many fish and insects use body fluids from inner organs to stabilize their structures. Wasps construct paper-thin walls and cells out of mud and wood made pulpy with their saliva—no polluting chemicals in these sealers.

Honeybees construct hives of perfectly symmetrical wax cells in which up to 30,000 bees live and work together. The six-sided cells are thin, yet sturdy; the wax comes from secretions of abdominal glands in the worker bees. They use their jaws to apply the wax in efficient teamwork. Each bee labors for thirty seconds and then is replaced by another. Some of the cells are used for holding pupae, others for storing food, which they make as well. In winter the bees cluster together for warmth.

These animal homes do not have negative impacts on their environment and even enhance it. If other animals can do so, and we're so smart, why can't we?

THINKING LIKE AN

ECOSYSTEM

Hibernation

One of the cyclical features of ecosystems is dormancy for both plants and animals. Perhaps dormancy evolved as a way creatures protect themselves from harsh climatic conditions when little food is available. They have adapted to dormancy in ingenious ways, so that this time is restful, nourishing, safe, and secure. Naturalist Hal Borland muses:

> The cold-blooded creatures are the hibernators—with only a few warm-blooded exceptions. Snakes, salamanders, frogs, toads, lizards, turtles, some fish—they all hibernate. Some insects hibernate, usually selectively. Queen bumblebees, mated and full of fertile eggs, hibernate, though all the other bumblebees die in autumn. . . . Some insects hibernate in caterpillar form, notably the little moth Isia Isabella which becomes the woolly-bear caterpillar.
>
> I wonder if there aren't stages of life and hibernation that we haven't yet properly classified. What would you call the germ in a bird's egg? Isn't it a form of life suspended, waiting proper conditions to rouse and grow? Or how explain the remarkable stages of any moth or butterfly? What is the seed of a pumpkin or a sunflower, say, but a form of life suspended—hibernating?—which needs only the right conditions of light, warmth, and moisture to rouse and become a thriving plant?
>
> —Hal Borland's *Book of Days*

ENERGY CHOICES

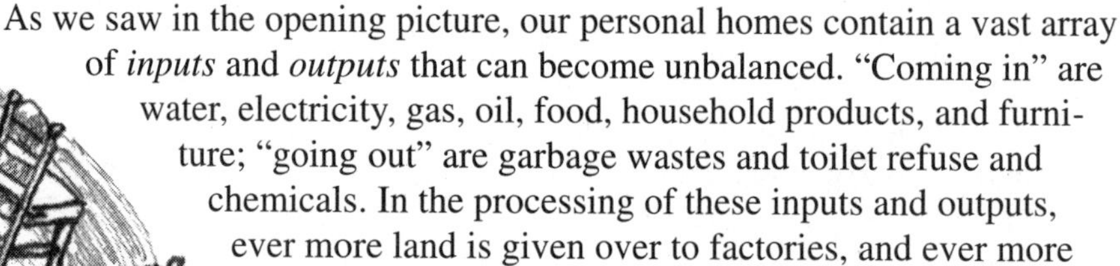

As we saw in the opening picture, our personal homes contain a vast array of *inputs* and *outputs* that can become unbalanced. "Coming in" are water, electricity, gas, oil, food, household products, and furniture; "going out" are garbage wastes and toilet refuse and chemicals. In the processing of these inputs and outputs, ever more land is given over to factories, and ever more quantities of energy are demanded by an exponentially increasing human population. It is imperative that we rectify our energy balance. We can't keep operating in the red. Being in the black means choosing energy resources that can be replenished at least as fast as they are used.

One of the saddest consequences of our predatorial raid on the Earth's energy resources has been the disruption of native homes, whose culture and livelihood are shaped by the ecosystem of a place.

Audubon scientist Jan Beyea assesses our energy options:

Because of the scale at which humans use energy, particularly fossil fuels, we are now a major player in setting the composition of the Earth's atmosphere. As a result, we affect the global temperature balance.

Climate protection requires a fundamental change in how we, as a society, cope with problems. It requires us to act now to stop destruction long into the future. We must act before the evidence of the destruction is actually visible. Those of us alive today have a special responsibility in history. If we do not reverse our course over the next decades, there will be little hope of later generations doing anything but surrendering to the effects of climate disruptions.

There is a key point that must be repeated about global warming, over and over again until it is widely appreciated. The gases we spew out today will take a long time before they are absorbed into vegetation and oceans. About 50–70% of our CO_2 emissions are recycled into the ocean surface waters rather rapidly. However, the residence time in the atmosphere for the remainder is of the order of hundreds of years. The fraction that stays up for the long haul will determine the climate of our descendants. Every time we drive our car, heat our homes with fossil fuels, or use electricity generated by fossil fuels, we ever so slightly narrow the options of future societies.

The analogy with nuclear wastes is very strong. Environmentalists have held it immoral to benefit from nuclear power while passing on the risks to future generations. Our descendants will be the ones who will have to deal with nuclear wastes escaping from repositories, not us. Similarly, they are the ones who will have to cope with fossil wastes, the wastes we will have left them in the atmosphere after having reaped the energy benefits for ourselves.

The United States contributes about 20% of world CO_2. Thus, the United States can't solve the greenhouse problem on its own. Yet, because we are the worst CO_2 polluters on a per capita basis, we must put our own house in order first. Only then can we expect other nations to put much effort into controlling emissions.

Herein lies the essence of our environmental crisis. Persistent trends in key ecological variables indicate that we have not only been living off the interest but also consuming our ecological capital. This means that much of our wealth is an illusion. We have simply drawn down one account (the biosphere) to add to another (material wealth).

—William Rees, *The Ecologist*

ENERGY OPTIONS

FOSSIL FUELS

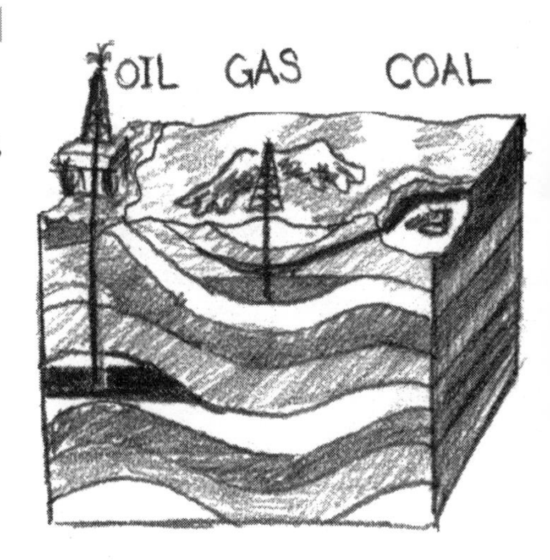

Natural gas emits the least CO_2 per unit of energy (about half that of coal). Oil is next in this regard, followed by coal. The worst CO_2 emitters are synthetic fuels from coal. Natural gas has promise as a short-term strategy, provided emissions of unburned natural gas, a greenhouse gas itself, are reduced during production and transmission.

The non-traditional fossil fuels, shale and tar oils, represent another enormous potential supply of energy, comparable to the world coal resources. But if we should ever tap significant amounts of these dirty fuels, any hope of controlling climate disruption will be lost.

We shouldn't write off fossil fuels completely. There may well be ways to improve the consumption of fossil fuels from the climate perspective. CO_2 removal is one example. It is possible to remove CO_2 from the exhaust gases of fossil fuels at power plants, such as those at utilities and large industries. Current estimates are that a 50% removal of CO_2 will double the cost of electricity, while a 90% removal will raise costs by six or seven times.

ENERGY EFFICIENCY

Using energy more efficiently is the cheapest and fastest way to reduce CO_2 emissions, while maintaining economic growth. Also it can be the most environmentally benign. I refer here particularly to eliminating energy waste by modernizing equipment in the home, office, factory, and transportation sectors.

SOLAR TECHNOLOGIES

Solar-related technologies derive their potential power at some point from the sun. Wind, for instance, arises from unequal heating of the Earth by solar energy. The success of technologies like wind turbines has shown that such alternatives can make a real contribution to the U.S. energy supply under the right regulatory climate.

SOLAR THERMAL

On the supply side, one of the most promising options is solar electricity. It can be derived from steam produced by high-temperature solar heat, assisted by the burning of natural gas. The cost of producing electricity, at least in the daytime, by this method is not too much greater than the cost of electricity from the latest nuclear power plants.

PHOTOVOLTAICS

The "hottest" form of solar electricity today is that which is produced directly when sunlight hits photovoltaic cells. The costs of these cells have been dropping dramatically, with the potential for producing electricity at costs well below that of nuclear power. Expanding research into photovoltaics would seem to be the most important energy research step that can be taken for the long term.

HYDROPOWER

Hydro in moderation is fine, but too much is a disaster. We have all too few crucial sites left for wildlife and river recreation. The natural flow of rivers is essential to ecosystems. For instance, in the spring when the ice melts, the resulting rush carries much-needed nutrients for wildlife far and wide. Dams change the natural rhythms by regulating the flow.

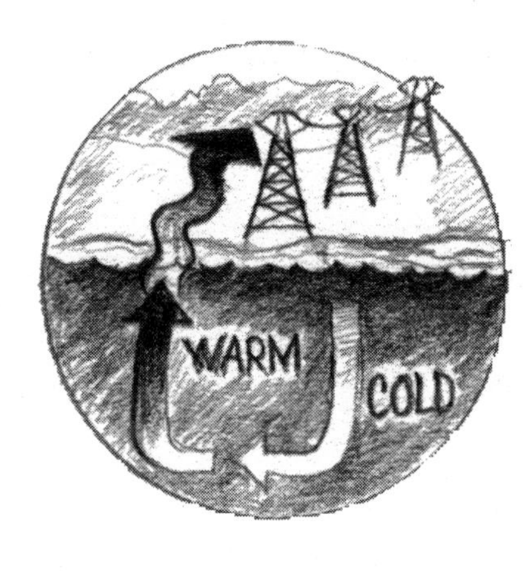

OCEAN-THERMAL (OTEC)

This option takes advantage of vertical temperature differences in the ocean to extract useful energy, but the environmental impacts of large-scale use of OTEC are largely unknown. For this reason, this technology is not a clear viable alternative to fossil fuels.

GEOTHERMAL

Geothermal energy is obtained from hot waters deep underground. The waters are heated as a result of radioactive decay deep in the Earth. The practical potential for long-term replacement of fossil fuels by this technology is not well known. However, the world's geothermal energy base is very large, comparable to the world's coal resource base.

BIOMASS

Biomass refers to biological matter that contains stored energy. Consider trees and plants, which collect sunshine and use it, along with CO_2 extracted from the atmosphere, to build up biological molecules, storing energy in the process. The stored energy can be extracted from these molecules for human use. (Note that biomass also includes living matter that eats plants or other living things. With very rare exceptions, all of the energy in biological systems can be traced back along the food chain to solar energy.)

Considerable biomass in the form of wood is already consumed in this country. The real potential for biomass, however, lies in its ability to be converted to a transportation fuel, e.g., alcohol. It is quite conceivable that the costs of producing alcohol from biomass will be dramatically reduced due to bioengineering, making ethanol the cheapest alternative to gasoline.

Once equilibrium is reached with biomass, it is an energy source that does not contribute to global climate disruption. The problem is that too great reliance on this energy source could have disastrous environmental impacts on land and habitat, unless agricultural and silvicultural practices are radically changed. Suppose, for example, the current U.S. transportation fleet were powered by alcohol. It could take 400 million acres of land to grow the necessary crops and wood. This is an area equivalent to our current crop base and would practically saturate our entire agricultural and silvicultural resource. The pressure to build so-called "biomass farms" would be enormous. If not properly designed, biomass farms could eliminate much of the wildlife habitat in the U.S. and put equivalent stress on the environment in other countries. Environmental biomass plantations should be designed to provide biological diversity that will allow wildlife to survive.

FUSION

Fusion energy, which powers the sun, would not produce CO_2. Hot, or normal, fusion technology attempts to produce temperatures as hot as the sun and hence tap fusion power. So far, success has been elusive. Current fusion cycles under study in the United States with government support are inherently radioactive, because they produce neutrons. When neutrons stop in the matter surrounding the reactor, they generate long-lived radioactivity, unless very pure (and expensive) materials are used. Another problem with neutronic fusion is that neutrons can easily be made to produce fissionable materials that can be used in conventional nuclear power reactors and for making nuclear weapons. In fact, the most practical use of neutronic fusion power will be as a fission breeder. As a result, it is doubtful that conventional fusion can compete with gentler technologies such as photovoltaics.

During 1989 there was a flurry of media reports about the possibility of fusion taking place at low temperatures inside a metal matrix. Most exciting was the apparent absence of neutrons and the possibility of cheap power. If true, the environmental implications would be disastrous. Why? If man can move mountains cheaply, the natural world will be transformed. As the saying goes, "Power corrupts and cheap power corrupts absolutely."

NUCLEAR

CONVENTIONAL

Nuclear power emits no carbon dioxide and little is emitted indirectly during the mining and transportation of the fuel, but the current technology has lost so much credibility that it is unlikely to be a viable alternative.

SECOND-GENERATION

There is the possibility of developing new designs of nuclear power that are meltdown-free. Research into them should be started, but we must recognize that they do not address other problems and other public concerns. They do not address the transportation of radioactive materials, their disposal, and the proliferation of weapons-grade material. Furthermore, it is questionable that the new designs can compete economically with other methods for avoiding carbon dioxide emissions, especially if they are really built to reassure the public on safety and quality assurance.

AN ENERGY POLICY FOR A GREENHOUSE WORLD

How can we ensure in a practical way that emission of CO_2 declines steadily, say at 2% per year over the next fifty years? (Note that a 2% per year reduction translates into about a 20% reduction by the year 2000.) The only realistic way to achieve this goal is by legislation, for instance, by placing a CO_2 limit-per-unit of energy on both new and existing plants, a limit that would tighten each year.

Over the long term, we'll have to pay more for energy to cut down CO_2 sufficiently, but the cost for our descendants of living in a deteriorated world would be much more.

A Carbon Dioxide Reducing Diet

The average American is responsible for the emission of 55,000 pounds of carbon dioxide and its equivalent in other greenhouse gases every year, mainly through the use of energy. Audubon wants everyone to go on a carbon dioxide diet and tells us how in a booklet called *The CO_2 Diet for a Greenhouse Planet*. The goal is to get us to cut our CO_2 emissions or "calories" by 2% a year (just 1100 pounds). If we did this (it's not as difficult as cutting down on ice cream or chocolate), by the year 2000 we would achieve a 20% reduction—a real contribution to solving the global warming problem.

On this diet we don't have to keep a daily diary. Once a year we list the global warming "calories" associated with our activities and the products we use. We set a goal of how many pounds of emissions we can sustain—in this case, for a healthy planet—and embark on a reduction plan.

Here are the amounts of CO_2 emitted, as figured by Audubon's scientists based on national averages:

Electricity—for every kilowatt-hour (kWh) of electricity we use, 1.5 lbs. of CO_2 is produced
Oil—for every gallon, 22 lbs. of CO_2
Natural gas—for every therm, 11 lbs.
Propane or bottled gas—for every gallon, 20 lbs.
Discarded trash—for every pound, 3 lbs.
Recycled items—every pound, 2 lbs.

First estimate the total use of fuels or products that contribute to greenhouse gas emissions. Then multiply the number of units (kWh, gallons, or whatever) of the item used by its CO_2 emissions factor (as given above, e.g., oil is 22 lbs. per gallon). This provides the CO_2 emissions for each product or activity. Adding these numbers gives the estimated total annual emissions.

To calculate CO_2 emissions from utilities, divide the amount of yearly payments for electricity, oil, natural gas, or propane by the unit price, which can be found on the bill. Apartment dwellers who do not get utility bills because the building is master-metered can ask the superintendent for information on utilities. A rough estimate can be obtained by dividing the building's total utility consumption by the number of apartments.

Once we know how many CO_2 calories we use, we can start a personal reduction plan by using efficient appliances and heaters, lowering water temperatures, and using less heat and air conditioning.

Fact: Compact fluorescent bulbs give the same light but use only 20% to 40% of the electricity of incandescent bulbs. Replacing a 100-watt incandescent bulb used four hours a day with a 22-watt fluorescent can save 114 kilowatt-hours and 180 pounds of CO_2 a year.

Eco Koan

A woman built a home that was the most environmentally advanced she could make it. She researched and chose the best building materials, the best furniture, the best paints. She made phone calls and wrote around the world to identify the most up-to-date products on the market.

Her sister in another city chose not to build a new house but to buy an old one. Each claimed that she was the better environmentalist. Who was correct?

The two sisters consulted the Eco Guru, who said: "Certainly Sister New House has made a great effort to reduce pollution in her new house. But Sister Old House has prevented the building of any new structures. On the other hand, with population pressure eventually someone is going to have to build a new house, who may not be as conscientious as Sister New House. Meanwhile, Sister Old House has delayed the day when old houses are torn down. After all, a major cause of buildings disappearing is demolition for new structures. But if someone doesn't tear down a structure, he or she will build on land that may have other good purposes. What shifts the scale for me is that Sister New House has set an example. There is the possibility that many other people would learn from her efforts and continue on improving down through time. On the other hand, I'm just a human and don't know much. You might go and ask a mountain."

ECO HELPER

Here is the story of one couple's desire to make a home and save a bit of nature at the same time.

In any season, the voyage to the Inland Island crosses suburban seas. Once it stood in open country, but today the tract houses reach within half a mile on one side. The Cincinnati circular super-highway is less than a mile away on the other. Drive past the billboards hawking home sites, then go a bit farther, passing a farm or two until the country road ends. A battered mailbox lurks in a lilac bush, and a lane turns off to the left. This is the port of entry to the island of which Josephine Johnson is sole owner, proprietor, monarch, chronicler, and leading citizen. . . .

There are no grand sights on the island, only very small ones. Those are rendered more perfectly and accessibly in a small book Josephine wrote eighteen years ago which is found today only in libraries and second-hand bookstores. . . . The Inland Island is magical because it

(Continued on page 61)

(Continued from page 60)

speaks so vividly about our place in the natural world, specifically of a self-proclaimed island on the land at the eastern edge of the Cincinnati sprawl. . . .

Josephine came to the island in 1956 with her husband, Grant Cannon, and their three children. When the Cannons bought the thirty-seven acres, it was in deplorable condition as "productive" land. But Grant and Josephine had other ideas. . . . They invited a state forester to walk through their new holding. He told them how to clear the woods of wolf trees, to build their value as timberland.

The Cannons had a different question. "What can we do to make a nature preserve?" we asked cautiously, the title belonging to vaster things such as the Serengeti Plain. . . .

"Sit back," he [a forester] said, "and watch the ecology develop. . . ."

That was all, and the tide of that odd word has come. . . .

The tide of forest succession has, indeed, come in on the island. The Cannons had horses for a while and the occasional worker to hold the creeping flood of trees out of the meadow. They cut paths back through the woods. They gardened around the house and sawed down a few trees to open the view. Mainly they let the ecology develop. The trees crowded in. The thickets grew denser. The bird population rose. The pond gained frogs, kingfishers, a tremendous bloom of duckweed, and a family of snapping turtles. The land returned to the pace of the old times before the farmers broke their hearts on such unsuitable terrain. . . .

The island was anything but primeval or innocent. The world always pressed in with its problems. The large creek was the weakest wall in the island's defenses. It was badly polluted by animal and human sewage when the Cannons first bought the land. With time, the water grew fouler. It was alive with dangerous bacteria. Suds and toilet paper were the chief wildlife. Finally the sewer district proposed a big concrete line through this part of the county, including a long reach beside Josephine's creek. She had no option but to agree. . . .

When the cistern pump burns out or vandals block the sewer with stones, Josephine wonders what on earth she is doing out here by herself. Living alone, her children scattered to the East and West coasts, she knows her tenure on the island will last only as long as her good health. When the dreary world news of war and want pours out of the television or the newspapers, she wonders if it is right for one woman to hold something as precious as an inland island—thirty-seven acres, two ridges, two streams, one pond, one house, one barn, and all the plants and creatures, great and small, that come with it. "I live on an island of sanity," she wrote, "the island of this place. I am fortunate. I no longer ask why."

—John Fleischman, *News From the Inland Island*

EcoQuiz

Q: What is household hazardous waste and why is it hazardous?

A: Household hazardous wastes are household products that are potentially dangerous to human health and the environment. They are most likely stockpiled in kitchens, basements, and garages (where they should be stored if possible). Since their disposal is not regulated by law, most find their way down drains or directly into the ground. When wastes that contain heavy metals, such as lead, mercury, and nickel, are incinerated, the heavy metals are released into the air. For these reasons municipalities have begun special collection programs to dispose of these wastes properly. But considerably more attention must be paid to reducing use of these products and improving the technology of their elimination. Get to know and watch out for the following substances:

Acids—found in household cleaners, pool chemicals, solvents; have a pH range of 0 to 5 and are corrosive;

Batteries, dry and wet—may contain lead, mercury, cadmium, and other corrosive chemicals;

Cadmium—a metal element found in paints, batteries; poisonous to humans;

Chlorofluorocarbons—man-made chemicals known in trade as Freon, Genetron, Isotron; they reside for years in the atmosphere and introduce chlorine into the stratosphere which destroys ozone;

Lead—poisonous when inhaled or ingested; found in old paint, pottery, pipes, dust, and some water supplies;

Mercury—a metal found in "button" batteries used in hearing aids and calculators; exists as a liquid at room temperature; poisonous;

Petroleum distillates—hydrocarbons produced in the refining of crude oil; found in lip gloss, fertilizer, pesticides, fuels;

Solvents—dissolve other substances; common toxic examples are aromatic hydrocarbons (e.g. benzene), alcohol, methanol, naphthas, turpentine, acetone, and some paints and cleansers.

ECOSYSTEM PATHWAYS

Most of us these days spend our waking and sleeping hours indoors, out of touch with the phases of the moon, the ebb and flow of tides, and the subtle seasonal shifts in the pressures on the flora and fauna outside. One reason for this is that we have allowed our lives to be built around electric appliances and fancy products, such as telephones, answering machines, stereos, radios, television, fax machines, and computers. We may think that many of the appliances in our homes are expensive in dollars, but their production and use are costly to ecosystems too.

- Factories use large amounts of energy, primarily fossil fuels—oil, gas, and coal—to power the manufacturing process. About half of our oil is imported. To reduce our dependence on foreign supplies, pressure is exerted to drill in wilderness areas such as the Arctic National Wildlife Refuge. However, we could reduce our needs considerably through efficiency measures. The amount of energy saved by the Energy Efficiency Law that went into effect in 1990 was more than what drilling in the Arctic National Wildlife Refuge would have produced.

Furthermore, because of the hundreds of millions of years it takes for oil, natural gas, and coal to form, these resources are now known to be nonrenewable on human time scales. Animal and vegetable matter in the sea was compressed by layers of sediment and very slowly underwent chemical changes, remaining hidden in large rock pockets until discovered by humans. Despite the fact that oil and gas will be exhausted relatively soon, very little money is invested by the government and industry in the development of renewable energy sources. Does it make sense to drill and mine until every last bit is extracted before thinking about what to do when the supply runs out?

- The United States has about 1300 coal-burning plants, generating 40% of our electriciity. Coal plants are by far the largest source of sulfur dioxide, nitrogen oxides, and other pollutants that are major contributors to acid rain. Carbon dioxide from coal plants is also a major contributor to the greenhouse effect. Air pollution from coal combustion has harmed the health of millions and in many parts of the country greatly exceeds federal air-quality standards.

- Let us look at a few aspects of the air cycle. Remember that the air that we, and other plants and animals, breathe in order to live comes from the photosynthesizing done by plants, mostly algae. The consequences of what happens if we cover over vegetation with blacktop or building projects are obvious and ominous.

Air is closely connected to the water, carbon, and nitrogen cycles that take many, many years to circulate. It is estimated to take 2 million years for water to circulate from being split by plant cells, evaporated, and returned to earth.

As air moves across the Earth's surface, it collects various chemicals produced by natural events and human activities. While air has hundreds of pollutants, the most common are carbon oxides; sulfur oxides; nitrogen oxides; volatile organic compounds, such as methane, benzene, formaldehyde, and chlorofluorocarbons (CFCs); solid particles such as dust, pollen, asbestos, lead, and liquid droplets including sulfuric acid, PCBs, and dioxins; photochemical oxidants, such as ozone; and radioactive substances.

As emissions of sulfur dioxide and nitric oxide are transported by winds, they form secondary pollutants such as nitrogen dioxide, nitric acid vapor, and droplets containing sulfuric acid and sulfate and nitrate salts. These chemicals descend onto earth through rain or snow or gases, commonly known as "acid rain." The lower the pH, the more acidic the solution. Each whole number represents a tenfold increase in acidity. A neutral solution is 7. Many problems experienced by water, marine life, plants, trees, and edifices occur when the pH level is below 5.5; it has been as low as 2.3—as acidic as lemon juice—in many areas.

- The materials used for manufacturing appliances—steel, metals, plastics, woods—are extracted or synthesized to the detriment of natural resources for all. A student reflecting on the appliances that surround him in his home wrote:

> The pipes and motors are made partly of metal. Metal must be extracted from iron ore. Bulldozers, fueled by gasoline refined from Saudi crude, remove piles of rocks blasted free from their ancient beds by Chilean nitrites. Rail cars transport the ore from the mine to a crushing plant, and later to a refinery. Searing furnaces melt down the oredust into a molten-bright soup, and floating impurities are swept away. The soup is set and cooled into solid sheets. Later the metal will be shaped into uncountable forms and shipped to lands near and far. I am supporting chemical plants, oil tankers, and the road and communication systems of much of North America, and the world.

- The manufacturing and operation of appliances interferes with

(Continued on page 64)

5

APPLIANCES AND EXPENSIVE TOYS

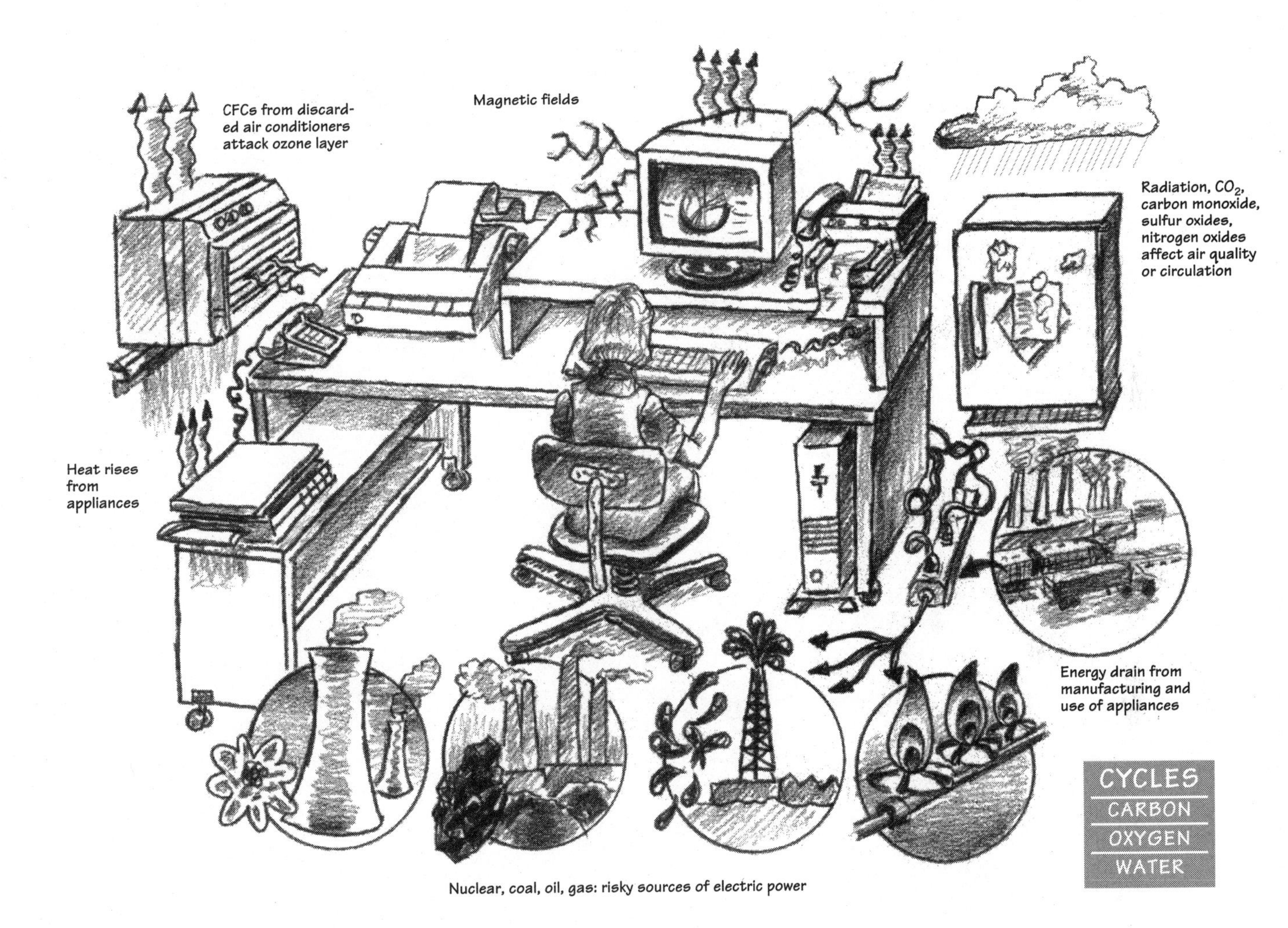

Nuclear, coal, oil, gas: risky sources of electric power

RECREATIONAL SHOPPING

Let's take a walk through our day with appliances. We get up in the morning to the tune of an alarm clock and perhaps turn on the radio for music or television for news. We turn on lights. We use the hot water heater. We brush our teeth with an electric toothbrush. We open the refrigerator to get something to eat. We use the can opener to open pet food. We use the toaster and/or microwave oven. Maybe the blender for juice. We turn on the coffee grinder and the coffeemaker. We are into the first half hour of the day and just beginning.

As we get dressed, we may use the electric shaver and hair dryer. We regulate air with humidifiers, conditioners, and dehumidifiers. Leaving the house, we press the button for the garage door opener. Or we

(Continued from page 62)

the carbon cycle. At the present time the carbon cycle has been altered by humans so much that it is feared that the temperature of the planet is being affected. While carbon dioxide ebbs and flows in natural processes, since the Industrial Revolution it has been released in excessive quantities by the burning of fossil fuels. Because human activities now emit more CO_2 than the plants and oceans can absorb, it accumulates in the atmosphere and has become the number-one contributor to global warming.

Improved appliance efficiency standards enacted in 1990 will save more than 30 billion kilowatt-hours in 1995, about 22 million tons of CO_2 emissions.

• An increasing population leads to greater demands for more products, which in turn leads to more drilling and mining, more burning of fossil fuels, more cutting of trees, more disturbance of soil, and more pollutants in the air with fewer trees to absorb them.

• Household batteries are little packages of toxic chemicals. They end up in landfills, where there is a danger that their mercury, cadmium, or nickel will leach into drinking water as the batteries degrade. If batteries are incinerated, the metals are concentrated in the ash, end up in the ground, or are released into the air.

• Where do our discarded appliances go? Most are not biodegradable and end up contributing to the mountains of garbage that overload our landfills, incinerators, and vacant lots. Moreover, most appliances come in excessive packaging that also adds to the garbage waste problem. Too bad. Appliances could be recycled and packaging made out of recyclable materials.

press the button in the elevator of our apartment building. In the course of the day we may do laundry in the washing machine and dryer, exercise on a treadmill, use a cellular phone, play a Nintendo game, write with an electric typewriter, listen to tapes on a recorder, and watch a movie on our VCR. Note that many of these items have separate motors and separate power converters. (A possible conservation option for the future is to have one motor source power for our electrical gadgets.)

All these items are ubiquitously advertised. You as a consumer are being constantly encouraged by producers to buy more and more of them. Furthermore, producers will make small changes or add on a convenience just to get you to throw away a perfectly good appliance in order to upgrade it. And you know that you just can't live without the latest!

In this way we and the Earth are mightily short-changed.

It must be clearly understood that while destitute countries destroy their environment by slashing and burning their forests or overworking fragile land until it becomes a desert, wealthy industrial countries, such as ours, destroy by over-consuming appliances and other goods. Shopping, not baseball, has become the major American pastime. Shopping centers proliferate along our roads. We are a nation with more and more people demanding, replacing, and discarding things.

A sustainable society is one in which the present generation meets its needs without jeopardizing the prospects of future generations. Each of us can make a difference by examining our lifestyle. Many of us owe a lot on our credit cards for appliances and expensive toys that we do not really need. We can ask how we can simplify and how we can get manufacturers to meet our needs for durable products. We can do this out of love for ourselves and our country.

Watch Out for Slipping CFCs

One way we can keep future generations and the fate of the ecosphere in mind is by being aware of how chlorofluorocarbons, or CFCs, used mainly in refrigeration and (formerly) in aerosol sprays, are eating away at the ozone layer, the atmospheric shield that protects us from the sun's lethal ultraviolet rays. Layers of ozone hover from the ocean's surface to as high as six miles.

When chlorofluorocarbons were first synthesized in the late 1920s, they seemed perfect as coolants and propellant gases in spray cans because they had no negative impact on health and are nonflammable. They consist of chlorine, fluorine, and carbon atoms and vaporize at low temperatures. In the past, they were used for foaming such plastic materials as Styrofoam. The problem is that they escape into the atmosphere, where they both increase the greenhouse effect and destroy ozone molecules. In the 1970s CFCs were banned in spray cans in the United States. Halocarbons include CFCs, halons, and other compounds that also deplete ozone. These compounds are generally released by discarded refrigerators, automobile air-conditioners being repaired, fire extinguishers, and air horns. Some of these materials can be collected in bags during servicing and disposal, thereby preventing them from entering the atmosphere, but best of all would be a ban on their use.

In the late 1980s the protective layer of ozone thinned to such a degree that a hole opened up over Antarctica. More recently, ozone levels have declined over the northern United States, Canada, and Northern Europe, exposing people to the risk of melanomas, cataracts, and DNA mutations. Children are especially vulnerable because they play in the sun a lot and have sensitive skin. In response to the "holes in the ozone layer," twenty-four nations, including the United States and the former Soviet Union, agreed to cut back on CFCs 35% by 1999 (Montreal Protocol Act). However, this act is not good enough, because even with reductions, some of the CFCs released today will still be in the atmosphere a century from now. The CFCs we have produced so far have yet to complete their damage to the ozone layer. When they do, the danger from the sun's rays is going to be more acute.

Corporations such as Du Pont plan to partly convert to hydrochlorofluorocarbons (HCFCs), which also destroy the ozone layer, although at a lower rate. Corporations can be pressured into utilizing alternatives, which do exist. Refrigeration, for instance, can be safely achieved with passive and active solar technologies, evaporative air-conditioning systems, water and alcohol, and helium.

Eco Koan

Two sixty-year-old men, each with three children, are arguing about who is the better environmentalist. Mr. Consume-a-lot says that he is, of course.

Mr. Consume-almost-nothing is outraged. He says, "How can you say that? You have bought every wasteful product that a wasteful society can produce. You have thrown things away when they could have been used over and over again. You laugh at pollution. You have voted against every candidate with environmental values."

Mr. Consume-a-lot responds, "Yes, what you say is all true. However, your children have felt so deprived by your zeal to be pure and not pollute that they are going to spend a lifetime wasting the environment. And my children have been so alienated by my excesses that they are following your example. So, on balance, I would say that I inadvertently have become the better environmentalist."

The two men took their dispute to the Eco Guru, who said, "Clearly, raising children cannot be controlled. What lesson do each of you draw from your experience?"

Mr. Consume-a-lot: "I learned that it really doesn't matter what you do to the environment because somebody else will take care of the problems."

Mr. Consume-almost-nothing: "I learned that I cannot force my values onto my children. I should have spent more time learning about environmental education. But, in any case, I hope to work for some legislation that will restrict what my children and his children can do to the environment."

Eco Guru: "There you have your answer."

Whistle Up a Breeze

Peter Steinhart

Fifty years ago, on summer nights, people repaired to their porches, board decks overhung with shading roofs, sometimes stretching around two or three sides of a wood frame house. There, they whistled up a breeze by moving slowly in rocking chairs or gliders. They talked quietly and listened to crickets and the far-off wailing of freight trains. Upstairs there were sleeping porches, screened to keep out the moths and mosquitoes but open to the wider currents of the summer night. The floors creaked, fireflies glimmered, and the scent of honeysuckle and hay floated dreams.

Thomas Jefferson brought porches to Virginia houses, believing that the Greeks invented them as a way to gather politicians and philosophers. But porches are part of the architecture of Malaysia, North Africa, and Japan, and American farmers who never heard of the Parthenon built farmhouse porches so they could work and rest outside in heat or rain. A U.S. Department of Agriculture study in the 1950s found that farm families still regarded them as a necessity. After the Civil War, vacationers brought the idea of the porch back from summer hotels and cottages and installed them in suburban houses. By the turn of the century, porch-sitting was a national habit. Today new houses have no porches. . . .

Few things have so transformed our culture as air conditioning. It opened up the arid Southwest and the humid South to retirement communities and new industry. Computers cannot work at high temperatures, and air conditioning was one of the prerequisites of the electronic age. It is essential to air travel and to the production of precision instruments and drugs which need exact manufacturing conditions.

But these benefits have not come without cost. Air conditioning uses considerable amounts of fossil fuel, which increases carbon dioxide in the atmosphere and raises global temperatures, leading us no doubt to demand more air conditioning. In the West, where hydroelectric power drives millions of the devices, we have dammed and dried out hundreds of miles of rivers for the sake of cool bedrooms. Fifteen percent of our electrical output—one-fifth of residential and nearly one-third of commercial electrical consumption—goes to space cooling. On hot summer days, a third of the electricity in the Northeastern states and a half in the South Central states goes to air conditioners. One-fifth of the chlorofluorocarbons (CFCs) we put into the atmosphere come from automobile air conditioners.

There are enormous opportunities to reduce these costs. In a study for the State of Arkansas, Amory Lovins of the Rocky Mountain Institute showed that a $5,000 retrofit on a single-family house in Little Rock would reduce the house's peak electrical energy demand by 83 percent and its annual electrical use by 77 percent. The solutions were simple: window-glazing to repel infrared light, insulation in walls, ceilings, and floors, compact fluorescent lights and more efficient appliances to cut down generation of internal heat, and lighter-colored walls and roof to reflect summer sunlight. Some of the savings came from a color television that used 50 watts instead of 150, and so generated less heat. Finally, having reduced the need to cool, the house could make do with a much smaller air conditioner that was also twice as efficient as the old one. . . .

In summer air we draw into our blood the sugars of pine and honeysuckle stream water, and new-mown hay. Who knows how their chemistries affect us internally? Perhaps these scents go deeper than our noses, into our blood, and thence into the core of feeling and being.

Porch-sitting offers us the chance to sit outside, to listen to owls and crickets and watch the stars glimmer through the exhaust of summer days. On such nights you bathe yourself in currents of air. Local eddies bring the scents of hay and honeysuckle, of pine and sage. Global currents carry the savor of the sea and the spice of distant meadows. The scent and sound of humankind, the babble of distant bazaars, and the murmur of quiet conversation from the backside of the planet float on the air of a hot summer night. Bathed in these currents, we are part of the plasma of life. We float in solution with the molecules of coyote howl, birdsong, and insect flutter, the whispers of sand, and the murmur of falling water.

But in the air-conditioned age, the swirl and eddy we sit inside is wholly domestic. We breathe in the noxious smoke of tobacco, the acrid vapors of formaldehyde, the ozone of muttering refrigerator coil, the stale draft of ductwork, the household odors of anxiety and regret. Our lights come from the cathode-ray tube of the television set, and the thoughts glimmering in that light are small indeed. We forget we remain human only to the degree that we are open to the wider world. Our capacity for inspiration is diminished. To escape boredom we turn up the TV sound or play loud music, and then we thicken our walls to shut out our neighbor's distractions. We draw the curtains, turn away from our neighbors, and mutter along with the machines.

To Air-Condition or Not

According to the American Council for an Energy Efficient Economy, 60% of U.S. households have air conditioners. Their use consumes the electric power output of seven large coal power plants, releasing approximately 100 million tons of CO_2 per year. Air conditioners are on the rise, with 75% new homes equipped with central air conditioning systems. Do we really need central systems which cool every part of the house regardless of who is in each room?

Granted, air conditioning has important uses. Especially in the hot South and Southwest! To list a few: cooling factories and hospitals, enabling food to be shipped without spoiling, and storing bread and milk so that poisoning is reduced. Design of air conditioners can make a big difference in electricity consumption and in CFC pollution.

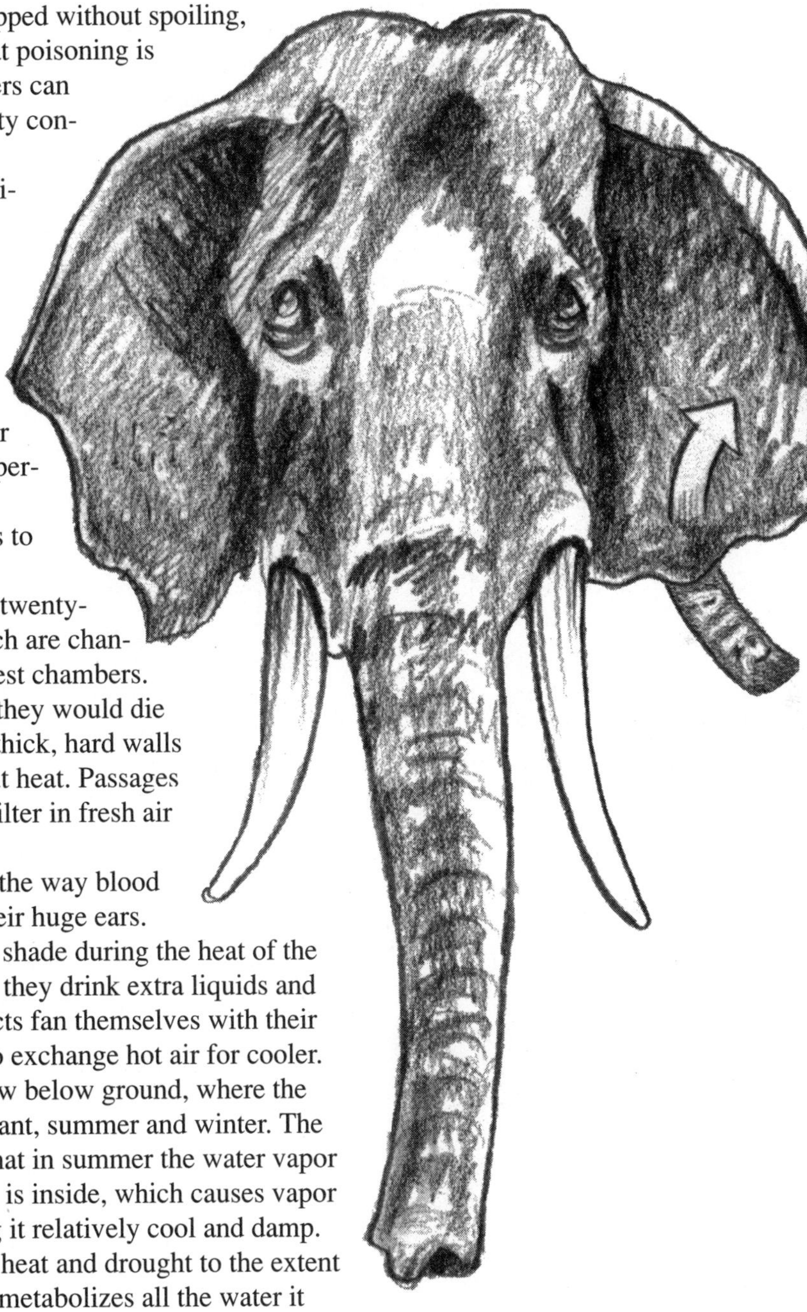

Before resorting to air conditioners in our homes, we can try reviving time-honored practices to block out the heat of the sun. Some alternatives include placing trees or shrubs or awnings in front of east-and-west windows, installing tinted or specially coated windows, and operating fans.

Animals have devised ways to cool themselves. Termites in hot Africa and Australia build huge, twenty-foot-high mounds, inside of which are channels and air ducts that cool the nest chambers. Without this ventilation system, they would die within hours. The mounds have thick, hard walls that seal in moisture and keep out heat. Passages are porous, with tiny holes that filter in fresh air and allow stale air to escape.

Elephants are kept cool by the way blood flows through the thin skin of their huge ears. Many animals wisely stay in the shade during the heat of the day and forage at night. Like us, they drink extra liquids and plunge in cool water. Some insects fan themselves with their wings. Panting allows animals to exchange hot air for cooler.

Many desert animals burrow below ground, where the temperature remains fairly constant, summer and winter. The real advantage to burrowing is that in summer the water vapor pressure outside is higher than it is inside, which causes vapor to move into the burrow, making it relatively cool and damp. The kangaroo rat has adapted to heat and drought to the extent of not ever needing to drink—it metabolizes all the water it needs from dry seeds.

THINKING LIKE AN

ECOSYSTEM

Communication

Animals outdo technology in their communicative and perceptive abilities. Sometimes it seems that our computer systems, which we use to create information networks, imitate insect antennae. We humans must develop sirens and alarm systems to ward off predators, whereas animals communicate warnings with their own "built-in" equipment.

In the animal kingdom, sophisticated or simple, methods and messages vary enormously from species to species. Humpback whales "sing" to each other in an elaborate language that can carry through 700 miles of open sea. Single-celled protozoa in water release attractants to signal readiness for breeding. In social groups, animals convey sex, location, strength or weakness, hierarchy, alarm, territory, and warnings. Signals, including subtle chemical and electrical nuances, are sent and received by every type of animal sense organ.

Touch is the most primitive and basic interaction. Many mammals groom each other in order to bond psychologically. Insects, crustaceans, spiders, worms, and mollusks use antennae or chemoreceptors for elaborate purposes.

Sight, sound, and smell are more effective for communicating over longer distances. Mammals often proclaim ownership of territory with scent marks made with droppings, saliva, or urine, very often with special scent glands near their eyes or other parts of their bodies. Sighted creatures use visual displays, postures, and gestures to indicate meaning. Birds are especially adept at showing special plumes and color patches and singing unique songs to attract only their own kind. Since many fur-bearing mammals evolved without color vision, and sport coats of gray, black, and brown, their visual displays are usually flashes of white—the alarm signals typical of deer—as well as complex gestures and postures, such as tail-swishing, teeth-baring, and ear-flattening. Color displays are used in the courtship of butterflies and fiddler crabs; fireflies and some deep sea fish use light that carries great distances. Tropical birds and primates use varied hoots and calls to travel through the distances and obstacles of a jungle.

Weather changes are known to affect the sounds made by grasshoppers, frogs, and crickets. Who knows what other factors can interfere with the communication of animals? We are not accustomed to giving animals such consideration, but their requirements for surviving and raising families depend on unhindered communication.

ELECTRICITY

—DO WE HAVE TO BE ITS SLAVES?

In 1879 Thomas Edison invented the light bulb, and in 1882 the world's first electric utility opened in New York. That is not so long ago in the scheme of things, and yet we cannot conceive a world without our electric lights and appliances anymore. In our total dependency on it, it has come to regulate us. For instance, night is no longer a time of darkness. All we have to do is turn on the lights and keep on going, without interruption. We can perpetuate daylight and work longer, which may be both good and bad for us. Regardless of the way electricity detaches us from the day-and-night cycle, it is also the central force of modern life. Producing electricity constellates a variety of crucial environmental problems: how to achieve efficient use, the limits of fossil fuels, the destructiveness of hydroelectric dams, safety of nuclear power, investment in solar technology, and how industry can contain environmental damage costs.

Efficiency Makes a Big Difference

Fact: In 1950 the U.S. demand for energy totaled 33 quads (40% oil, 18% gas, 37% coal, and 5% other). By 1990 the demand increased to 81 quads (41% oil, 24% gas, 24% coal, 7% nuclear, 4% other).

Here are dollar costs to the consumer of traditional options:

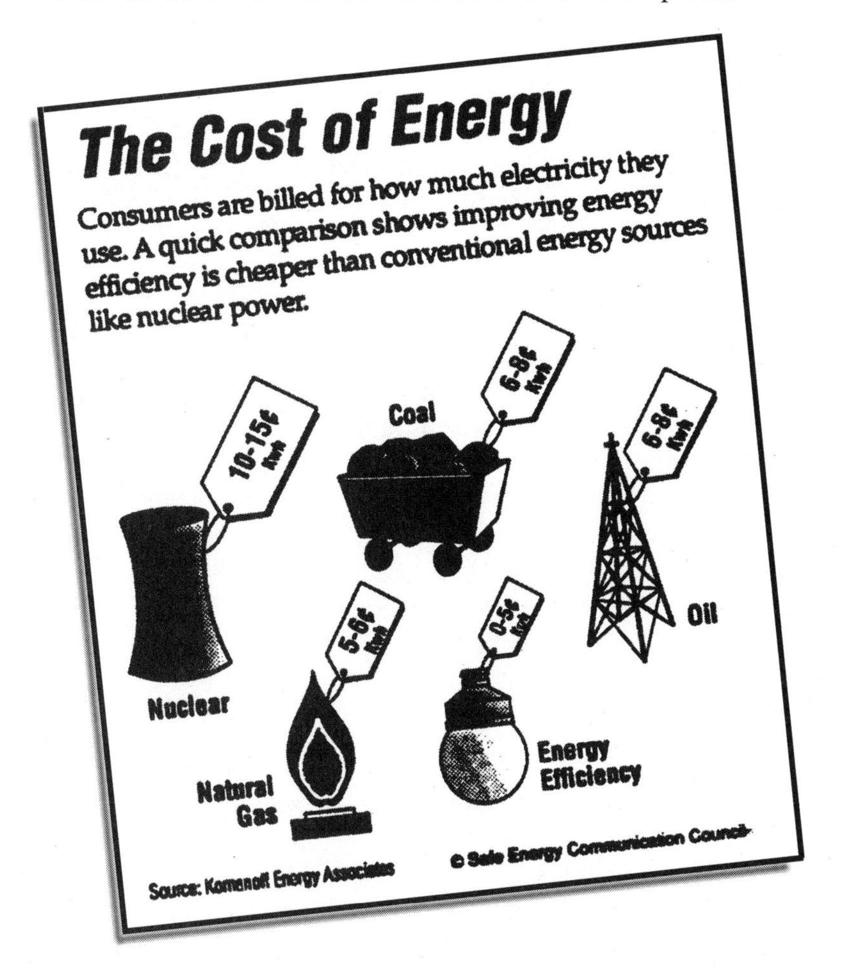

Energy efficiency can make a big dent in bills to oneself as well as to the Earth. (The American Council for an Energy Efficient Economy has published *The Consumer Guide to Home Energy Savings*, which carefully spells out how, when you purchase, to make energy-efficient choices in heating, water, lighting, cooling, food storage, and laundry appliances.) Everyone knows that ideas for energy-efficient devices abound, but there hasn't been enough commitment in the marketplace or legislatures to get them operative.

The current mix of washers, dryers, refrigerators, and dishwashers in this country draws nearly one quarter of the nation's yearly electricity output. Yet this fraction stands to fall sharply because appliances manufactured today are twice as efficient as those built in 1972. Some high-efficiency models require only 10% to 20% the energy of their older counterparts.

Efficient appliances were given a boost by Congress in 1987 with passage of the National Appliance Energy Conservation Act. The law established minimum efficiency standards for most household appliances and required manufacturers to place energy guide labels on their products. The yellow label displays the expected annual energy cost in dollars (based on the average U.S. electricity cost) of a given model, flanked by the equivalent cost for the most-efficient and least-efficient machines in the same class (such as similar-sized refrigerators). The cheaper the cost of running a machine, compared to others in its class, the greater its efficiency. When buying an appliance, think also in terms of its expected lifetime. The life-cycle cost equals the initial cost plus lifetime operating costs. Increased efficiency of many new appliances means savings for the consumer.

The net efficiency of the entire energy-delivery process includes each energy conversion step in the process, from extracting the fuel, purifying and upgrading it for use, to transporting it, and then using it.

Today's American family uses an average of 3200 kilowatt-hours of electricity a year to power its appliances, accounting for emissions of nearly three tons of carbon dioxide, the leading greenhouse gas, and close to sixty pounds of sulfur dioxide, a major cause of acid rain. On average, appliances alone count for 3700 pounds of CO_2 emissions per person in the United States. Every kilowatt-hour of electricity you avoid using saves over one and one half pounds of carbon dioxide that would otherwise be pumped into the atmosphere. Replacing a typical 1973, eighteen-cubic-foot refrigerator with an energy-efficient model will save about 1160 kWh and over 1700 pounds of CO_2 emissions per year. That same refrigerator will save over twenty pounds of sulfur dioxide emissions.

Fact: Electricity generated by fossil fuels or nuclear reaction requires about three times as much energy at the power plant than actually reaches the home.

Our world must move away from carbon dioxide–producing fossil fuels for most of its energy needs. Natural gas emits the least CO_2 per unit of energy, approximately half that of coal. Oil is in between gas and coal. The worst CO_2 emitters are synthetic fuels from coals; very bad also are nontraditional fossil fuels, such as shale and tar oils. In the long run it might be possible to remove CO_2 from the exhaust gases at power plants. For instance, methods for scrubbing the CO_2 from coal in the burning process may hold promise, albeit they are expensive. If so, fossil fuels need not be written off completely.

Solar or nuclear power plants do not generate greenhouse gases during operation. However, nuclear's second-generation, hypothetically "safer" designs do not eliminate the possibility of nuclear waste or sabotage. The best and most practical option for the long term is to commit ourselves to a solar future, viewing nuclear power and fossil-fuel combustion with CO_2 removal as insurance policies to be deployed only if the promise of photovoltaics, biomass, and other forms of solar energy runs into unexpected obstacles.

Solar is particularly attractive because it produces no waste during operation and is inexhaustible. Not all solar power comes directly from the sun: both wind and hydroelectric power are solar, since wind is created by the sun's uneven warming of the atmosphere and since the water that collects behind dams was originally rain, which in turn was water vapor evaporated by solar heating. But hydroelectric dams can damage the environment.

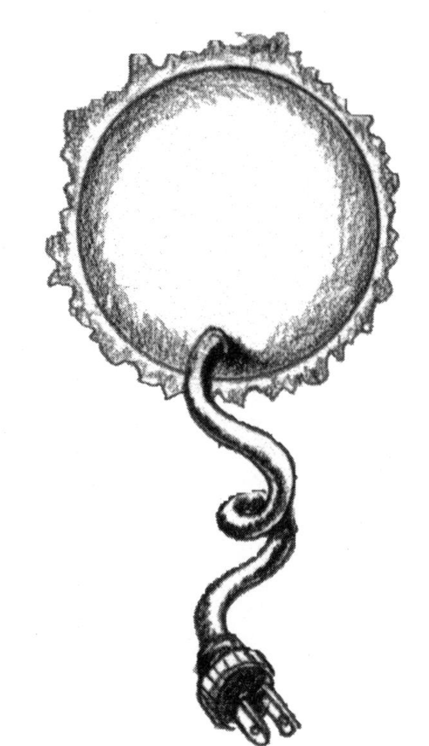

The Toll of Hydroelectric Dams

The high price of big dams includes disrupting water flows that many living organisms depend upon, the uprooting of people, an increase in diseases, more risk of earthquakes, and sterile fisheries. Rivers and their tributaries are the lifeblood of ecosystems. As water cycles are altered, fish and birds are disoriented by the changes.

Thousands of rivers have been dammed. The Platte, for instance, which flows through Nebraska, Wyoming, Kansas, and Colorado, has been diverted since the 1800s. With more than 70% of its flow changed, the formerly wide, unvegetated, braided (of several channels) river is now choked with plants, and the river is squeezed into a narrow, twisted path. About 95% of the habitat preferred by breeding sandhill cranes—a wide riverbed with vegetation-free sandbars interspersed with shallow, open water—has been lost. Hydropower also carries risks of mercury and methane contamination.

The argument of those against long-range environmental assessment and planning is that each new dam is only a small fraction of the total, so the latest dam can't be that bad. The refusal to look ahead to the final results is reminiscent of how in the past the United States forged ahead with a new technology without thinking of the consequences—for example, the automobile and nuclear power plants. For decades our society has ignored the gradual deterioration of much of its natural resources and seen small losses mount up to disaster. Then the argument becomes, "Well, the damage has been done, so the next dam or highway, or land conversion won't make things much worse." James Bay in Canada is a prime example.

Listen to remarks of the Grand Chief of the Crees of Quebec—Matthew Coon Come—as he describes what his people have undergone as the result of Hydro-Quebec's proposed development of forty-seven dam sites and 667 dikes for electrical power at James Bay (commonly known as James Bay I and II), where Crees have sustained themselves for 6000 years.

> I represent 10,000 Crees. We are still hunters, fishermen, trappers. We occupy an area of 144,000 square miles of land. We are people of the land. We have always lived on that land. We harvest what the land provides us and

The Platte Pretzel

Suzanne Winckler

Every spring some 500,000 sandhill cranes—80 percent of the species' world population—gather for a month or so along the Big Bend of the Platte to feed in the cornfields and remnant wet meadows and roost in the shallow river channel. It is here that the cranes fatten up in preparation for the arduous business of nesting that lies ahead on their breeding grounds in Canada, Alaska, and the former Soviet Union. Nowhere else on their migration route do the cranes linger so long as they do on the Platte, because nowhere else on the continent provides what they need when they need it.

This staging of sandhill cranes is one of the last remaining wildlife spectacles in North America. . . . Besides this assemblage of sandhill cranes, the Big Bend of the Platte sustains millions of migrating and wintering geese and ducks, including virtually all the midcontinental white-fronted geese and impressive numbers of pintails and mallards. In summer, the last remaining sandbars on this stretch of the river provide nesting grounds for two diminutive Great Plains birds—the interior least tern, which is endangered, and the piping plover, a threatened species in the Great Plains.

Occasionally during their migration between Alberta and Texas—sometimes in the fall, sometimes in the spring—several of the 133 whooping cranes in the Wood Buffalo flock will stop off for a few days along the Big Bend of the Platte. Their appearance always causes a stir. Like gemstones, the most dazzling and rarer they are, the more we cherish our wild creatures. Consequently, the endangered whooping crane is often cited as the ne plus ultra of Great Plains species that frequents the Platte. And because our federal and state wildlife laws have come to reflect this peculiar gemstone mentality about wild things and wild places, the whooping crane has had to bear the onus of proof in public hearings and courtrooms that the Big Bend of the Platte is worth saving.

It is the long-held opinion of National Audubon Society that the Big Bend of the Platte is worth saving not for one but for countless reasons, not the least of which is the river's importance to the sandhill crane, a species that is relatively stable at present but that could be devastated by further degradation of the river. Saving the Platte explores a view that is all but lost in environmental debates these days: that we should save places not for that which is rare and endangered but for that which is common and vulnerable. . . .

Occasional high flows still come coursing down the main stem from the South Platte, or as a result of heavy rains that fall outside the embrace of any of the reservoirs, or in years of widespread and massive snow in the Rockies and on the Great Plains. But for all practical purposes, the Platte is now subdued, like a person on tranquilizers. Additional dams and diversion projects on the various stems and tributaries of the Platte would hold back more sediment and erase the remaining peak flows. The Platte would then be like a person lobotomized.

maintain a relationship with the land, which has allowed the animals and the birds to grow and flourish for thousands of years. One of the Cree elders said, "My father and grandfather hunted here and after all their lives on the land, they did not leave a trace that they passed this way." This was the accomplishment of my people.

The impacts of James Bay I are only gradually unfolding. To my people who live off the land, it has become a nightmare. It has contaminated our fish with mercury. It destroyed the spawning ground of the fish. It destroyed the nesting ground of the water fowl. We have nowhere else to go. All we want is to transfer the knowledge we have to our children, to teach them a way of life that they can use to support their family. They can then live peacefully and live in harmony with nature as our ancestors did.

Our territory is very flat. We live in a place where the last ice age started, a land of low relief, striated lakes and wetlands, with a low cover of spruce shrubs. If you build a dam on one of our rivers, the water quickly envelops an enormous territory and eventually finds another outlet to James Bay. The dam builders must erect enormous structures, scouring the land for sand and gravel, tearing away at the fragile plant life to locate the vast quantities of material they need to confine the waters on such a level surface.

[The project] rearranged the migration road of all animals. You see it destroy the small and big game. The muskrat, otter, and beaver are no longer there. When you do this, you drive us off the land because we depend on those animals and have taken care of them. They've made the rivers to flow backward. The seasons are backwards. The rivers flow between November and March, and in the springtime, when they're supposed to flow, they don't because they collect the water in the reservoirs.

If man finds another way to use energy, these dams will be a monument to his stupidity. They will be an embarrassment to our children and their children's children. Those will be the only monuments you will see on Cree land, because we don't build them. Unlike James Bay I, James Bay II is not built yet. We can stop a major ecological disaster and cultural genocide for our people before it happens.

Solar Enthusiasts

The technology for solar has become more efficient and cheaper since the 1970s. Photovoltaic cells, which produce electric current when bathed in sunlight, dropped from $50 per peak-watt to $5. If they drop to $1, they could be very competitive. Investment by business and government in research could help. Solar enthusiasts see vast tracts of photovoltaic collectors transmitting cheap electricity over long distances. The proliferation of the technology has already created new jobs and markets.

Replacing fossil fuel units with solar collectors could be done over fifty years. The land requirement for solar would be equivalent to an interstate highway system or 5% of our farmland. It would be important to use lands that are not ecologically sensitive or especially productive. Electricity from photovoltaic cells installed on top of commercial buildings (as well as possibly residential buildings) could feed power into the grid during the daytime, where it would be stored by utilities for nighttime use. Compressed gas, pumped hydro, and hydrogen could be used to store it. Electricity at night would cost twice as much and lead to the development of technologies for storing heat and the maintenance of cool temperatures. Cleaning of the solar collectors would have to be done carefully to avoid water pollution. Electricity could be transported, following the sun from east to west, across the country and along the transmission network.

In order to jump-start utilities into using more solar energy, the National Audubon Society launched a nationwide Solar Brigade to get thousands of people to write a note to their electric company when they paid their bill, that said, "We want 10% solar by the year 2000." If utilities would make this conversion, the cost to consumers should be no more than a 1% a year rate hike. Such a conversion would also eliminate 170 million tons a year of CO_2 from the Earth. Just as public pressure forced companies to put environmentally friendly products on supermarket shelves, the public can make its wishes known to electric utility companies.

The Solar Brigade is part of a long-range goal to get our country on the road to an environmentally sound, sustainable energy program.

Paying the Damage Costs

Electric utilities and other industries need to be held responsible for the extensive ecological damage they cause. We shouldn't destroy wildlife, wilderness, and native cultures out of greed. We must think of generations to come. When we resolve to protect nature, engineers are challenged to come up with better solutions. They can do it, too, once they know what we won't stand for. Thus, including environmental damage costs in the price of electricity would lead to a more accurate sense of how expensive resources are relatively and inspire industries to engineer more efficient uses. The price should include not only damage by the facilities, but also pollution damages from the production of materials themselves, such as the steel in nuclear plants, wind turbines, and computers. Damage to buildings, clothing, and health also needs to be considered. One way engineers could be challenged is to automatically include environmental costs. If damage cannot be eliminated, at least cleanup is to be provided for. Such an ethic would stimulate industry to do less harm to the environment.

One example of the need to take precautions is the siting of energy transmission routes because of the possible risks to health from electromagnetic fields. A number of epidemiological studies have found associations of time-varying magnetic fields with cancer.

Electric and magnetic fields are invisible lines of force that surround any wire conducting electricity, including power lines, house wiring, and wires in appliances. An electric field is associated with the voltage or force that moves electricity through wires; a magnetic field is produced by current, a measure of the movement of electric charge through the wires. The strength of such fields drops off with distance from the source, so power lines present a potential risk primarily to people who live or work near them. Generally, if you live more than about 200 yards from transmission lines, magnetic fields from household sources will be stronger than those from the lines. Appliances, such as electric blankets, toasters, clocks, shavers, and can openers result in field intensities greater than those from transmission lines.

Industry can come up with ways to shield unwanted magnetic forces, although that step alone may not be enough because as yet scientists do not understand the exact connection between these fields and illness.

Remember, an ecologically planned world is not too much to expect from professionals and engineers.

Our businesses and industries can utilize the many products and technologies that already exist in order to make the transition to a sustainable society. We can expect them to make better appliances that last longer and to not extract more materials from the earth than they are willing to keep circulating and cycling in the economy.

And we likewise can examine our own lifestyles and not just wag our fingers at others. We can invest in appliances that work well and not be seduced by advertising for the "latest" model. We can restrain impulse shopping and live more efficiently. Not only will our bank account benefit, so will the natural "safety deposits" of our world.

EcoQuiz

Q: Why do utilities offer rebates on air conditioners, freezers, washing machines, and microwave ovens?

A: Utility companies and then regulators realize that it costs less for them to give you a rebate for buying an energy-efficient appliance than to build new power plants to meet higher demand. They want to pay you to save money for them. We all win in the long run.

Q: Are current computers a threat to health?

A: Toxic chemicals, such as glycol ether (a solvent used in circuit boards), arsenic, and lead in microchips have been linked to birth defects, miscarriages, and reproductive problems.

Q: What can you do about the CFCs and HCFCs swirling in your home or car air conditioners?

A: You can make sure there are no leaks in your automobile's system. It is the leaks that are ozone-deadly. Check annually.

Q: Do you have to be dependent on your electric utility?

A: No, enterprising souls can buy or make solar devices to operate equipment. A good source of information is the American Solar Energy Society, 2400 Central Avenue G-1, Boulder, CO 80301.

ECOSYSTEM PATHWAYS

When we step outside our home in the city, suburb, or country, do we know what ecosystems surround us? Do we see how our neighborhood is dependent on nature cycles that extend far beyond its borders? Flora and fauna, from microorganisms to our pet dog, share this habitat and play a role in its future. In ecosystems each member of a community of organisms depends on the others and perpetuates the poem of life.

Ecosystems are like a piece of fabric; one thread pulled can unravel the whole. They are full of paradoxes. They can be fragile and yet enduring, contain beauty and violence, and demonstrate astonishing examples of cooperation as well as competition. Life goes better for all when the sound of the frog can be heard, when the stream gurgles, when the reed sways in the breeze.

How do the cycles of nature operate in our open spaces? The picture at the beginning of this chapter provides some clues:

- Our yards have their place in the midst of the circulating global cycles of water, carbon, oxygen, nitrogen, and minerals. For these to function successfully, they cannot be overly stressed.
- Decomposers, such as fungi and bacteria, cause the decay of dead bark, leaves, and other debris, which in the process releases carbon dioxide, water, and heat. If decay is blocked, the food chain may be broken, with profound effects on other ecosystems.
- A community of plants and animals keeps itself alive through forms of cooperation, and in check through competition for the same food, water, minerals, and sunlight. Destruction may be caused by disease, poisoning, predators, storms, or overcrowding. Pollination and fertilization is carried on by wind dispersion, birds scattering seeds, and insects seeking nectar.
- Symbiotic activity is key. For example, some beetles carry on their backs a small community of mites dependent on them for survival. In the domain of ants and aphids, aphids get their food by sucking juice from plants. Ants will stroke aphids with their antennae, prompting the aphid to release a drop of sweet liquid from its abdomen, which the ant then drinks. As a consequence, ants protect aphids and will occasionally carry them to good sites for feeding. The barn owl is the most effective controller of rodents. Loss of even one could alter the local predator-prey balance.

We do not know enough about the needs of insects and vegetation to decide which components can be left out. We do know that plenty of diversity is essential for the survival of all.

- Trees prevent soil erosion as their roots hold soil firm. Their leaves decrease the concentration of carbon dioxide in the atmosphere. Because of the breathing they do for us, they have been called our lungs.
- Trees, brush, and flowers provide food and habitat for birds and insects. The "ecotones," or edges between communities existing in lawn and shrub, pond and marsh, are particularly rich because they contain an overlap of biotic organisms.
- Each yard has a particular microclimate—temperature range, humidity level, amount of light and shade—that determines which microorganisms, plants, and animals can inhabit it.
- A compost pile converts food and yard debris into fresh rich soil.
- A marsh, pond, or bog is full of diverse wildlife and plants that filter out sediment and break down waste, purifying groundwater, protecting land from flooding, and preventing erosion.
- A woodpile is home to rodents and insects that balance the local insect population and help decompose dead matter.

Sue Hubbell, in *A Country Year*, writes about her attitude to such creatures:

> **In truth, I don't mind the wood cockroaches that come in on my firewood. Their digestive system and mine differ enough so that we don't share the same ecological niche; they do me no harm, we are not competing, so I can take a long view of them. There is no need to harry them as a bee would, or to squash them as a housewife would. Instead, I stoop down beside them and take a closer look, examining them carefully. After all, having in my cabin a harmless visitor whose structure evolution has barely touched since Upper Carboniferous days strikes me, a representative of an upstart and tentative experiment in living form, as a highly instructive event. Two hundred and fifty million years, after all, is a very long view indeed.**

Humans have deeply disrupted outdoor spaces. Some of the pressures that can easily upset the delicate balance of ecosystems are:

- Roads. Pavement covers and destroys natural activity above and below the surface. It fragments ecosystems. It attracts cars and more development, which generate noise and fear and fumes and is highly threatening to plants, animals, and insects, if not humans.

(Continued on page 78)

6
OUR YARDS

Our backyard may be a grassland, desert, park, wetland, forest, lawn, or vacant lot flanked by sidewalks. In the United States most humans live in a city apartment or suburban house, within 200 miles of an ocean, lake, or river shore. Development of human communities, through covering land and waterways with concrete, steel, stone, asphalt, wood, brick, and glass, obliterates local ecosystems and has an impact on larger ones. A current unresolved debate is whether it is better for humans to live closely and vertically packed in cities, leaving broader expanses of land open to plants and wildlife, or in the single-unit homes that characterize horizontally sprawling suburbs. Wherever we live, we can make our outdoor environments more congenial for ourselves as well as for the rest of nature.

Architecture-philosopher Lewis Mumford once said, "Forget the damned motor car and build cities for lovers and friends." While his statement neglects nature, it does suggest irritation at how much development has accommodated automobiles more than anything else. In many European cities efforts have been made to "calm" traffic by strategically placing trees, bushes, and flower beds along the road, gentle inducements to go slowly

(Continued from page 76)

- Chemical fertilizers, pesticides, and fungicides can be deadly to the food web of plants, animals, and insects; poisonous to the water below ground as well as to soils; and are often transported to neighboring land.
- Imported exotic flowers or shrubs disrupt or displace native plants and the communities dependent on them.
- Fumes from the unregulated lawn mower and barbecue pollute the air, the water, and the soil.
- The construction of a building foundation erodes soil.
- A garage full of paints, oils, pesticides, and other toxic chemicals is a time bomb—its contents are all too likely to be dumped into the ground or to volatilize into air and interact with sunlight to produce smog.
- The hose wastes water if left running when cars are washed or gardens watered, and may cause soil erosion if used heavily.

THINKING LIKE AN ECOSYSTEM

Pull of the Moon

Peter Steinhart

There are times at night when I am driving or hunched at my desk over some beetle of purpose, and I look up to see the full moon looming large and yellow over the horizon. It is a commanding presence. Something unhitches my mind from its distant errand and draws me outside into the night. If I can find the stillness inside me, the silence of the evening slips over my imagination. I become aware of the vast distances of space, the undeflected flight of ancient starlight, and the freshness and improbability of earthly life. It makes me feel young and free. At such moments a surge of joy leaps through me.

There was plenty of reason for the ancients to worship the moon. The moon exerts a profound influence on the Earth. As it passes overhead, its gravitational force pulls a bulge of seawater, causing the high tides. As it lines up with or opposite the sun—at the times of the new and full moons—the tides are highest. It even pulls a bulge in the Earth as it passes. Moscow rises twenty inches twice a day. And at times the Empire State Building is sixty-three feet closer to the Eiffel Tower. Because it tugs harder on whatever is closer, it pulls the Earth more than it pulls the water on the far side of the Earth. So there is a bulge of water on the backside of the globe too—a second daily high tide. . . .

Creatures of the sea are governed by the pull of the moon. Many of them feed or reproduce in lunar cycles. Oysters open their shells when the moon is high. Shore crabs, mud snails, and mussels are most active at 12.4-hour intervals. Fiddler crabs change color on 12.4-hour schedule. The chambered nautilus forms a new chamber in its spiraled shell every lunar month. California grunion come ashore to deposit their eggs in the sand of California beaches on the high tides of the spring full moons.

The palolo, a worm which burrows into the coral reefs of Fiji, bursts to the surface in wriggling masses of green and brown during the October full moon. The luminescent marine worm Eunice fucata surfaces at night in the Caribbean on a lunar cycle. . . .

Though we are distantly descended from aquatic creatures, we keep some of the old timepieces. The human menstrual cycle is twenty-nine days, the length of a lunar month, and human gestation is nine lunar months long. Births are slightly more common when the moon is full. Human blood chemistry changes regularly on a twenty-nine-day cycle. The human heart beats faster under a full moon. Bleeding from wounds is greater at new and full moons. Some people gain and lose weight in lunar cycles. . . .

The modern world is moonblind. Outside our walls and windowshades the same silvery moon shines. But in the house-habit of the modern age, we no longer see the world in moonlight. The glare of streetlights and the dust of pollution veil the night sky. And the ringing of telephones, the mutter of prowling automobiles, and the errands of an interlocked humanity turn our eyes away from the heavens. Our age favors precision, narrow calibration, gridwork dependency, the kind of discrimination we perfect in daylight. Though men have walked on the moon and science has published its secrets, we know it less. Few of us can say precisely what time the moon will rise tonight.

and yield to pedestrians, cyclists, and children. One reason for the popularity of shopping malls is the opportunity they afford to be away from traffic in a place with shade-giving plants and sunlight. Many urban communities replace barren lots with communal gardens. In congested Newark, New Jersey, for example, there are 1500 small-scale gardens producing $750,000 worth of fruits and vegetables. Compost from the citywide yard-waste program feeds the gardens. In one year over 3000 trees were planted. With community leaders, landscape architects, and city/town planners working together, cities can be made greener.

Perhaps one of the prime causes of human crime and despair is separation from nature. When we live surrounded by concrete and traffic, we become alienated from our environment. Often such people assault their impersonal neighborhoods. Nature appears to be a soul need as well as a biological one.

How Flowers Changed the World

Loren Eiseley

A little while ago—about one hundred million years, as the geologist estimates time in the history of our four-billion-year-old planet—flowers were not to be found anywhere on the five continents. Wherever one might have looked, from the poles to the equator, one would have seen only the cold dark monotonous green of a world whose plant life possessed no other color. . . .

Flowers changed the face of the planet. Without them, the world we know—even man himself—would never have existed. . . . Their appearance parallels in a quite surprising manner the rise of the birds and mammals.

Slowly, toward the dawn of the Age of Reptiles, something over two hundred and fifty million years ago, the little naked sperm cells wriggling their way through dew and raindrops had given way to a kind of pollen carried by the wind. . . . Once fertilization was no longer dependent on exterior water, the march over drier regions could be extended. Instead of spores, simple primitive seeds carrying some nourishment for the young plant had developed, but true flowers were still scores of millions of years away. After a long period of hesitant evolutionary groping, they exploded upon the world with truly revolutionary violence. . . .

The true flowering plants (angiosperm itself means "encased seed") grew a seed in the heart of a flower, a seed whose development was initiated by a fertilizing pollen grain independent of outside moisture. But the seed, unlike the developing spore, is already a fully equipped embryonic plant packed in a little enclosed box stuffed full of nutritious food. Moreover, by feather-down attachments, as in dandelion or milkweed seed, it can be wafted upward on gusts and ride the wind for miles; or with hooks it can cling to a bear's or a rabbit's hide; or, like some of the berries, it can be covered with a juicy, attractive fruit to lure birds, pass undigested through their intestinal tracts, and be voided miles away.

The ramifications of this biological invention were endless. Plants traveled as never before. They got into strange environments heretofore never entered. . . .

The explosion was having its effect on animal life also. Specialized groups of insects were arising to feed on the new sources of food and, incidentally and unknowingly, to pollinate the plant. The flowers bloomed and bloomed in ever larger and more spectacular varieties. Some were pale unearthly night flowers intended to lure moths in the evening twilight, some among the orchids even took the shape of female spiders in order to attract wandering males, some flamed redly in the light of noon or twinkled modestly in the meadow grasses. . . .

Across the planet grasslands were now spreading. A slow continental upthrust which had been a part of the early Age of Flowers had cooled the world's climates. The stalking reptiles and the leather-winged black imps of the seashore cliffs had vanished. Only birds roamed the air now, hot-blooded and high-speed metabolic machines. . . .

Grass has a high silica content and demands a new type of very tough and resistant tooth enamel, but the seeds taken incidentally in the cropping of the grass are highly nutritious. A new world had opened out for the warm-blooded mammals.

Every yard is a theater of biodiverse activity, in which millions of organisms are caught up in acts of murder, mating, self-sacrifice, feast, and birthing. As Diane Ackerman has written:

> There are marauding bands of hunters; sculptors of leaf and bark; some of the first papermakers on Earth; rustlers and ranchers with their own stockyards; gatherers of fruit and petal; architects working in clay; nurseries for the young; catacombs, lean-tos, and towers; weavers of silk and cotton; builders of cities; practitioners of order; many different tribes and society.

BIRDS

Every year Audubon members and many other bird lovers around the world conduct bird censuses. They do it in part because birds are the first groups of animals to be affected by adverse threats to habitats, including pollution. The bird count gives much-needed databases in quantifying bird populations across our continent, from the Arctic shores to tropical rain forests and beyond into Latin America, where many North American breeding birds spend their off-seasons.

The findings have been discouraging: waterfowl populations have been reduced because of the loss of wetlands to development; so have threatened bird species, such as the marbled murrelet and the spotted owl, because 400,000 acres a year of ancient forests are logged; likewise, songbirds because of the millions of acres of rain forest lost each year and, equally important, because of habitat fragmentation on the breeding grounds.

How Much Room Do Creatures Need, Anyway?

Among the many seasonal nature cycles is the migration of species. If we observe closely the comings and goings of the birds and other creatures in our yards, we will realize how vital our yards are to their feeding, nesting and breeding needs.

From June to September, *American Birds* reports that migrating males of many species of songbirds precede the females to stake out a breeding territory. Hawk migrations reach peak intensities on warm sunny days when thermal updrafts are well developed. Northbound migrants along the western Gulf of Mexico often encounter severe headwinds that force millions down into the spare vegetation along coastal Texas, Louisiana, and Alabama. Waterfowl tend to be confined to a narrow corridor determined by the availability of suitable habitat. Many kinds of waterfowl begin to move northward as soon as the lakes and ponds are released from the grip of ice.

It must be remembered too that many mammals migrate. Gray whales migrate from the north Pacific to the coastal bays along Mexico to care for their young in these warm waters, and then return northward. In the Arctic

Hope is the thing with feathers that perches on the soul.

—Emily Dickinson

caribou migrate from one range to another in search of fresh supplies of lichens and other plants. Salamanders migrate to breeding pools in the woodlands to lay eggs and then return to the surrounding woods. In the fall, monarch butterflies migrate from our meadows to mountains in Mexico where they rest until returning north in the spring.

Such creatures are truly citizens of the world: they don't recognize boundaries between states or nations, your property or your neighbor's. Birds migrate in order to find food and warmer places to raise their young. The North Temperate Zone embraces fifty times more land surface than the South Temperate Zone. Space is important because each breeding pair needs to obtain sufficient food for themselves as well as for their voracious and rapidly growing young. In addition, the explosion of flowers, fruits, and insect populations during the northern summers enables birds to have more young and allows those young to grow faster than is the case in the tropics. Migrating to the Arctic for the summer, for example, provides twenty-four hours of daylight. This exposure greatly reduces the amount of time that nest-bound young are vulnerable to predation. Nonmigratory birds, it has been found, have smaller families, and the young tend to grow at a slower rate. Food gathering in the tropics requires more work in just twelve hours of daylight.

Among migrants, the urge to fly occurs when the required amount of sunlight combines with weather conditions to create chemical changes in their bodies, causing them to become restless with the urge to move in a certain direction. Sandpipers spend their winters at the southern tip of South America, then fly to raise their brood north of the Arctic Circle. Hummingbirds double their weight before taking off from Central America to reach the eastern United States. Many birds fly at night only. Most fly at low speeds, meaning that their journeys take weeks or months.

The more rapid flyers, such as plovers and falcons, go from the Arctic to Patagonia in a series of overwater flights. Orioles shuttle between South America and our northern states at a speed of about 25 mph. Barn swallows take two minutes to go a mile. Robins, crows, pelicans, and sparrows fall in the 20 to 25 mph bracket. Hummingbirds and golden eagles go 50 mph. The altitude at which most birds fly depends on that of the most favorable winds, but is seldom more than 5000 feet. Smaller birds seem to hold to below 1000 feet. On the other hand, mountain ranges force migrants much higher. Cranes and storks in the Himalayas have been seen at above 20,000 feet.

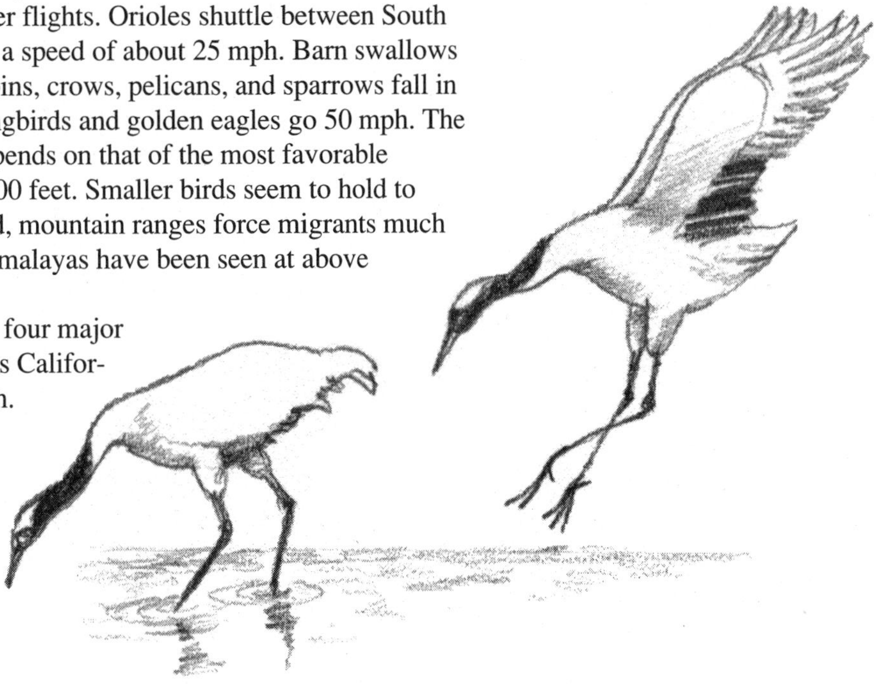

In the United States there are four major flyways. The Pacific extends across California, Nevada, Utah, and points north. The Central extends across Texas, Colorado, Wyoming, Montana, and Canada. The Mississippi extends from Louisiana northward to Minnesota and Nebraska and across Illinois, Indiana, and Ohio. The Atlantic extends from Florida north through states along the Atlantic coast to New York, Michigan, Wisconsin, and eastern Canada.

Here are some of the best sites to seasonally witness migratory birds, according to *American Birds*:

SPRING

Bonaventure Island, Québec
Baxter State Park, Maine
Elk Lake, New York
Scherman-Hoffman Sanctuaries, New Jersey
Delaware Bayshore, New Jersey and Delaware
Huntley Meadows County Park, Virginia
Dry Tortugas, Florida
Whitefish Point, Michigan
Point Pelee, Ontario
Indiana Dunes National Lakeshore, Indiana
Great Smoky Mountains National Park, Tennessee/North Carolina
DeSoto State Park, Alabama
Churchill, Manitoba
Lostwood and Des Lacs National Wildlife Refuges, North Dakota
Cheyenne Bottoms, Kansas
Big Bend National Park, Texas
Lost Maples State Natural Area, Bandera County, Texas
Grand Teton National Park, Wyoming
Pawnee Grassland, Colorado
Glacier Bay National Park, Alaska
Okanagan Valley, British Columbia
Malheur National Wildlife Refuge, Oregon
Point Reyes, California
Morongo Valley, California
Cave Creek Canyon, Arizona

SUMMER

Grand Manan Island, New Brunswick
Block Island, Rhode Island
Bombay Hook National Wildlife Refuge, Delaware
Cape Hatteras, North Carolina
Jekyll Island, Georgia
Bill Baggs/Cape Florida State Recreation Area, Florida
St. Marks National Wildlife Refuge, Florida
Crane Creek/Magee Marsh, Ohio
Lake Calumet, Illinois
Saylorville Reservoir, Iowa
Grand Isle State Park, Louisiana
Last Mountain Lake/Big Quill Lake, Saskatchewan
Great Salt Plains National Wildlife Refuge, Oklahoma
Falcon Dam, Texas
Dempster Highway, Yukon/Northwest Territories
Deer Flat National Wildlife Refuge, Idaho
Bear River Migratory Bird Refuge, Utah
Rocky Mountain National Park, Colorado
Madera Canyon/Florida Wash, Arizona
Percha Dam State Park, New Mexico
Leadbetter Point, Washington
Klamath/Tule National Wildlife Refuges, Oregon/California
Monterey Bay, California
Point Loma, San Diego, California

AUTUMN

Hartlen's Point, Nova Scotia
Newburyport, Massachusetts
Cape May, New Jersey
Montauk Point and Jones Beach, New York
Brigantine Division, Forsythe National Wildlife Refuge, New Jersey
Chesapeake Bay Bridge-Tunnel, Virginia
Huntington Beach State Park, South Carolina
"Ding" Darling National Wildlife Refuge, Florida
Niagara River, Ontario/New York
Reelfoot National Wildlife Refuge, Tennessee
Pascagoula River Marsh, Mississippi
Hawk Ridge at Duluth, Minnesota
Morton Arboretum, Illinois
Table Rock Lake, Missouri
Garrison Dam, North Dakota
Lake Tawakoni, Texas
Laguna Atascosa National Wildlife Refuge, Texas
Bonny Reservoir, Colorado
Bosque del Apache National Wildlife Refuge, New Mexico
Corn Creek, Nevada
American Falls Reservoir, Idaho
Sulphur Springs Valley, Arizona
San Juan Islands, Washington
Sauvie Island, Oregon
Palo Alto Baylands, California
Salton Sea, California

WINTER

L'Anse-aux-Meadows, Newfoundland
Provincetown, Cape Cod, Massachusetts
Sandy Hook, New Jersey
Blackwater National Wildlife Refuge, Maryland
Bodie Island, North Carolina
Merritt Island National Wildlife Refuge, Florida
Everglades National Park, Florida
Amherst Island, Ontario
Presque Isle State Park, Pennsylvania
Kalamazoo Nature Center, Michigan
Land Between the Lakes Recreation Area, Kentucky/Tennessee
Wapanocca National Wildlife Refuge, Arkansas
Sabine National Wildlife Refuge, Louisiana
Platte River, Nebraska
Marais des Cygnes Wildlife Management Area, Kansas
Lake Hefner, Oklahoma
Bentsen–Rio Grande Valley State Park, Texas
Jackson Hole, Wyoming
Bitter Lake National Wildlife Refuge, New Mexico
Buenos Aires National Wildlife Refuge, Arizona
Davis Dam, Nevada/Arizona
Clover Point, Victoria, British Columbia
Tillamook Bay, Oregon
Gray Lodge Wildlife Management Area, California
Upper Newport Bay, California

Our feathered friends obviously experience a host of dangers on their flights. One of them is the tendency to regard the disposal pits of oil companies as sources of water, when actually they are full of toxins, oil, and brine. But the most significant problem is the loss of habitat. The lives of birds hinge on the presence of migratory stopovers being at the right place at the right time. With increasing numbers of humans developing more land and clearing out forests, essential territory is disappearing. Over the last decade bird-watchers everywhere report declines in migrants.

The National Audubon Society has a program called Birds in the Balance, whose goal is to maintain and restore bird populations at/to historical levels of abundance throughout their ranges in the Americas and the Caribbean, emphasizing the protection and management of avian habitats. The time to save a species is while it is still common—not after it is critically endangered. One urgent reason for not letting a species become endangered is that its genetic structure may be essential to our future in ways that we cannot comprehend at present. Therefore, species and habitats must be saved for their very uniqueness. Extinct means gone forever, as we know.

Because birds, as well as other animals, need "living room" in order to feed and nest successfully, the fragmentation of landscape is a major contributor to population declines. Some birds spend their lives in a quarter of an acre; others roam for miles. Habitats have carrying capacities based on the soil and plant life in them. When one or more bird populations are greater than the ability of the land to support them, the surplus will die or move away. Humans greatly affect populations by disturbing the plant life and the corresponding food chain on the land.

Imagine birds—or rabbits or deer—attempting to get from one place to another in your neighborhood. Are they confronted with huge dwellings, fences, roads with deadly cars, wires, or other interferences? It has been observed that when forests are cut back, migrant birds spill over into the edge of brush. They no longer have a canopy of high branches or trees of different ages to forage around. One need of birds is to be able to safely forage in early sunlight, which a forest permits more than unsheltered areas. They also seek to breed in areas where others of their kind live successfully and where there is more reproductive potential. Thus, when a range cannot host enough members of their species, they will not survive. As yet, a great deal is unknown about the precise needs of creatures, but we do know that we must be more sensitive to their needs in order for them to survive.

Birds dream of territories enclosed by singing.

—Margaret Atwood

A case in point is the white-crowned pigeon—not the pigeon we see flocking in the city, but a rare breed that resides in the Caribbean and the Florida Keys. It is a slate-gray bird with a white crown and a red bill with a white tip. Its legs are also red. Such pigeons are extremely shy of people, feed exclusively on fruit, and nest on isolated mangrove islets. They depend on the fruits of tropical hardwood trees, and the trees depend on them for propagation through excretion of undigested seeds. But the continued destruction and fragmentation of hardwood forests threatens the maintenance of the white-crowned pigeons' population. It is listed as a threatened species. The birds seem to require at least twelve acres to attract young pigeons to arrive, then disperse to feed and breed. They also need to be safe from raccoons. To protect these birds we can make sure that a network of protected habitats is provided so that the birds at least have "stepping-stones" from one to the next. Also, people can plant more hardwood trees, such as poisonwood, in their neighborhoods.

Steve Kress, an Audubon biologist, has diligently done some remarkable restorative work with his "Puffin Project." Over a century ago puffins, having been slaughtered for meat and feathers, had disappeared from islands off the Maine coast. Ever since Kress had been an ornithology instructor at a camp on Hog Island, he'd thought that the area was impoverished by the loss of its original birdlife. Since the mid-1970s puffin chicks have been brought from their Newfoundland hatching ground in hopes that they will later return to breed. Kress's crew experimented with building adequate nesting burrows. Decoy puffins were perched atop rocky cliffs. Recordings of puffin sounds blared. Now there are 150 nesting pairs on Matinicus Rock and fifteen on Seal Island. Similar methods have been used to restore the endangered roseate tern to Maine sanctuaries.

In the Galápagos archipelago it was observed that humans had introduced many predators—rats, dogs, cats, pigs—which had devastated the dark-rumped petrel breeding colony. Kress and his team tried to encourage the birds to nest in areas with few or no predators. They hand-dug 160 burrows within an extinct volcano in the Santa Cruz highlands. In four years, they attracted six pairs, which nested successfully, and others appeared ready to come. This exemplary work can be applied throughout the world to lure wildlife to recolonize areas, but it takes persistence, patience, and care.

In fixing habitats, we have to keep firmly in mind that

> *Everything is connected to everything else.*
> *We cannot do "just one thing"*
> *to solve "just one problem."*

How to Make Our Yards Wildlife-Friendly

Our yards are under siege. An obsession with neat and highly controlled landscaping, for example, can create havoc in backyard ecosystems. We may weed out strong healthy native plants and use our yards to show off foreign ones. We prune in fanciful shapes and ruthlessly spray a leaf or stem that dares to creep around a patio. At the sight of a gypsy moth cocoon, we panic and call in the chemical sprayers, causing destruction to insects and birds. Our yards are hardly private oases either. Pollutants and wastes flow out of our homes into the soil, air, and water of those who live near us or farther away.

A way to lessen the threat to the ecosystems in our yards is to learn which organisms are present and to follow a program of integrated pest *management*, as opposed to elimination. For instance, we can cut our lawns high so that weeds are shaded out. Many environmental and gardening books are full of excellent advice on organic gardening, composting, and recycling.

Audubon cannot recommend *any* chemical treatment of plants or insects because the hazards to living creatures are simply not known, despite the fine print on the labels. If you have been using pesticides or herbicides, you can

What is a garden pest? Someone who thinks: Shoot it, cut it down, or kill it with chemicals.

gradually reduce use, making a goal, for instance, of using 50% less than the year before. Meanwhile we all can try other ways to manage or control the insects and groundhogs that devour our veggies.

The best-known safe methods are: (1) plant selection, e.g., basil interplanted with tomatoes wards off worms; (2) habitat modification, e.g., removing damp wood to deter slugs and rats; (3) physical controls, e.g., handpicking; (4) barriers, e.g., screens, fences, safe traps, and sticky goos; (5) cultural controls, e.g., mulching; (6) biological controls, e.g., birds, toads, snakes, owls, milky spore disease; and (7) building healthy soil with compost.

We can learn to appreciate bugs and animals as allies instead of drowning them with chemicals. Shrews, moles, and hedgehogs eat invertebrate pests, such as woodlice, millipedes, and slugs. Birds feed on grubs, caterpillars, and aphids. Frogs and toads are predators of many small insects. Ladybugs feed on aphids. Ground beetles and centipedes eat eelworms, cutworms, and insect larvae. Earthworms, of course, are vital for aerating the soil and carrying organic matter underground. They all help to ensure the continuation of life.

With a little ingenuity our yards can be made into friendly habitats for birds, rabbits, butterflies, or flowers, not to mention ourselves. Stability in nature is often dependent on the diversity present.

To get a sense of the possibilities in a yard, make a map. It does not have to be very precise, but should give a picture of the layout of the yard and what's in it. Show groupings of large and small trees, shrubs, annual and perennial gardens, as well as types of bird feeders, sources of water, and hiding places, such as brush, log, or rock piles.

It may help you to know that birds make their nests at different heights. For instance:

- In high treetops are found: warblers, vireos, orioles
- In saplings or lower branches: robins, bluejays, grosbeaks, redstarts
- In low bushes: sparrows, cardinals, catbirds
- On the ground: ovenbird, junco, towhee, quail, whippoorwill

Notice how the property is bordered—by woods, a neighbor's backyard,

Saving Seeds

Jon Luoma

Kent Whealy had been reading warnings from geneticists about the erosion of diversity in America's seed stock. Commercial vegetable seeds were, increasingly, hybrids developed from limited parental lines with traits like tough-skinned shipability or uniform ripening, favored by growers. But in the nation's backyard gardens were uncounted other varieties that had been preserved among families and friends. Many had traits the commercial growers shunned—too small, not uniform, late ripening, too delicate to be machine-harvestable. But the vegetables had been grown all those generations for good reason. Chiefly, and often in sharp contrast to supermarket varieties, they had flavor.

So in 1975, the Whealys set out to locate a few other savers of seeds, with a notion that a little network might be set up, whereby seeds could be exchanged among gardeners and thus better preserved and shared. The first year, by "advertising in garden and back-to-the-land" magazines, they found six heirloom gardeners.

The network, called the Seed Savers Exchange, based in Decorah, Iowa, has burgeoned and blossomed since then. It has purchased a small farm as a headquarters and as a site for its vegetable gardens, for an heirloom apple orchard, for a collection of rare chickens and cattle, and for a series of seasonal festivals. Its winter catalog was 272 pages long, with its nearly 800 members offering, for the cost of postage to other members, more than 6,000 varieties of seeds.

or other houses or buildings. Try looking at the area from a bird or mammal's point of view. Does it offer the essentials for survival—food, water, shelter, and safety for raising young?

A good rule of thumb in planning a wildlife habitat is to provide a variety of vegetation. Start by assessing how much lawn area is really needed. Some lawn is desirable for play or sitting areas, but much of it is more costly to maintain than it is worth in value to wildlife. Instead of lawn consider a meadow or prairie of perennial native wildflowers and grasses. A varied ground cover can make it easy for grasshoppers or songbirds to forage.

An uncut lawn may seem unkempt, with its dandelions, thistles, and other so-called weeds. However, this anarchy is a haven for many species of wildlife. Chickweed produces a multitude of seeds that attract many ground-feeding birds. Propagating black-eyed Susans, sunflowers, goldenrod, milkweed, pink fleabane, butter-and-eggs, wild strawberries, and asters will create a beautiful space. Prairies are enhanced by grasses, such as switchgrass, little bluestem, Indian grass, and broom sedge.

Butterflies are attracted by tithonia, zinnias, asters, and butterfly weed. The caterpillars need leafy foods, such as pussytoes, pearly everlasting, and fennel to complete their growth cycle. Butterflies also like water in hidden places where the young can be safe. It's entirely possible to count at least two dozen different species of butterflies in a backyard if they are provided for.

You can make a substantial contribution to ecosystems by cultivating plants that are native to your region. You can even re-create entire native plant communities, whether prairie, desert, or deciduous forest. In many instances these restore decimated flora and give hope to habitats. While nothing we do can substitute for the complexity found in nature, we can learn to work with natural systems and help heal damaged areas.

Here are the reasons why it is worth perpetuating native plants that are indigenous to a region:

1. They maintain natural diversity. Aliens take over space and nutrients, resulting in a monoculture.
2. They maintain the genetic integrity of plant populations.
3. Because they have adapted to a particular landscape, they tend not to need pesticides or fertilizers.
4. Encouraging natural diversity is good for the wildlife inhabitants too.

"Succession" is the term for the stages of growth that inevitably take place, if unimpeded, on a piece of land. The larger the area, the more stages are likely to be visible. The first stage is often what we see in vacant lots. In five years' time an unmowed meadow will enter the shrub stage. Brush is essential for giving birds and mammals shelter, nesting spaces, and a generous supply of berries. Some land should be allotted to young trees, shrubs, and vines.

Land let go for ten years will develop woody vegetation and enter the sapling stage, usually a poor habitat for wildlife. The trees are too high for ground-browsing animals and produce few nuts or seeds. Thin, stalky saplings don't make good dens or nests.

By contrast evergreens are excellent for wildlife. Thick and bushy, they shield animals from cold winds and provide nesting. Junipers are one example, cotoneasters another. These provide fruit or seeds desired by birds, and flowers pleasing to butterflies and hummingbirds. They don't attract pests or seek to spread out and dominate surrounding vegetation.

The final stage of forest growth is beneficial to wildlife in other ways.

Fruits and nuts are abundant. Holes can be made in tree trunks for squirrels, raccoons, opossums, and songbirds to dwell in. Old, dying trees attract insects that woodpeckers and other bark-probing birds like. Hawks and owls roost on high branches.

Varying levels of slope and vegetation invite a wider variety of wildlife to fill its niches. A pond or stream will attract waterfowl, propagating fish, marshes, mosses, and much else, as you can see in the article below: "A Pond Is Born."

While "biosphere" refers to the way all of the Earth's ecosystems work together; a "bioregion" consists of the operative ecosystems in a particular area. Knowing about your region will help you comprehend your yard and its place in local ecosystems much better. Furthermore, knowing how your region

A Pond Is Born

Glen Martin

The big D8 Caterpillar had not been gentle in gouging the pond out of the hillside of decomposed pumice and serpentine. The banks were raw and eroded. Little vegetation grew on them. But the Cat operator had known his business, and the site showed it. The overflow pipes were correctly situated; the earthen dam had been layered and tamped correctly. The pond was full from the winter runoff, which was the conclusive proof of its soundness—not a leak anywhere.

The water was pellucid; it looked as sterile as a Precambrian bay. No algae or higher plants, no fish or amphibians: just some mosquito larvae and the odd hellgramite or water bug. It was a young pond, no more than a year or two old.

The site for the pond was formerly a rocky mountainside pasture that had grown little save star-thistle, tarweed, and ripgut brome. Few creatures, wild or domestic, had found much in the way of shelter or sustenance on the meadow.

Yet it soon became apparent that the pond was serving other, subtler purposes. From the beginning, it was a favored nocturnal venue for black-tailed deer. . . . As the summer deepened and water grew scarce elsewhere on the mountain, we marked other spoor: skunk, opossum, raccoon. Pickings must have been slim—if there were any frogs or salamanders around, they were scarce.

I took a walk one morning, when the sky was marked by scudding clouds fat with sleet and rain . . . and the air exploded with whistling wings: a big raft of lesser scaup, goldeneyes, and buffleheads. . . . All winter I watched the waterfowl sally through the pond, something I've done each winter since. . . . Scaup and mallards typically dominate these seasonal populations, but there are often fair numbers of wigeons, canvasbacks, and shovelers. Blue-winged, green-winged, and cinnamon teal usually log annual visits.

By spring it was apparent that something else had been using the pond besides ducks: bluegills. . . . And so [grew] the population of things that liked to eat fish. First I saw the great blue herons. . . . Common egrets came later. . . . Most secretive of all our local Ardeidae are the green herons. . . . Not so shy are the common mergansers. . . . Several times we have seen common terns keening and dipping over the water. . . . We've seen black-necked stilts and avocets prod in the rich mud, and we've found killdeer nests along the rocky shingle.

Over the years, the perennial vegetation along the pond has grown up. Sedges and watergrass are thick along the water's edge. . . . The cover has brought birds, the various species ebbing and flowing according to the seasons. . . . The pond and its vegetation have encouraged rodents. . . . Both the rodents and the songbirds have drawn raptors. . . . And then there are the bullfrogs. They first showed up in the pond five years ago, and now the evenings of spring and summer resound with their vocalizations. . . . Gray foxes also come down for the frogs.

The pond gets along without us. In fact, the less we deal with the pond, the more it thrives. There's something humbling about that. The pond has established its own exquisite biological equipoise, independent of—even in spite of—our meager efforts at enhancement. It is a consummately balanced wheel that was originally set in motion by crude human endeavor, but is now driven by its own internal mechanism. Its genesis was that of some cosmic cake mix: just add water, and presto!, instant habitat. To walk around it is to sense the natural world straining at its bonds, waiting for the smallest niche or opening, demanding to flower, to express itself wildly and fecundly and in infinite diversity. Nature is fragile, but it's also paradoxically tough, flourishing in unexpected ways at unexpected times. I would not have thought that digging out a simple reservoir for winter runoff would yield such stunning dividends.

has changed over the years can tell you about its health status and suggest ways you can be more environmentally sensitive. Finding out the answers to the following questions will guide you.

EcoQuiz

1. Trace the water you drink from precipitation to tap and down the drain in order to know what happens to this vital resource on its journey.

2. Where does your garbage go? Take responsibility for how it is composted, recycled, burned, or left to accumulate.

3. Compare how much rain fell in your area in the past years. Your yard's ecological community will vary if the climate becomes drier or wetter.

4. What were the primary subsistence techniques of the culture that lived before you? How has the land been used?

5. What geological event/process influenced the landform where you live?

6. What species have become extinct?

7. Name five edible plants and their season(s) of availability. What spring wildflower is consistently among the first to bloom? This awareness informs you about your yard's growing season and native plants.

8. Name five grasses. Are any of them native?

9. Name five resident and five migratory birds.

PART III
Community Ecology

ECOSYSTEM PATHWAYS

In ecological terms our society "produces" and "consumes" but doesn't have technological decomposers at work to handle all the old appliances, plastic packaging, and metal and glass products that are manufactured for use in our homes, businesses, and factories. Furthermore, our current disposal methods use up fossil fuel energy sources and emit pollution. Reducing materials at the source, reusing, recycling, and composting products are practical solutions, although not panaceas. The picture shows the problems our waste matter makes for ecosystems.

- In every mountainous landfill lies a valuable mine of resources—metals, reusable glass containers, recyclable paper and plastics, and food and yard wastes high in soil nutrient value. Through recycling we can reuse much of this waste before it gets to a landfill.

- Paper comprises 36% of our solid waste; about 50 million tons of paper waste are generated each year in this country. Incineration of paper sometimes gives off air pollutants; recycling it not only reduces our waste load, but also saves valuable forests. If we recycled all of the newspapers printed for one Sunday edition of *The New York Times,* for example, we would save 75,000 trees. In addition, the reclamation process uses only between one fourth and three quarters of the energy required to produce paper from raw materials and less than half the water. The recycled paper product can be used to make cereal boxes, newsprint, corrugated containers, tissue paper, wallboard, and writing pad bases. Even so, the deinking and bleaching process used to clean paper for recycling generates pollutants.

- Aluminum cans are extremely energy-intensive products when made from scratch. Manufacturing aluminum from scrap instead of bauxite cuts energy use and air pollution by 95%. Each soda can thus produced saves the energy equivalent of half a can of gasoline.

- Steel cans weighed into U.S. landfills at 2.5 million tons in 1986. The tin from the outer coating of these cans is expensive and imported. When recycled, the tin can be recovered and resold, or used to make new cans. The steel can be reclaimed for manufacturing.

- Glass bottles make up 8% of our waste stream. There are two types—refillable and nonrefillable. The reusable types require a thorough cleaning: they are heavier because they can be used up to thirty times and then melted for reuse. The lighter, nonreusable glass can be remelted with raw materials to create new glass bottles, jars, and fiberglass, and can be recycled almost indefinitely. This last option requires a great deal of energy, however, so the more durable bottles are preferable.

- Plastics comprise 7% of our waste stream by weight—but a whopping 32% by volume. They may take hundreds of years or longer to decompose, so they essentially become a permanent part of the landscape when thrown out. There are at least forty-six kinds of plastic, all currently made from petroleum or natural gas. A plastic ketchup bottle may be made of six different layers. In the future, plastics are likely to come from plant and animal products.

- Yard wastes and organic kitchen wastes make up one fifth of the waste stream—28 million tons per year—that could be composted and used to create nutrient-rich soil.

- Metal cans may take hundreds of years to decompose. But aluminum cans, if recycled, can come back to you in only three months.

- Toxic chemicals get into water, air, and ultimately creatures' bodies. Hence the need for toxics reduction and separation of household hazardous wastes, such as shellacs, paints, batteries, and drain cleaners. This fact has not yet penetrated deeply into the waste management profession.

- Incineration is as threatening to public health as burning coal. This form of pollution could be reduced only if certain kinds of paper or plastic were burned and hazardous wastes were kept separate. PVC plastics should be separated out. Co-firing with natural gas should be instituted to ensure proper combustion at all times, which can dramatically reduce the emission of toxic substances.

- At landfills toxics leach into the water system. If the air is stuffy and the right bacteria and moisture are present, methane, a dangerous greenhouse gas, forms. Landfills should be aboveground to allow for easy monitoring and collection of leaking liquids. Air emissions should be controlled. Methane can be collected instead and used for energy.

- Garbage is imported and exported by municipalities without state-of-the-art recycling, composting, and hazardous-waste-separation programs.

- Illegal dumping by individuals and private waste haulers to avoid paying disposal costs adds toxics uncontrollably. This problem could be addressed by having a fee added to the purchase price of virgin materials so that disposal

(Continued on page 94)

7

OUR GARBAGE

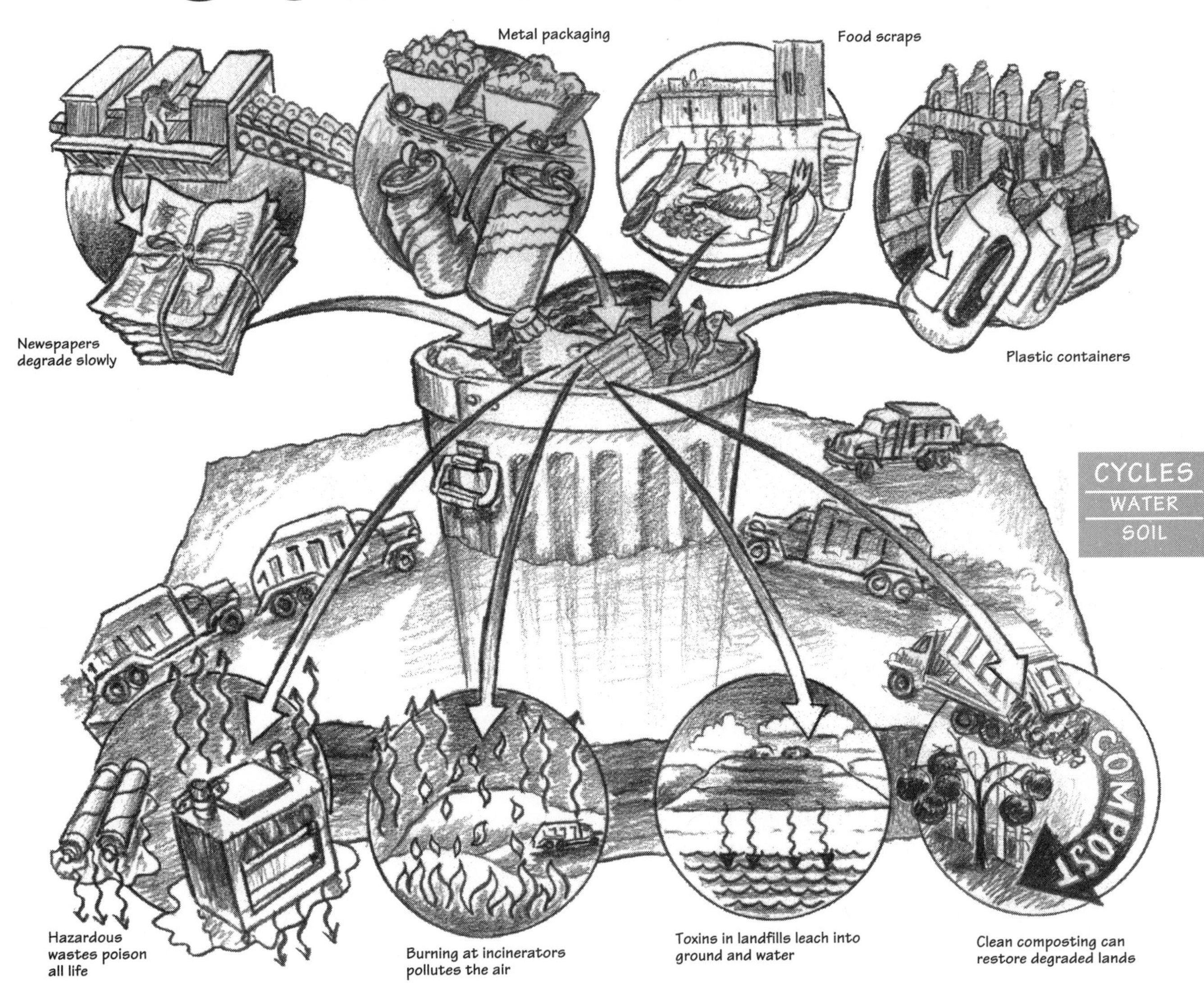

TAKING THE LID OFF GARBAGE

In the United States we generate more garbage than any other country in the world—in 1988 a heaping 160 million tons of commercial and residential waste, or three to four pounds per person. We fill 63,000 garbage trucks every day—lined up, they would stretch from San Francisco to Los Angeles (about 400 miles). The waste disposal problem has reached this magnitude at least partially because many consumer products and most packaging materials are designed to be used once and then thrown away.

Most of our garbage—about 80%—is dumped in landfills, but now we are running out of cheap space that meets permit standards to accommodate this flood of rubbish. The Environmental Protection Agency predicts that soon half of all U.S. landfills will be full. As old sites close, new ones become increasingly difficult to start up because of public opposition.

Communities are facing tough choices. More and more, they are placing a premium on open space and environmental quality. Some community planners are proposing huge, costly incinerators—often used to generate

E C O S Y S T E M   P A T H W A Y S — c o n t ' d

(Continued from page 92)

costs would be covered. This would go far to eliminate illegal dumping.

- The oil industry generates billions of barrels of toxic waste every year.

- Garbage, either thrown overboard directly or run off indirectly, ends up in our lakes, rivers, and oceans. We have mistakenly acted as though water would magically make it disappear. Out of sight, out of mind has been our operating principle so long that the result is showing up in inedible fish, unswimmable beaches, and dead birds and marine life. The contents of bird and turtle guts are found to be filled with shreds of plastic bags, filament line, bottle caps, combs, toothbrushes, and toys. Fish and birds are caught in plastic rings and netting. Unable to move, they slowly starve. Frequent oil spills coat the feathers, fins, and nests of these creatures, causing them to die helplessly.

- Our water receives chemical additives from many sources. Most public water systems are dependent on surface water—lakes, rivers, and streams. The rest are dependent on groundwater, which has filtered down through layers of sand, gravel, and porous rock. Surface water is vulnerable to acid rain, industrial sludge, and harsh cleaning agents as well as to eutrophication (a proliferation of algae from too much phosphorus and nitrogen, which ends up depleting the oxygen needed by plants and fish). At the far-from-perfect processing plant, water is chlorinated, fluorinated, and disinfected with other compounds to keep pipes from eroding.

power. Because of their potential to emit air pollutants, and because they reduce the garbage to an ash that could be toxic or at near-toxic levels, incinerators are usually controversial and difficult to site. Many communities opt to have the stuff hauled away—sometimes even to developing countries—in order to keep it out of their backyards. This is a very shortsighted approach, however. Garbage never really goes away. As we've learned, it just becomes a problem for someone else, and eventually for the entire world.

Some communities are using a strategy that combines waste reduction, reuse, recycling, and composting to reduce (though not entirely eliminate) the waste that must be incinerated, placed in landfills, or otherwise disposed of—an approach that seems promising.

In Audubon's vision of the future, product designers will work closely with recyclers and composters to facilitate better methods for handling solid waste. Producers would be responsible for disposal and hence the costs would be built into their products. Thus, they would be more inclined to find efficient disposal methods. Audubon wants to see regulations, economic incentives, designs, and consumers be sensitive to the environmental impacts of products entering the waste stream, as well as to the pollution caused by the extraction of raw materials during processing. Products should be labeled as to their recycled content and recyclability.

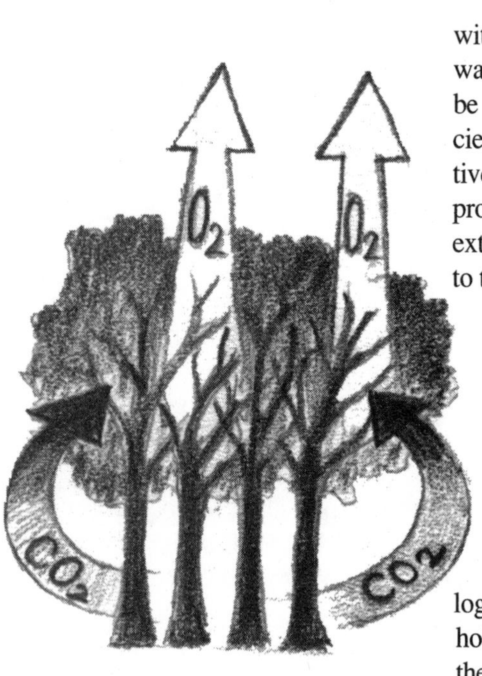

Loss of Precious Trees

Let us recall the importance of trees and the tragic consequences of removing them. One of the huge consequences of our industrial age and its manufacturing of products has been deforestation. Replacement, where it occurs, has often been monocultures, intensively managed. Ecologically, forests are like sponges, slowing down runoff and absorbing and holding water. They reduce the severity of flooding. As the Earth's lungs, they eliminate much of the carbon dioxide buildup. They are the principal home for biodiversity. Aside from their incomparable beauty and value as food and shelter to wildlife, we use the fruit, foliage, sap, bark, stumps, logs, and roots of trees for many products that benefit us.

Retaining our trees is vital to our future. Yet around the world millions of trees are cut down annually. We need to correct the impression, however, that South Americans and other persons in developing areas are the only ones to blame for putting the world at risk by removing their CO_2-absorbing trees. It is a shock to realize that our land, which was once mostly covered with forests, has been converted to field or pasture.

A people without children would face a hopeless future; a country without trees is almost as hopeless.

—Theodore Roosevelt

As more forests are cleared, a region's climate gets hotter and drier. Eventually the soil loses its fertility and becomes a desert. Regeneration can take hundreds of years. Old-growth forests, containing trees that are often hundreds of years old (e.g., Douglas fir, western hemlock, giant sequoia, redwoods, loblolly pine, and most of the tropical forests) are home to a greater diversity of plant and animal life than secondary forests. *American Forests* magazine reports that deforestation condemns at least one species of bird, mammal, or plant to extinction daily. The more we destroy, the less flexibility we have for future survival and genetic material to create new food sources and medicines. In considering how our public forests are used commercially, first priority should be given to their ecological functions and diversity of life.

Explaining the problem, Brock Evans, Audubon's Vice President for National Issues, testified in Congress as follows:

> [The problem is] the application of various "scientific" forestry principles, as brought over from Europe by Gifford Pinchot, the founder of the U.S. Forest Service, about 100 years ago. The basic principles of "scientific forestry" that Pinchot brought over from Switzerland were the principles of what we now know as intensive agriculture: monoculture plantations, thinning, and extensive pest and fire suppression. Due to the force of his powerful personality, these principles were applied nearly universally as governing tenets of what came to be known as "scientific forestry," as taught in all U.S. forestry schools and practiced by the wood products industry and the Forest Service.
>
> For a long time, these techniques seemed to work. . . . But what we did not know, or avoided facing up to until recently, is that Mother Nature has her own ways of dealing with fires and insects, and that the forests of our country have evolved over the *ages*; a "natural forest" is there precisely *because* it is *adaptable* to conditions of pests, disease, and fire. Scientists have long been telling us that dead trees are not just fit to be "salvaged" as if that were their only use; they also play a crucial function in the health of the forest ecosystem, in providing homes for woodpeckers for example, which in a healthy forest keep the insect population down. . . .
>
> Everywhere across the Northwest, especially in its east side forests, we now see the consequences of these applications of "scientific forestry": the salmon fishing industry—a billion dollar industry which provides 60,000 jobs—is in serious decline, due to over-harvesting in too many spawning watersheds. Both the spotted owl and the marbled murrelet are now listed under the terms of the Endangered Species Act, again because of over-harvesting on the west side; other interior forest-dependent species such as lynx, marten, fisher, vaux' swift, are felt to be in serious decline also. The reduction in population of these "canaries in the coal mine" is a fact that is telling us we must be much more careful in the future about how we manage forests, and we have to listen to nature much more carefully than in the past. . . .
>
> *Specific wildlife aspects of "salvage" programs.* The National Audubon Society has conducted a very successful Adopt-a-Forest program in the Blue Mountains of eastern

Oregon and Washington state, an area now quite hard hit by the massive declines in forest health due to wrongful application of "scientific forestry" principles mentioned above. Hundreds of our members and staff have spent literally thousands of person-hours on the ground, walking through the forests, identifying wildlife habitat and wildlife species, inventorying groves of ancient forests, noting damage, interacting with agency and industry officials, members of local communities, Indian tribes, fish and wildlife experts, and many others. We are now in the process of working with all interested parties in this area in an effort to devise the best possible forest health restoration program, as opposed to a mere salvage program, so that we can have our forests back again.

Salvage programs should stay entirely out of roadless areas and riparian zones. The scientists of the Blue Mountain Institute, who have studied the forest health situation extensively, have concluded that, while it is indeed necessary to "salvage" portions of the dead and dying trees, no program can be justified which enters into either roadless areas or riparian zones. We have already mentioned the riparian zones and the adverse impacts which would occur from salvage logging there. The roadless areas are, in almost every case, the last remaining refugia and best habitat for birds and mammals. Here you find the healthiest trees—because they haven't been high-grade logged yet; here you find the most shade, the most cover, the least disturbance by humans, and the best possibilities for regenerating a healthy forest later on, by preservation of the gene pool.

Thus, the recommendation of the National Audubon Society is that, where maintenance of healthy fish and wildlife populations is concerned (a) there must be no salvage logging in riparian zones; (b) there must be no salvage logging in roadless areas; and (c) any salvage logging of the extensive remaining areas should be carefully designed to preserve sufficient dead and dying *standing* trees for cavity nesters and for game cover.

Plastics vs. Paper

In the debate between whether a paper or plastic product is better to use, the focus from an environmental point of view should be on the manufacturing process, the weight of the product, and the extent to which it will be reused or recycled. Reducing the toxicity of additives is important for both products. Because the manufacturing of paper also involves the heavy use of chemicals, as well as releasing them into the air, paper is not always a better choice than plastic.

William Rathje, an archeologist, has spent recent years excavating America's landfills and has found that less than one tenth of 1% of the total was made up of fast-food packaging, diapers less than 1%, plastics about 12%, and that the greatest part was paper—above all, newspapers. Moreover, some of the newspapers were forty years old and still readable. Contrary to popular belief, paper can take just as long as plastic to decompose in landfills. Since both paper and plastics do not compact much, they shorten the lives of our landfills and when burned give off toxic chemicals.

When we consider the manufacturing of paper, we see that aside from the destruction of trees and habitat, it involves (1) the burning of fossil fuels that emit sulfur dioxide and nitrogen oxides (leading culprits in acid rain), (2) the discharging of caustic chemicals into streams and rivers, (3) chlorine bleaching that results in the escape of dioxins into air and water. Dioxins are linked to cancer, birth defects, and suppression of the immune system. According to the American Paper Institute, 70 million tons of raw paper are produced in the United States each year. Half the trees used for our paper are taken from Canada. In meeting our huge demands for paper, trees need to be managed in a sustainable manner that supports a great deal of diversity. And the industry should be moving away from fossil fuel use and molecular chlorine bleach in paper production.

In the case of plastics, the emphasis should also be on shifting the use of polymers away from those that are built up from toxic monomers, such as PVC and PVDC, toward those that are not, such as polyethylene and polypropylene. If a plastic product is constructed for long-term use, or is very light, it is sometimes a better choice than paper.

The plastics industry needs to plan to recycle more of its product, to develop ways of using recovered plastic. Unfortunately, plastics recycling is not well developed because there are so many different types of plastic resins in use, making them difficult to efficiently sort, collect, and reprocess. Currently, we recycle only a minor percentage of our plastics, with soft drink and milk containers among the most frequently reclaimed. Recycled plastics can be used in flowerpots, drainpipes, toys, traffic barrier cones, carpet backing, and fiber fill for pillows, ski jackets, and sleeping bags. Since much of the paper and plastic found in our trash comes from packaging designed to be used once and thrown out, controlling this problem at its source is key.

One mass-marketed item for which the packaging could be revised to eliminate much waste is the compact disc. Most CDs come in a package within a package. The disc itself is housed in a plastic "jewel box," which in turn fits into the cardboard longbox, roughly twice its size. A customer typically junks the longbox and its cellophane wrap. Environmentalists have long viewed the longbox as a prime example of unnecessary and excessive packaging. And with CDs selling at a clip of 250 million a year (they recently surged ahead of cassette tapes), that adds up to a lot of garbage—more than 23 million pounds a year. Some record companies and inventors have experimented with different forms of packaging and have printed with vegetable-based inks. Surely the day of using all-recycled material cannot be too far off.

For both paper and plastic products, environmentalists agree that no product should be called "recycled" unless it contains more than 50% recycled material. The following are suggested guidelines for their use.

Guidelines for Paper

Recycled paper is defined as that made with "postindustrial fiber" (that is, leftover from paper mills or printers) or "postconsumer fiber" (that is, paper collected after use from businesses, institutions, or individuals). Some 85% of postindustrial fiber is already recycled. Further guidelines are:

1. 100% recycled fiber, of which at least 10% must be postconsumer. After 1994 the percentage of postconsumer fiber required should increase to 15%, rising another 5% every few years:

2. Or: 51% recycled fiber, of which at least 30% must be postconsumer.

Any purchase of paper bleached with molecular chlorine or mixtures of molecular chlorine and other bleaches should be accompanied by a written justification for the necessity of the purchase.

Guidelines for Plastic

Purchase of plastic products must favor polyethylene and polypropylene over all other polymers. In an office setting, purchase of plastics containing polyvinylchloride (PVC) must be accompanied by a written justification of why this purchase is absolutely necessary. This requirement is particularly important in areas where PVC ends up being incinerated.

All paper or plastic products should be accompanied by a certificate indicating that they have not been manufactured with the intentional use of the four heavy metals, lead, cadmium, mercury, or hexavalent chromium. This certificate is particularly necessary when it comes to printing inks and pigments used in paper, and coloring agents used in plastics.

Recyclability Guidelines

Efforts should be made to purchase products that are recyclable. To qualify, a product must meet one of the following criteria:

1. At least 75% of consumers must have access to recycling facilities for that product in at least 75% of the communities in the jurisdiction where the guidelines are being applied;

2. Or, a manufacturer of a particular product, such as film spools, must recycle more than 50% of its products;

3. Or, 50% of the products in an entire industry, such as aluminum cans, are actually being recycled.

Responding to the garbage crisis, McDonald's recently converted from its polystyrene hamburger containers to paper-based packaging. Polystyrene releases the carcinogen benzene in its making. But the new material is coated with plastic and thus fails the recycling test. Such mixing of materials makes recycling impossible, at least at the current level of affordable technology. The coated paper does have less volume and therefore will take up less space in landfills, but it is only slightly lighter in weight and thus likely to be as much of a polluter as before. *Solving the packaging problem has to take into consideration the whole cycle from production of the material to collection, marketing, and processing of the waste.* The pros and cons are not simple to sort out. Other steps McDonald's, with the help of the Environmental Defense Fund, has taken to improve its products, such as use of recycled paper, are much more clearly an improvement.

There is no perfect solution to the garbage problem. Even if all Americans cut their wastes in half and recycled the other half, we would still produce some 40 million tons of garbage annually.

Some activists suggest you unwrap a product at the store and leave the package there or mail it back to the manufacturer, thus forcing the makers to deal with it. At any rate, if you can get one law passed to control waste or get one company to change its ways, you can overcome a lifetime of pollution. For instance, a recycling program for 100 employees in one year can save 100 years of office waste.

Facts: Every ton of material that is reused saves from 1.5 to 3 tons of new materials.

—Recycling your newspapers for one year could save four trees, 2200 gallons of water, and fifteen pounds of air pollutants.

Perhaps our problems can give us more respect for nature's decomposers. These feeders on waste do their best to transform the dead and toxic into life-supporting nutrients. Microorganisms not only break down and make ammonia, nitrogen, and phosphorus but also convert industrial chemicals and pesticides into simple compounds that plants can absorb. Without decomposers the entire world would be knee-deep in plant litter, dead bodies, wastes, and garbage. In some cases microbes mutate in order to become more efficient at digesting toxins.

Some creatures keep water clean and have come to be utilized in water purifying systems. A bed of mussels can filter a million gallons of water a day. Shrimp and crabs scavenge and recycle the dead, shred plant material, and filter mud and sediment through hairlike combs of their mouth parts. In the process they return minerals and nutrients to water.

Just as individual species have their key roles to play, our communities can be organized more efficiently to manage their production, consumption, and decomposition. Humans have so overloaded nature's system of waste decomposition that the system has not been able to keep up. We need to remember the principles of ecology in order to rectify the balance.

THINKING LIKE AN ECOSYSTEM

Niches

The species found in ecosystems can be described as:

1. Native, which normally live and thrive there;
2. Immigrant or alien, such as those introduced by humans. Some are beneficial, some run riot;
3. Indicator species, which serve as warnings that a community is being degraded. Examples are songbirds and frogs, both of which are big insect-eaters. They are declining in numbers, probably because of loss of habitat. If things go bad for them, we can be sure we are next.
4. Keystone species, which are important to a number of organisms in the community. An example is the alligator in the Everglades. Alligators dig deep holes, which collect water during dry spells. These are refuges and feeding grounds for aquatic life and birds. Alligators also eat gar, a predator fish, and in so doing maintain a balance in fish populations.

Some species are specialists; that is, they can live in only one type of habitat. An example is the tiger salamander, which can breed only in fishless ponds. Others, such as mice and deer, are generalists because they can live in a variety of habitats. Many species interact in some way—as competitors, predators, or symbiots. A community has a wide stratum of specialized niches, which minimizes competition and enables many species to survive together. It is this marvelous organization, still so mysterious to us, that enables all of our lives to continue.

The population level of species depends on temperature, on the ability to tolerate pollution, and on water, light, and soil nutrients. If populations are placed at risk, we humans may be next. To ensure the continuity of the human species, we must protect our biodiverse communities.

SCARAB

Natalie Angier

In the vast world of beetles, they have the stamp of nobility, their heads a diadem of horny spikes, their bodies sheathed in glittering mail of bronze or emerald or cobalt-blue. The ancient Egyptians so worshipped the creatures that when a pharaoh died, his heart was carved out and replaced with a stone rendering of the sacred beetle.

But perhaps the most majestic thing about the group of insects known romantically as scarabs and more descriptively as dung beetles is what they are willing and even delighted to do for a living.

Dung beetles venture where many beasts refuse to tread, descending on the waste matter of their fellow animals and swiftly burying it underground, where it then serves as a rich and leisurely meal for themselves or their offspring.

Each day, dung beetles living in the cattle ranches of Texas, the savannas of Africa, the deserts of India, the meadows of the Himalayas, the dense undergrowth of the Amazon—any place where dirt and dung come together—assiduously clear away billions of tons of droppings, the great bulk of which comes from messy mammals like cows, horses, elephants, monkeys and humans.

Scientists have long appreciated dung beetles as nature's indispensable recyclers, without which the planet would be beyond the help of even the most generous Superfund cleanup project. But only recently have they begun to understand the intricacies of the dung beetle community and the ferocious inter-beetle competition that erupts each time a mammal deposits its droppings on the ground.

Researchers are learning that every dung pat is a complex microcosm unto itself, a teeming habitat not unlike a patch of wetland or the decaying trunk of an old redwood, although in this case the habitat is thankfully short-lived. For scarabs, it may be said that waste makes haste, and entomologists have discovered that as many as 120 different species of dung beetles and tens of thousands of representatives of those species will converge on a single large pat of dung as soon as it is laid, whisking it away within a matter of hours or even minutes. . . .

The diversity of beetles that will flock to a lone meadow muffin far exceeds what ecologists would have predicted was likely or even possible, and scientists are being forced to rethink a few pet notions about how animals compete for limited goods and what makes for success or failure in an unstable profession like waste management. They are learning that beetles have evolved a wide assortment of strategies to get as much dung as possible as quickly as possible, to sculpt it and manipulate it for the good of themselves and their offspring, and to keep others from snatching away their valuable booty. . . .

As beetles go, scarabs are exceptionally sophisticated. In Africa and South America, where some species are the size of apricots, the beetles may couple up like birds to start a family, digging elaborate subterranean nests and provisioning them with dung balls that will serve as food and protection for their young. And these dung balls, called brood balls, are not slapdash little marbles. With a geometric artistry befitting the sculptor Jean Arp, the beetles use their legs and mouthparts to fashion freshly laid dung into huge spherical or pear-shaped objects that may be hundreds of times the girth of their creators. Some beetles even coat the balls with clay, resulting in orbs so large, round and firm they look machine-made. . . .

Still working as a duo, the beetles then roll each ball away from the dung pat and down into the underground nest. The female lays a single egg in each brood ball; among the largest species, there may be only one ball and thus one baby per couple.

Safe within its round cocoon, the larva feasts on the fecal matter. As the infant develops over a period of months, the mother stays nearby and tends to the brood balls with exquisite care, cleaning away poisonous molds and fungi and assuring that her young will survive to emerge from its incubator as an adult. That sort of maternal devotion is almost unheard-of among beetles, which normally lay their eggs in a mindless heap and lumber away. . . .

All of which means that little dung will go to waste. . . .

The incentive to move quickly is great. Not only does every beetle want to get away with the biggest slice of the pie, but while they are scavenging in an exposed heap of dung they are extremely tempting to many insectivores.

Community Solid Waste Programs

Because the United States urgently needs solutions to its mounting garbage problems and is seeking better techniques to reduce the volume of waste, collect and sort it, develop site disposal facilities, and manage hazardous toxins, the National Audubon Society, with a grant from Fuji Photo Film, has packaged a multicomponent program that enables communities to implement environmentally sound, workable solutions to their present solid-waste-management problems. The program has been tested and refined in ten pilot communities through the stages of fact-finding, setting goals, educating constituents, and building consensus among concerned parties.

COMMUNITY CHECKLIST

Here are some questions to ask about your own community:

1. Has it commissioned a study of solid waste options that includes recycling, composting, and collection of household hazardous waste?
2. Does it have a source minimization program such as minimizing the use of products and packaging that are difficult to recycle with current technology?
3. Does it have a curbside or other recycling program? If so, is its rate high enough—at least over 50%?
4. Does it have a plan to develop municipal composting facilities to handle source-separated food, yard waste, and other compostable wastes?
5. If the community incinerates garbage, does it (a) have an auxiliary energy source to maintain temperatures, (b) have pollution scrubbers, c) treat the ash separately from regular municipal waste?
6. If there is a landfill, is it double-lined and monitored for leaking toxics?
7. Does it import garbage from other communities?
8. Has it established a forum for dialogue and consensus-building among business leaders, government officials, environmentalists, and other citizens concerned with solid waste?

In order to facilitate the joining of local governments, business leaders, and citizen groups in coming up with strategies that work best for all, Audubon has prepared a two-volume Guidebook that shows step-by-step exactly how communities can establish a solid-waste program.

Audubon's vision is a nation—and world—with a sustainable waste management system powered by renewable energy. To get there, Audubon would like to attain a reduction of 97% of the amount of household toxic waste going into the municipal waste stream, with a diversion of 90% of household trash through waste reduction, recycling, and composting.

Composting Is Hot

Many homeowners have kept compost piles in order to enrich their gardens with the "black gold"—rich humus—that emerges from the debris. More recently, towns and cities are collecting compostable waste, processing it in large plants, and bagging and selling it as valuable fertilizer. Enthusiasts can be found among employees of Procter & Gamble, Walt Disney, McDonald's, Nynex, and among members of Congress.

The composting cycle consists of the decomposition, under controlled conditions, of organic materials by microorganisms. During aerated composting, the microorganisms consume oxygen while feeding on organic matter. A great deal of heat is generated, and carbon dioxide and water vapor are released into the air. CO_2 and water losses can amount to half the weight of the initial materials. Composting reduces the volume and mass of the ingredients, transforming them into a friable mass of nutrients, which incidentally kill most weed seeds and disease organisms. The most important conditions that encourage the speed of transformation include a balanced supply of carbon and nitrogen, oxygen required by the microorganisms, moisture that permits biological activity without hindering aeration, and suitable temperatures.

Since some composting materials, such as sewage sludge, can contain human pathogenic organisms that persist if the high temperatures reached in the composting process are not maintained, it is important that systems take appropriate safety measures. Odors can be contained by enclosing compost facilities. The advantage of composting is that waste is reduced without dirtying the air as incinerators do, while also creating a nutrient-loaded product that can be used in landscaping, agriculture, nurseries, golf courses, cemeteries, and land reclamation.

While composting is a biologically simple process, execution on a community level is more complicated. Although increasing numbers of communities are reducing the amount of waste stream that ends up in a landfill or incinerator, unfortunately a lot of materials that could be processed as compost never get separated out. Communities need well-developed plans for composting food wastes, kitchen scraps, and non-recyclable wastepaper, which make up roughly half of household trash. Yard trimmings, sludge, horse manure, and building materials can also be composted. Los Angeles is trying to do this in order to attain a 50% reduction in its waste stream by 2000.

In a groundbreaking partnership, Audubon and Procter & Gamble conducted a month-long experiment in "source-separated" community composting. In source-separated composting, individuals separate compostable materials from their trash so that they will not be contaminated by household hazardous wastes and non-compostable materials. The experiment was conducted to show how much and what kinds of household materials are compostable and to gauge the level of public participation in such a program.

Five hundred volunteer households in Greenwich and Fairfield, Connecticut, were asked to set aside their compostable wastes, such as coffee grounds and food scraps, as well as materials such as soiled paper, cardboard food containers, and even diapers. The waste was picked up and taken to Fairfield's composting plant. The plant screened out residual plastic and processed the compost, which was then transported to the Connecticut Agriculture Experiment Station to test its quality.

Everyone who participated was excited by the results of the project. They were able to show that by composting residential wastes alone, 30% of household waste can be kept out of landfills or incinerators. Combined with traditional recycling, that figure jumps to between 70% and 80%. Many found separating compostables as easy as separating recyclables.

Demonstration projects elsewhere around the country are raising expectations for composting—and bringing together environmentalists and business. Such programs provide valuable data for others to use. The cost of building a facility is 4.5 to 8 million dollars per hundred tons per day. A publicly owned plant would have to charge a dumping fee of 50 dollars a ton to operate, while a privately owned enterprise would have to charge 80 dollars. Compost, low in contaminants and free of shards of plastic and stones, can be sold for 5 to 20 dollars per ton.

ECO HELPERS

WATERWAY WATCHDOGS

(Citizen keepers enforce laws the government can't)

Peter Steinhart

The Baykeeper glides along the shoreline of San Francisco Bay. The high-rises of San Francisco's business district gleam in the morning sun. At the wheel of the twenty-six-foot motorboat is Michael Herz, a small-boned man with a salt-and-pepper beard and a look of calculation in his eyes. He has just passed downwind from a huge Liberian-registered oil tanker—one of the 1,200 tankers that stop in San Francisco Bay each year—and encountered an invisible but throat-constricting cloud of petrochemical vapors. He has circled the tanker, looking for a slick on the water or a fresh discharge from the scuppers. A five-gallon bottle is ready to take a sample for laboratory analysis. A deckhand walks defiantly along the tanker's tail, four stories above the tiny Baykeeper, challenging Herz with his stare. But Herz cannot find a sheen on the water. He does not take a sample. . . .

Herz is part of a growing number of citizen enforcers patrolling the air and waterways of the nation. There is now a Hudson Riverkeeper, a Long Island Soundkeeper, a Delaware Riverkeeper, a Puget Soundkeeper, a New York–New Jersey Harbor Baykeeper. The Izaak Walton League has more than 3,000 streamwatchers, monitoring the individual streams and demanding that regulatory agencies act when someone pollutes the water. There is even a group of people interested in starting a Desertkeeper in Arizona. They are moved by the conviction that state and federal agencies aren't adequately enforcing environmental laws. . . .

The hook upon which the watchdog's hat is hung is a series of provisions in the environmental laws that allow citizens to file suit if government agencies are not pursuing polluters with due dispatch. Such suits are allowed under the Clean Water Act, the Safe Drinking Water Act, the Superfund law, and a number of other laws dealing with hazardous chemicals. Some of these laws provide for paying the legal costs of the watchdogs that bring the suits. Since 1982, by which time it had become clear that the Environmental Protection Agency was not going to enforce water-quality laws, the Natural Resources Defense Council and the Sierra Club Legal Defense Fund have filed more than two hundred suits. Environmental-law clinics at Oregon, Tulane, and other universities have also filed suits.

Superfund Sanctuary

Frank Graham, Jr.

Jim Rod manages this National Audubon Society sanctuary in Garrison, New York. The setting, 50 miles above New York City, on the eastern shore of the Hudson River, is spectacular. The U.S. Military Academy at West Point and Storm King Mountain dominate the opposite shore, and when snow isn't falling the hills behind them are silhouetted in the blue distance. More than a century ago, this stretch of the river inspired the haunting paintings of the Hudson River School. But the river's beauty masks a nasty truth—for Rod is managing the only Audubon sanctuary that is also a Superfund site. . . .

Cadmium, an element that poisons living organisms when it accumulates in the open environment, has infiltrated the muck among the cattail roots. The animals most obviously affected are muskrats, ordinarily one of the commonest of marsh dwellers. . . .

Rod found cadmium levels of up to 15 parts per million in the kidneys of muskrats in the marsh. Almost all of the "rats" in the contaminated area had liver lesions. Females gave birth at about the same rate as those in uncontaminated marshes, but juveniles made up 86 percent of the female population in cleaner areas and only 17 percent in the cadmium hot spots. In other words, young muskrats die soon after they begin nibbling on poisoned cattail roots. A follow-up count proved that there are simply a lot fewer muskrats in Constitution Marsh than in compatible marsh upriver. . . .

So what is cadmium doing in a nice place like Constitution Marsh? Humans have been leaving their imprint on the marsh for a long time, but originally in a more benign way. A 19th-century owner built a dike at each end of the marsh, dug a network of channels inside, and planted wild rice. Things began to deteriorate further when the railroad came through on its way from New York's Grand Central Terminal to Albany. Saltwater infiltrating the marsh, along with competition from cattails, filled off much of the rice. Yet enough persists to attract flocks of ducks and geese that stop here each autumn and spring. . . .

In 1969 Laurance Rockefeller and Lila Wallace bought the 270-acre Constitution Marsh and turned it over to New York State, with the stipulation that Audubon manage the property as a sanctuary. The benefactors must have assumed that they had secured the marsh and its inhabitants for all time. But the marsh was already in trouble. From 1952 until it closed in 1979, the Marathon Battery Company of nearby Cold Spring regularly discharged cadmium from its factory into Foundry Cove, just to the north of the sanctuary. . . .

The U.S. Environmental Protection Agency designated Foundry Cove, Constitution Marsh, and the nearby Hudson River a Superfund site in 1983 and ordered extensive testing for cadmium in the area. Enter James P. Rod. . . .

But sitting back and basking in the wonders of the marsh is not Rod's style. He goes out and fights for its survival. When the EPA was slow to deal with the abandoned battery factory and the yards around local homes that were still tainted by cadmium, he convinced the state to pressure the EPA and to chip in additional funds. He realizes his marsh is no healthier than the streams that flow into it. . . .

"Recently I learned about plans to discharge a high volume of sewage into one of the marsh's tributaries," he said. "I went to the Department of Environmental Conservation and asked: 'When was the last time you did a survey of the brook?' '1936,' they said.

"So I said, 'Let's do one now.' They did, and they found brown trout in there. They upgraded the brook's rating to 'trout spawning stream.' Now they are revising the sewage-discharge plan." . . . Such persistence has brought measurable improvement to the Hudson and the five large tidal marshes surviving below Albany. . . .

"This may be the most viable large river left in the entire North Atlantic Basin," he pointed out. "Few striped bass have reproduced in the Chesapeake area in the last twenty years. The Connecticut River has lost its salmon and the Susquehanna its shad. Goodness knows what's going on in European rivers. But the Hudson has all the species that were here when the Europeans arrived, and it now produces more than 50 percent of the striped bass off the Atlantic coast."

EcoQuiz

Q: At the grocery store, if you are offered a choice between a paper and plastic bag, which one do you take?

A: If you picked plastic, you are wrong. Plastic bags contaminate the oceans. They degrade very slowly, they are nonrenewable, and their production results in pollution. So is paper the answer? Nope. Brown paper bags used in most supermarkets are made from virgin paper, without contributions from recycled paper. Papermaking pollutes the water, can release dioxin, contributes to acid rain, and costs trees. Much of our paper comes from so-called "superior" trees grown with nonrenewable fossil fuel fertilizers in intensively managed and sterile environments. It's not at all clear that papermaking as it is practiced today is even a sustainable enterprise. The right answer is to pull out your personal carrier, just as Europeans have been doing for decades. You don't consume either paper or plastic.

ECOSYSTEM PATHWAYS

Our buildings and workspaces are man-made (open) ecosystems, subject to natural laws and processes. The brilliant, towering structures regulate temperature, transpire, and breathe. They consume energy and resources to house, support, and feed the productivity of their inhabitants.

Just as decisions regarding our forests determine the quality of air for all living beings, decisions that go into designing and constructing office buildings, shopping centers, and industrial buildings have impacts on the environment that extend far beyond the immediate area, well into the future. The products that go into making, and the energy used to run, our offices and buildings contribute to air pollution, water pollution, and habitat destruction worldwide.

With an estimated 4 million commercial buildings in the United States, there are nearly 35 million office workers who spend seven to twelve hours a day at work. How effective we are at keeping our buildings and cities functioning in a healthy way determines to a large extent how healthy we are.

As the picture shows, the materials and energy that go into and out of a single building are drawn from all corners of the world. Supplying this demand exhausts labor and resources. Consider this:

- It takes tons of steel, aluminum, concrete, stone, and wood to construct simply the shell of a building and the trucks, planes, boats, and trains to get these materials from source to site.

- Building materials—sand and gravel, subsurface metal ores, clay, marble, limestone, and granite—are also the building blocks of our Earth. They are being extracted faster than the Earth is replenishing them.

- Energy—mostly oil, gas, and coal—comes from underground deposits. These fuels are also used to extract the raw materials that go into making walls and building façades, furniture, and office equipment, as well as to power lighting and regulate air temperature.

- Trees are chopped down in the forests of Brazil, Canada, and the American Northwest—among other places—to supply offices with furniture, paper, and more paper. Most of the paper we use comes from forests in Canada. When it is used once and thrown away, the Earth is robbed of a vital resource.

- Aluminum ore is taken from mines in Jamaica for a building in Kansas City, Missouri; marble for steps leading to a San Francisco building is brought from Italy; oil from Arabia is running a New York office building. The farther away these materials and energy sources are transported from source to building site, the more energy is consumed.

- Heating, ventilation, and air-conditioning systems break down, windows and light bulbs need replacing, interiors are redesigned, and outer walls erode and break away over time. That means even more energy and resources are poured into the building on a continuous basis to maintain and repair it. More durable building and office products mean fewer resources have to be extracted from outside ecosystems.

- Transportation of employees to and from a place of work takes more energy and materials to provide and power our cars, trains, subways, buses, and commuter helicopters and planes. Some people commute on a regular basis to and from their workplaces as much as three hours each way.

- Billions of magazines, fact sheets, pieces of mail, membership or advertising appeals—made from paper, clays (for the glossy look), glues (for lick-'em-and-stick-'em envelope backs and labels), and petroleum (a component of most inks)—come into or go out of offices, direct-mail warehouses, and the like on a daily basis.

- Offices and employees who recycle paper are acting as the office ecosystem's decomposers, and they bring the system full circle by supplying producers with a needed resource that would otherwise be taken out of another ecosystem.

- Once a building deteriorates, renovating it or recycling it (reusing the existing structure) utilizes less energy and primary resources than does demolishing it and building a new one.

8

OUR OFFICE BUILDINGS

Guest author: Mercédès Lee

CITIES

An office building is part of a larger, more extensive and complicated infrastructure encompassing other office buildings, supermarkets, stores, warehouses, roads, parking lots, train tracks, and more. Cities were created out of a human, and practical, need to bring people together. Such agglomerations create an organized structure to provide such basic services as clean water, sewage systems, communication links, and transportation for communities. Cities are one of the fastest-growing segments of society—the Worldwatch Institute predicts that by the year 2000 more than half of the people in the world will live in cities.

The urban ecosystems in which we have built our office complexes, department stores, and factories cannot in themselves provide what is needed to sustain such intense activity and population densities. They are open ecosystems. Therefore, energy, materials, and food are taken and transported from ecosystems far away. In our workplaces, energy is used to run the multitude of fans, pumps, motors, air conditioners, heaters, refrigerators, lights, elevators, computers, copiers, and industrial equipment. Multiply that by the number of other buildings in a city, and by the number of cities around the world, and the amount of energy consumed reaches beyond our ability to comprehend. This energy does not come from a socket, it comes from the Earth, at a cost most of us never see:

- Wild rivers—home and feeding grounds for fish, birds, and other creatures—are dammed and wedged with turbines.
- The crust is probed and poked with steel straws and machines that suck out oil and gas. Left behind are debilitating scars and ponds filled with poison.
- The living skin of our Earth is scraped and gouged for coal, altering rock and soil cycles.
- Nuclear power plants produce the most lethal substance known to humans—radioactive waste. How to dispose of it safely defies even our best-intentioned political institutions.

Islands of Heat

The temperature is always warmer in the city than it is in the surrounding suburbs or countryside. This "heat-island" effect occurs for several reasons. In a city, the sun's energy is not converted and used in the same way as it is in open landscapes with trees and vegetation. Concrete, asphalt, and stone absorb heat. Hours after the sun has gone down, trapped heat continues to radiate from the city's streets and building façades.

Early societies knew the value of the sun's energy and depended on its warmth for sheer survival. Early cliff-dwellers—Greeks and Southwest Indians, for example—built their stone and adobe villages oriented toward the south to get the sun's full effect. During cool nights and cold seasons, they were kept warm by the sun-heated stone that housed them.

FACT: Nearly a quarter of all ozone-layer-depleting chlorofluorocarbons (CFCs) are emitted by building air conditioning and the manufacturing processes used to make building materials.

The factors listed below not only contribute to the heat-island effect of modern cities, but also generate heat and energy waste that in turn leads to poor air, which threatens the health of urban dwellers.

- Lighting systems generate heat that becomes trapped indoors by sealed windows and well-insulated walls.
- The heat generated from excessive indoor electrical lighting puts greater demands on air conditioning.
- To improve air quality indoors, good ventilation—bringing fresher air in and flushing indoor air out—increases the demands on a building's heating and cooling systems.
- Air conditioners remove and produce extra heat and moisture, spewing it into the local air, which adds to the heat and humidity of the urban environment.

In nature, air is kept healthy by a circulation system with checks and balances, as long as it is not overloaded with noxious pollutants. We can apply our knowledge about how ecosystems function in our work sites.

Sick Buildings, Unhealthy Employees

A moist, fetid pocket inside an old ventilation duct becomes a breeding ground for potentially harmful bacteria. . . . Nearby, a worker's hacking cough persists inexplicably. . . . Down the hall, a colleague wonders about the vague chemical smell she notices every time she sits at her desk. . . . A maintenance engineer discovers that it's been nine years since the ductwork has been inspected. . . .

All are harbingers of sick-building syndrome, a condition that, ironically, was exacerbated by the need to make buildings more energy efficient. The better-insulated buildings get—with walls being insulated and window leaks filled—the more stagnant the air can become inside, allowing indoor air pollutants to build up. Studies estimate that from 33% to 50% of all commercial buildings in the United States may be afflicted with sick-building syndrome.

The following story recounts the plight of National Audubon Society staff. It is a story that could be told by thousands of people, with only a few details changed here and there.

> Personnel were stuffed into four floors of a modern, glass-faced office building, sharing air space with a perfumery. To save on lease money, walls and cubicles were erected in a hodgepodge way. As a result, most Audubon employees never saw the light of day. But light deprivation wasn't a problem; artificial light splayed everywhere, with rows and rows of fluorescents constantly humming and flickering. In this space, where most staff spent a minimum of eight hours daily, 241 days of the year, it was hard to know how to dress. In the winter it could be either hot or cold, but space heaters were nearby just in case, with extension cord underfoot at every step. In the summer, some areas were stuffy, others were overchilled; that is, unless you stayed in the office past six o'clock, when the air conditioner was shut off.

Two components contribute to sick-building syndrome: poor ventilation systems and toxic off-gassing by building materials, furniture, and office products. Sick-building syndrome reveals itself through a host of human health problems, including malaise; headaches, irritated eyes, nose, and throat; distractedness; irritability; respiratory ailments; liver damage; and suppression of the immune system.

FACT: Sick-building syndrome costs America an estimated 6 billion dollars annually in lost work and medical expenses.

—*Newsday*, Nov. 15, 1992

Air-circulation systems that are poorly designed, installed, operated, or maintained can create a breeding ground for dangerous levels of bacteria and fungi. According to the Environmental Protection Agency, indoor air can be 100 times more polluted than outdoor air. There are, however, ways to improve ventilation in offices, stores, and industrial plants.

- On-site heating and cooling units can be linked with a circulation system designed to deliver regular infusions of filtered air, drawing heavily on outside air (where it is fresher than indoor air).
- Larger ducts to allow lower velocity can be installed to prevent moisture buildup.
- Access to fresh air can be improved through windows that open, are not sealed shut.

Indoor air pollution is one of the greatest threats to public health of all environmental problems.

—Lance Wallace, Environmental Protection Agency scientist

Common building and office materials contain harmful chemicals and solvents whose gases continue to be released over a long period of time. The main culprits are formaldehyde, benzene, xylene, toluene, and a few others. Inside air consists of more than 1,000 chemical compounds: a veritable witches' brew. These chemicals are hidden and silently released from carpeting, plywood, furniture veneer and fabric, and other materials. Workers can also be exposed to toxics in the office through photocopy and laserprint toner, liquid paper, and cleansers. There are alternative, low-toxic materials available, however, as well as things that can be done around the office that help reduce indoor air pollution and odors.

- Lead-free, water-based latex paints can be used instead of toxic, oil-based paints. Some are even free of smelly drying agents.
- Formaldehyde-free carpets can be installed over 100% jute pad.
- Surfaces of glued wood products can be sealed with plastic laminate to minimize harmful chemical releases.
- Furniture systems, non-paint wall coverings, finishings, and fabrics can be lab-tested to ensure minimum levels of formaldehyde off-gassing.
- Photocopy machines and laser printers can be placed in well-ventilated areas.

We need to begin thinking about the health of a building the same way a medical professional thinks about the health of a person.

—James E. Woods, Virginia Tech University

Building with the Environment in Mind

Most people think that environmental preservation is in direct conflict with development and economic vitality. Such does not have to be the case. There is tremendous opportunity to lessen our consumption of natural resources, such as water, energy, and minerals, and reduce the stresses we impose on living systems, including ourselves. At the same time, we can provide jobs that impart a sense of contributing to the well-being of our planet and fellow citizens. How? By adopting an alternative philosophy of development: rebuilding our infrastructure using environmentally sound principles and techniques. This includes using local resources when appropriate; applying energy-efficient technology; using less toxic, superior products; recycling the materials that we use; and using postconsumer recycled products.

Building projects by the Natural Resources Defense Council, the Rocky Mountain Institute, National Audubon, and others show that this new awareness has begun to penetrate the construction industry. Architects, engineers, and interior designers are beginning to construct buildings with the goal of reducing impacts on the environment by using energy-efficient technology and environmentally sensitive building and furniture products.

The Audubon House: A Model of Ecologically Conscious Design

Between April 1991 and September 1993, National Audubon renovated a century-old building in New York City for its new headquarters. Architectural critic Brendan Gill described the building as it was before work was begun:

> . . . a tawny fortress fashioned of brownstone, glazed brick, terracotta, and cast iron, one of the handsomest loft buildings in the city. . . . It was built by members of the Schermerhorn family in the early eighteen-nineties, to designs by the celebrated George Browne Post. . . . Like many another building, as it aged, [it] suffered declining revenues and diminished maintenance; its upper floors were filled with illegal sweatshops, and its ground floor was occupied by retail shops that came and went.

Every decision made by Audubon's architectural team, Croxton Collaborative, relating to building design and construction took into account two considerations: minimizing impacts on the environment and cost accountability. Designing in a way that is sensitive toward the environment does not depend on charity. It does mean placing the environment as a high priority, and it takes commitment on the part of all the design team members—building owner, architect, interior designer, engineer, consultant, contractor.

In the Audubon House, the extra costs of the advanced building materials, electrical systems, and furnishings will be recovered within three to five years of use. In other words, they'll pay for themselves. All systems and materials being used dramatically reduce pollutants associated with global warming, ozone depletion, and acid rain. Building products—from post-consumer recycled materials where possible—will help prevent sick-building syndrome. A comprehensive recycling program, including a state-of-the-art recycling center, is an integral part of Audubon's continuing effort to lessen its administrative load on the environment. The eight-story building was also "retrofitted"— reusing an existing structure—with advanced but readily available technologies for lighting, heating, ventilation, and cooling, to lower total energy consumption by more than 50%.

The Audubon approach has resulted in major environmental benefits and economic savings:

- Energy costs have been slashed approximately $100,000 per year.
- Acid rain emissions and ozone-depleting CFC refrigerants from air conditioning have been virtually eliminated.
- Energy-related greenhouse-gas emissions have been dramatically reduced by using gas rather than electricity for heating and cooling (the burning of coal and oil to generate electricity results in massive amounts of global warming gases).
- Electricity consumption has been cut by more than 50% (compared to the same building renovated the traditional way), placing less demand on heavily polluting power plants.
- Forty-two tons of paper will be recycled annually, rather than tossed into landfills or contributing to deforestation.
- Use of recycled post-consumer building and office materials will result in

conservation of a host of natural resources.

- The ventilation system provides six changes of filtered air per hour and prevents the buildup of moisture in ducts. The circulation of more fresh air, combined with the use of low-toxic office materials, will help prevent sick-building syndrome.

"The Audubon model," says Audubon President Peter Berle, "could be adapted to any building or renovation project. The Natural Resources Defense Council environmental building project provided the technical background and inspiration for Audubon's endeavors. We hope that our having successfully gone through the process will provide inspiration for others—building owners, architects, interior designers, engineers, and government policy makers—to integrate sustainable practices in their workplaces. We see NRDC and Audubon as mere stepping-stones to a more sustainable future for our cities."

The value of the Audubon model is not that Audubon did it. It's that others can do it, too. The next sections explain the rationale behind the steps Audubon took to accomplish its conservation goals.

The Hidden Costs of Energy

Because there are environmental costs to everything we do, we have to make decisions cautiously. There are two ways to power workplaces: (1) use an energy source at the building site (such as a gas-fired machine to run the heating, air conditioning, and ventilation systems), or (2) get energy exclusively from electricity provided by a regional utility. Environmentally enlightened building owners, their engineers, and architects take care to evaluate the environmental consequences of various energy sources when planning a building project. Deciding which energy source to use is not easy when you're aware, and concerned, that the health of ecosystems—and people—is at stake.

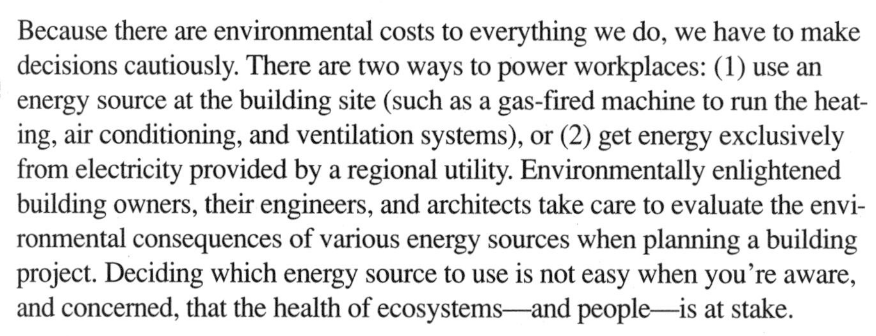

FACTS: Office buildings account for a third of the nation's peak electricity consumption, a figure exceeding the output of all U. S. nuclear plants.

By the year 2020, their appetite for power will double, requiring fifty more nuclear power plants to satisfy their inefficient use of electric power.

Each year, buildings and the power plants that supply them emit 14% of the gases associated with global warming and the greenhouse effect, and seed 15% of acid rain.

Selecting an energy source to power Audubon's new building came at a time when the organization was battling against New York State contract negotiations to obtain hydropower from Québec. Québec was planning to dam nearly every river flowing into James Bay. Utilities throughout New York State were considering buying this hydropower to provide electricity to Audubon and other customers. Hydropower was once touted as a "clean" energy source. When we see the record of ecological devastation caused by the thousands of mega hydroprojects around the world, hydroelectricity no longer looks so innocent. Audubon wanted to minimize their direct contribution to the demise of the important wilderness ecosystem of James Bay in any way it could. Audubon was also concerned about carbon dioxide emissions and the gases that are responsible for the greenhouse effect.

As a result, Audubon chose an on-site heater/chiller to supplement power to the building. That, combined with energy-efficient features to create a superior "thermal shell," enabled Audubon to scale down the size of its heating and cooling unit by half.

- The outer walls were insulated with materials that were free of CFCs.
- Windows and skylights were double-paned and incorporate "heat mirror" sheets—a transparent, wavelength-selective material that allows light (but little heat) to penetrate in summer and retains heat in winter.

If the Audubon approach of integrating efficient systems were adopted by all new commercial buildings, from now until the year 2020 the United States could save as much energy in the commercial sector as it currently consumes—100,000 megawatts per day.

One megawatt is a million watts, or the power needed to light 10,000 100-watt electrical bulbs. A suburban shopping mall with 100 stores or so draws roughly 3 megawatts, the towers of the World Trade Center use about 32 megawatts each.

—Sam Howe Verhovek
The New York Times Magazine

Lighting: Where You Want It, When You Want It

Lighting is one of the costliest components of the American office. But that problem is also one of the most easily remedied, providing substantial paybacks: electricity costs can be reduced and the quality of light can be enhanced for those who utilize natural light more effectively and integrate readily available high-performance lighting technology in their building design.

The key to efficiency in lighting systems is to put light where it's needed, when it's needed, as opposed to wasting it by lighting everything. In addition to the practical approach of placing offices and partitioned workspaces to take advantage of natural light, energy-conserving lighting systems include fluorescent light fixtures, desk lamps using compact fluorescent light bulbs, and occupancy light sensors. The location of the photosensitive devices is key to controlling the electrical lighting relative to the sun's intensity. The brighter the natural light, the dimmer the overhead electrical lights, so that an even amount of light is maintained in work areas. In office rooms and corridors the lights go on only when a person is occupying the area. They automatically shut off when the area is vacated.

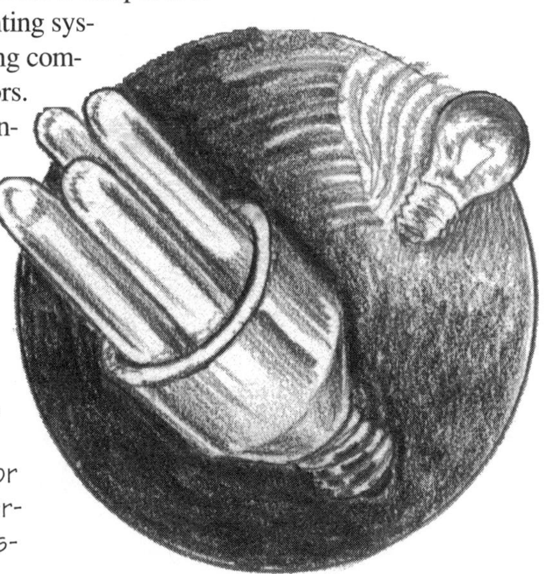

FACT: In 1986 alone, American buildings used about 321 billion kilowatt-hours of energy just for lighting. If all incandescent bulbs used in commercial buildings were replaced with compact fluorescents, close to 30% less energy would be used.

THINKING LIKE AN

ECOSYSTEM

The Sun

The sun is a fireball composed mostly of hydrogen (72%) and helium (28%), with pressures high enough to release enormous energy in the form of electromagnetic radiation. Our Earth's rotation around the sun is one of nature's most basic rhythms, setting the seasons and timing of ecological clocks. But many humans, shut away in homes or offices with artificial lighting, lose touch with these cycles.

Our Earth receives the sun's energy in the form of light of many wavelengths, both visible and invisible. All life on Earth depends on the sun's energy. Yet only plants can tap into it directly to manufacture and store their own energy foods, the carbohydrates. This store is the basic energy supply for all food chains. It also provides all our fossil fuels. The sun's energy translates into mechanical energy—the ability to move and think. Its energy works in so many marvelous ways that many people have considered it as god.

Sunlight also consists of an electromagnetic spectrum of different types of kinetic energy measured in wavelengths. The wavelengths of visible light extend from violet, the shortest, to red, the longest. We use these waves in distributing color to enhance our environments. Beyond the visible spectrum are the invisible wavelengths—the shorter ones of ultraviolet, X rays, and gamma rays and the longer ones of infrared, microwaves, and radio waves. We have learned to use the way objects absorb and reflect these waves in many ways.

FLOOR PLANS: WHERE IS THE SUN?

One value of the sun is in its gift of light. There is qualitative value in natural light—in its variability, in how it marks the change of time, and how it affects our moods. Natural light is a basic need; it gives us a sense of orientation and of the passage of time. The sun can be used to save commercial energy through what is called "passive solar technology"—essentially by using the sun as a source of light and heat, supplemented by electrical lighting and heating only when more is needed.

The basic floor plan of an energy-efficient office building like the Audubon House is configured relative to shadow movement through the day. The "democratization," or overall sharing, of sunlight is fundamentally important to maximize energy efficiency. In the Audubon building, where there was sun coming through the windows, the design team avoided placing partitions and walls up to the ceiling, which would essentially have closed out the natural light. Instead, they used an open-floor plan and grouped lower-partitioned workspaces nearest the windows along the southern wall, which has the longest exposure of daylight. No enclosed offices were put along this southern wall at all. Intermediate and high partitions for privacy were placed the next tier away from the windows. Enclosed offices were situated far back against the solid, windowless north wall. These are designed with interior windows so that staff within these work areas have a view of the natural light. Along the west side, which has a lesser degree of sunlight coming in, they put large windows on the interior faces of these offices so that any available natural light can penetrate the interior open offices.

FACT: The typical office uses 2. 8 watts of power per square foot; Audubon uses well under 1. That will translate into an approximate electrical savings of $40,000 annually. This energy efficiency required an initial investment of $100,000 more than the cost of a conventional lighting system. That expense will be recouped through energy savings in less than three years, and it will provide savings that will grow through the years as energy costs climb.

SOLAR ENERGY

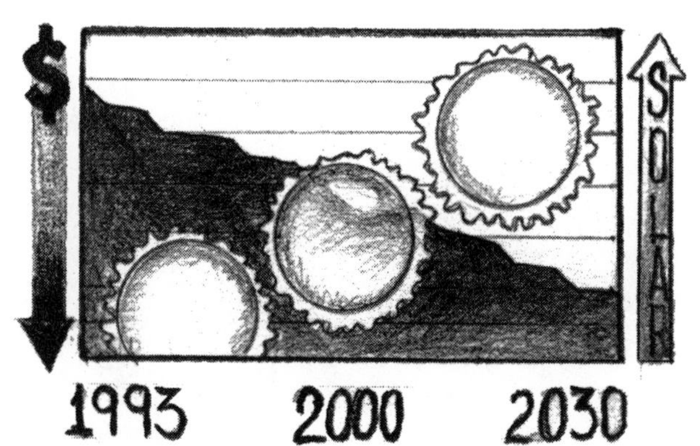

Although passive solar energy, via clerestory windows and an open floor design, is considered an integral part of energy-efficient design, active solar technology—transforming the sun's energy into electricity—is not used on a wide scale. It is still considered too expensive to implement. For example, even though Audubon sought the least-damaging source of energy to run its retrofitted headquarters building, active solar energy wasn't economically practical.

Solar energy costs, however, are coming down. Photovoltaic electricity costs dropped by a factor of four in the 1980s to 30 cents per kWh. Nevertheless, solar electricity costs are still considered too high to compete with the lower cost of electricity obtained from utilities. The outlook looks promising, however: the Worldwatch Institute predicts that solar electricity will become competitive in the next decade. We must anticipate these future possibilities in our commercial-building designs.

CONSERVING MATERIALS: RECYCLING AND REUSING

Businesses contribute in many ways to the waste stream, not the least of which is through demolition and construction of new building complexes. Retrofitting saves energy and resources that new construction would otherwise require. These savings include the energy to extract, manufacture, and transport the building materials. It also alleviates the environmental impact of disposing of huge quantities of demolition materials.

FACT: Audubon chose to retrofit, or recycle if you will, its headquarters building rather than demolish it and start from scratch. Doing so preserved 300 tons of steel, 9000 tons of masonry, and 560 tons of concrete.

Any renovation project is not without its share of waste materials from demolition. Yet steps can be taken, as was done with the Audubon House, to keep as much out of landfills as possible. Typical materials separated for recycling include concrete, glass, wallboard, bathroom partitions, masonry, and roofing felt. Today there is a growing selection of recycled building materials to meet the most rigid environmental, cost, and aesthetic standards.

THE AUDUBON GARBAGE TEST

In addition to its Garbage Test for home use (see chapter 4), Audubon designed a test to help determine what kind of system was needed in its offices to help make recycling easier and less of a hassle. The Garbage Test was tested in the former headquarters building prior to relocating to establish what percentages of the waste stream consisted of recyclable paper, compostable food, and recyclable aluminum and glass. This information was then used in designing a recycling system for the new headquarters building.

For the test each department was asked to select leaders who would supervise the collecting and sorting of trash. Boxes were placed in central locations and the waste was weighed at the end of one week. The containers were marked and the trash sorted as follows:

A. *Hazardous* or other waste that should not be mixed with the general waste stream. Examples: Wite-Out, fluorescent tubes, batteries, laser printer and photocopy toner cartridges, paints, glue containers, polishes.

B. *Newspapers, magazines, glossy brochures* (material that is now, or soon will be, recyclable into newsprint, but not higher grades of paper). Examples: all dry newspapers and magazines.

C. *White office paper.*

D. *Mixed paper.* Examples: colored office paper; clean, dry paper from takeout restaurants; manila folders; interoffice envelopes; envelopes without labels; phone books; junk mail.

E. *Other paper.* Examples: Post-Its or other sticky paper, take-out cardboard, fax paper, paper with labels.

F. *Corrugated cardboard.*

G. *Compostables.* Examples: lunch and food waste; soft, wet paper; tissues; coffee grounds.

H. *Mixed materials.* Examples: ballpoint and other disposable pens, pencils, paper envelopes with bubble wrap; liners; reports with plastic covers not easily removed; damaged computer disks.

I. *Plastics.* Examples: Styrofoam packing material, wrapping, takeout food containers that have been washed.

J. *Glass, cans, metals.* Examples: clean or washed containers, broken paper clips.

K. *Electric or other heavy equipment.* Examples: broken clocks and fans.

L. *Miscellaneous.*

Many employees could be seen at the end of the day, lingering at the sorting boxes, trying to figure out what items were actually made of and where they should go. It wasn't easy. They quickly saw that everyday objects, such as envelopes, were a combination of things, such as paper and glue, and therefore couldn't be recycled. By the end of the week, they saw garbage in a very different light.

Highlights of the results were that for an office of 150 employees, 1135 pounds of waste were thrown out during the test week, more than 80% of that in the form of paper. That translates to 227 pounds of garbage being generated a day, or 1.8 pounds per person per day. Here's the breakdown by category:

PERCENT	MATERIAL
0. 1	Shiny cardboard
0. 2	Electrical or heavy equipment
0. 3	Hazardous materials
1. 8	Mixed materials
1. 9	Plastics
4. 0	Corrugated cardboard
4. 1	Compostables
4. 7	Glass, can, metals
7. 2	Other paper
7. 5	Miscellaneous
9. 3	Mixed paper
13. 7	Magazines, glossy brochures
13. 7	Newspapers
31. 6	White office paper
Total: 100	

The data showed that Audubon already recycled 35% of its trash (white paper, cans, and bottles). Adding newspapers to the recycling list could get the total to 50%. If Audubon could find businesses to take and recycle glossy magazines/paper and mixed paper, as well as make it easier for employees to separate such waste at their desks, they could recycle as much as 73% of their waste.

If food scraps and soiled paper could be composted, the amount of material being removed from the office waste stream could go as high as 80%.

The point of Audubon's undertaking the Garbage Test was to determine what kind of waste was generated in the office in order to install an advanced recycling system tailored to employee needs. The system works as follows:

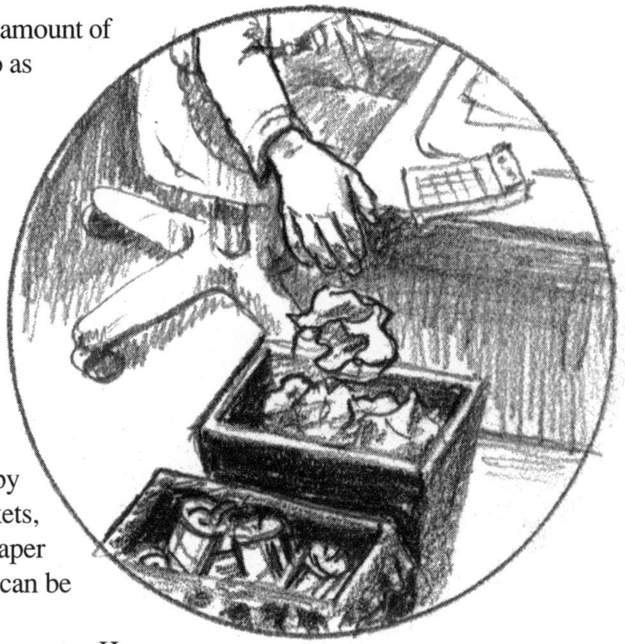

- Each floor in the building is equipped with four waste chutes to enable convenient separation and disposal. Each chute is dedicated to a particular type of waste: high-quality paper, aluminum/plastics, mixed paper, and food waste. Other materials, such as newspapers, glass, and hazardous items, are placed on shelves for collection by custodial staff. Each desk is equipped with dual wastebaskets, one for mixed paper, one for residual trash. White office paper is collected in dual desk trays, with one slot for paper that can be reused for notes.
- The disposal chutes terminate in the subbasement recycling center. Here waste is processed and packaged for transport to the appropriate recycling outlet. Cans and plastic bottles are crushed; newspapers and cardboard are

bound; hazardous waste is securely packed for pickup. This is where food and other biodegradable waste is to be mulched in the center's special composting equipment.

- An on-site composting center will be used to turn food waste (organic materials) into rich humus. Compostable waste material from the office will be shredded and placed in insulated, ventilated garbage bins in the basement of the building. Temperature and moisture level during the curing period is particularly important and will be monitored closely to achieve fast, aerobic composting conditions, and to reduce odor problems. The final compost product will be used for potted office plants and the rooftop garden.
- Audubon's purchasing guidelines urge that incoming products be recyclable. Audubon developed purchasing guidelines with an understanding of the current recycling market. Whenever practical, and whenever costs are below 10% extra, it buys only those products that contain recycled content and that can be recycled after they are used. Everything from stationery, file folders and Post-Its to the packaging of food products must pass muster before it is ordered.

Audubon's unique recycling system and room required a one-time investment of $185,000. Eventually, markets for recyclables and government regulations may add economic incentives for building owners to install similar comprehensive systems.

It is not enough to think that we are environmentally enlightened if our purchasing decisions take into account impacts of a product only after it is discarded (the "downstream" concept). For example, when considering a particular product and analyzing its environmental "signature," it is important to consider: (1) Where will it end up—in a landfill or incinerator? (2) Is it to be disposed of as hazardous waste? (3) Or will it degrade or can it be recycled? But we must also be aware of "upstream" impacts, which include the pollution created during the extraction of raw materials or processing. Paper, for example, has very high upstream impacts.

Busy as Ants

There are striking similarities between workplaces and ant colonies. In both cases the agglomeration of individuals compartmentalize their territory and work at a fast pace. Ant colonies and workplaces become highly structured organisms.

Communication is an essential component in this flurry of activity. For example, while humans have phones, faxes, and the postal service to keep in touch with the outside world, ants have antennae. As Lewis Thomas reflects, "Somehow, by touching each other continually, by exchanging bits of white stuff carried about in their mandibles like money, they manage to inform the whole enterprise about the state of the world outside, the location of food, the nearness of enemies, the maintenance requirements of the Hill."

Ants are quite enterprising and are one of the most successful of the insect families. The number of individuals seen in an area of half an acre may be astronomical, for colonies number from a few members to several hundred thousand. More ants are present in any one locality than can actually be seen, as a large proportion of any nest is always busy in the galleries, and there are some species in which the workers never appear above ground.

These insects have a highly organized and cooperative social order. They seem to thrive under modern conditions every bit as well as they did before cities and farms replaced fields, woodlands, and meadows. And they are industrious, taking on projects much grander in scale than themselves. The Allegheny mound-building ant, for example, builds structures four or five feet in diameter and two feet high. A single colony may occupy several mounds.

The 8800 species of ants are essential to the survival of the Earth. They disperse seeds and make soil. If they were to disappear, most fish, amphibian, reptile, bird, and mammal populations would crash for lack of food. Dead vegetation would pile up and dry out, closing nutrient cycles, causing other vegetation to dry up, as well as the last remnants of the vertebrates.

Ants are in tune with the rise and fall of the sun. Ant nurses take care that the proper temperature is maintained in the nest. During the heat of the day they move the young to chambers deep in the ground. When the coolness of evening descends, they carry them up nearer to warm stones or pavements.

Urban Wildlife

Urban areas are not only the centers of economic activity for millions of humans, they are home to a host of wildlife, some revered (like the peregrine falcon), some the object of disgust (like the common cockroach). The diversity of wildlife is less in cities than in natural ecosystems, but nonetheless the creatures that inhabit our heavily populated areas are amazingly resilient and versatile. They have to be in order to adapt to and survive the drastic changes in their world.

Our city neighbors include assorted birds, mammals, reptiles, amphibians, and fish, as well as insects, spiders, and countless other species of invertebrates. Sometimes they occur in greater numbers in the city than in the wild. They thrive in parks, vacant lots, lawns, fields, street trees, sidewalk cracks, cemeteries, landfills, and waste areas.

How did these critters become city slickers? While some may think many of them, such as cockroaches and rats, are revolting and a nuisance, they are wildlife, part of the ecosystem we have fashioned for ourselves. Some wildlife species that once occupied natural habitats were forced out, became extinct, or were pushed into smaller and smaller patches when the land became urbanized. But some animals came to flourish in the infrastructure of our cities because they were able to adapt easily to changes in their food source and because there was less competition for food and living space.

In Praise of . . . Starlings?

Michael Harwood

I first began to admire *Sturnus vulgaris*, the European starling, about twenty-five years ago on a morning after a snowstorm. I lived then in a frame apartment house halfway up a hill on Staten Island, New York City. The snow-mounded roofpeaks of the houses on the next street below were at eye level, and I noticed a small troop of starlings on one of the roofs, playing in the fresh powder. For all of you who wince at the first hint of anthropomorphism, let me say that playing is just what I mean. These birds had created a little schuss-track from the roofpeak down to the ledge, and again and again they flew up to the peak, sat back on their tails and undertail coverts, and slid down the track. . . .

Our immigrant starlings are common partly because they have taken superb advantage of the fruits of human civilization. Many of them live alongside us, which makes them seem—by a warp in our perspective—not truly wild. More than that, their remarkable success makes them downright pests sometimes, particularly from late summer through winter, when they collect in night roosts, often with blackbirds, in aggregations that may number in the hundreds of thousands. . . .

More important, starlings are remarkably resourceful, versatile, and flexible. They glean grubs from our lawns and farms, like robins; and they stalk barnyards, pastures, and the backs of cows, like so many dark, miniature cattle egrets. Hopping and striding along with their purposeful, stiff-legged gait, they haunt our fast-food restaurant parking lots and our garbage dumps, like gulls, rummaging under the top layers of trash as if they were turnstones on a beach. They hawk insects in flight, like flycatchers and martins; they comb evergreens for insects, like titmice, and berry bushes for fruit, like waxwings. . . .

The best example of its adaptability is the way it took hold on the foreign turf of North America. After several attempts at introduction had failed, a hundred starlings released in New York City in the early 1890s established a successful colony. Barely more than a half-century later starlings had appeared as far north as James Bay, as far south as Florida and Mexico, and as far west as California. . . . Whenever they found a suitable place and began to multiply, starlings competed with native birds for food and housing, and their persistence and versatility made them winners more often than not.

Imagine those brassy little pioneers on the frontiers, wandering great distances, exploring new terrain, braving blizzards and deserts to find their versions of the promised land. If we weren't so anthropo*centric*, we would recognize that what we admire so much in the recent history of our own immigrant species in North America—the ability to overcome all sorts of difficult environmental challenges—is the very characteristic that distinguishes the immigrant starling. And if *Sturnis vulgaris* has faults, they are unsettlingly like those of *Homo sapiens*.

E c o Q u i z

Q: What can you do in your office to help reduce your contribution to pollution and waste?

A: Here are some simple habits you can work into your everyday routine:

Reduce paper

- To make it easier to recycle at your desk, have separate bins for recyclable items vs. nonrecyclable trash.
- When photocopying documents and memos, use both sides of the paper.
- Use routing slips to circulate your internal memos and reports instead of making multiple copies of the same document.
- On wastepaper that has copy only on one side, use the other side for scratch paper. Even envelopes can be used for notes.
- Reuse envelopes when sending letters, documents, and materials out. Just put your label over the old address.
- When shopping for stationery or paper for printing, whenever possible choose paper that has at least 10% postconsumer recycled content. Unbleached, non-de-inked, postconsumer recycled is the best.
- Fill the paper feeder of your copier or laser printer with one-sided, already-used paper.

Ride a bike to work, take mass transit, or carpool

Conserve materials

- Donate old furniture or equipment to charities or nonprofit organizations.
- Use a mug for coffee instead of disposable cups—even when you go to the deli. They are usually happy to oblige.
- Recycle your toner cartridges (Canon and Hewlett-Packard have recycling programs for their copier and laser printer equipment) or go to your nearest business supply store and ask them to refill the cartridge.

E C O S Y S T E M   P A T H W A Y S

How to control and manage transportation in this country has people either tearing their hair out or throwing up their hands. Visions of brave new worlds abound, but so far the political and economic will to carry them out has been weak. With better understanding of the issues, perhaps we can move forward to implement the solutions that are at hand.

As shown in the picture, the negative ecological impacts of our transportation system are that

- Present systems depend almost exclusively on oil.
- Roads cut through habitats and destroy species. Wetlands and forests are increasingly encroached on. Noise disturbs breeding and nesting requirements.
- Our transportation vehicles interfere with the carbon and nitrogen cycles. They are major CO_2 producers and accelerators of global warming as well as a large source of health-threatening air pollution and acid deposition. The major pollutants are carbon monoxide, nitrogen oxides, sulfur oxides, volatile organic compounds (e. g. , hydrocarbons), suspended particulate matter, ozone, and lead (from leaded gasoline). Whereas upper-atmosphere ozone protects us from ultraviolet radiation, unhealthy ozone is created at the street level as nitrogen oxides and hydrocarbons interact with sunlight as a result of vehicle exhaust combining with other pollutants. Stultifying smog occurs when a lid of warm air traps currents of air that would normally carry away some of these pollutants.
- High car use has a detrimental effect, but car pools or van pools, walking, and bicycling can greatly reduce it. Although teleconferencing and computer networking reduce the need for transportation, they may lead to increased development of roads and homes in rural areas.
- Health threats are high. About 500,000 persons die in car wrecks around the world each year and animals are slaughtered by cars in even greater numbers. An estimated 150 million persons in our nation's urban areas breathe unhealthy air or suffer respiratory problems. Carbon monoxide, mainly from motor vehicles, interferes with the blood's ability to absorb oxygen, and can cause fetal damage as well as heart problems.
- Car-induced air pollution also hurts vegetation and reduces soybean, cotton, and other crop yields by 5% to10%.

9

TRANSPORTATION

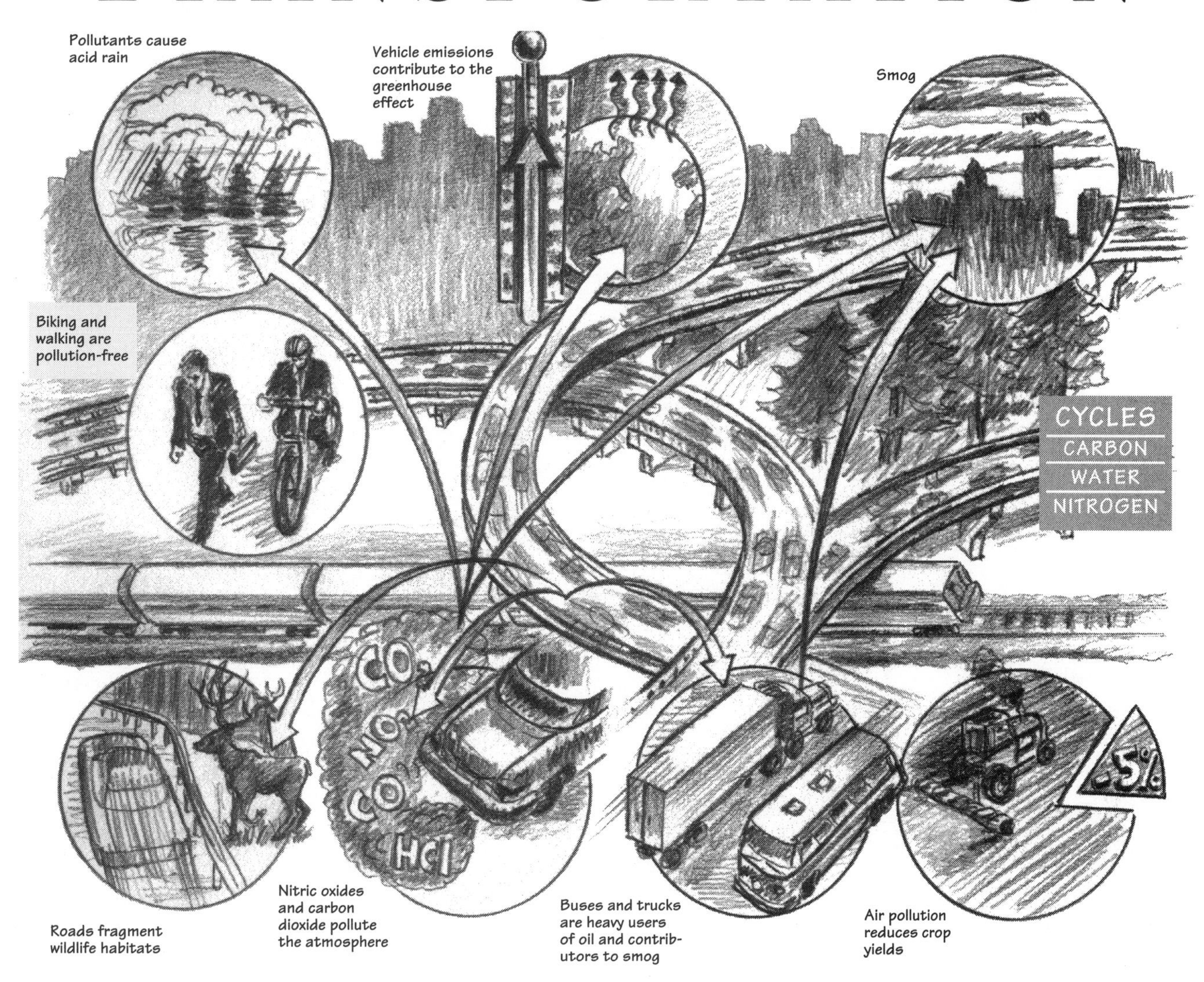

Countless Americans have a romance with cars. Is it because they are, indeed, remarkable machines? Or because they are advertised along with glamorous women and wild animals, suggesting sex and power? It is highly ironical when commercials show automobiles amidst pristine nature, the very settings that their presence most disturbs. Ads also promote a sense of being in control, which consumers seem to fall for. Many people use driving as a means to ease depression or to escape from disappointment or sagging hopes.

Listen to Lesley Hazleton:

> Something happens to my mind when I drive fast. Speed changes the way I think. The power of the engine becomes confused—or simply fused—with my own personal sense of power. At speed, I feel as though I own the road. Even as though I conquered it.

Your car is your second home. Your private space. If you are a teenager, maybe it's freedom from home, freedom to drive fast and go where you want to go. Having a car means you can take a date somewhere. You have status. If you are an adult, cars are still bought for their status value. You want more than just a functional

Eco Koan

Student: How much pollution am I responsible for?

Guru: For your own and your descendants' up to the seventh generation.

machine, you want luxuries—plush seats, push-button conveniences, air conditioning—and you are willing to pay thousands of dollars for a car that depreciates the moment you walk off the sales floor. Most adults buy cars on credit, just as they do houses, and are willing to be in debt for years for a product that is not built to last very long.

Maybe you are a mother who lives in the suburbs and has to chauffeur young children around to various after-school activities. Most likely you are an adult who drives at least part of the way to work. You get in your car and take local roads to the main roads. You will probably get lined up behind many other single-occupant cars, and you will wait . . . and wait . . . and wait. Your blood pressure mounts, you become frustrated, your temper rises. Meanwhile your motor is idling and exhaust is entering the atmosphere.

Here is where our population growth and consumption habits enter the picture. Because the United States is a growing industrial nation, we have more and more cars on the road. Building more highways, it's been learned, just leads to more congestion. Since the Federal Highway Act of 1921 the building of roads took a quantum leap; highways networked the landscape, along with a super rise in the number of cars on them. In 1990 there were 123 million cars, many more than found in much more populated India and China. Along with more traffic has come disturbance and disappearance of wildlife habitat, more animals killed on the road, and more people injured and killed in accidents. To a large extent our lives are dominated by watching out for traffic and looking at road maps.

During your car's lifetime, you will pay dearly for maintenance and repairs, aside from regular gas and oil. Do you ever wonder what happens to your car when you sell it? Or, what happens to useless cars when hauled away? Here is a description by the U. S. Office of Technology Assessment of how the parts of a car are handled.

> When an old car is junked, it is often first sent to a dismantler, who removes any parts that can be resold, as well as the battery, tires, gas tank, and operating fluids. The hulk is then crushed and sent to a shredder, which tears it into fist-sized chunks that are subsequently separated to recover the ferrous and nonferrous metals.
>
> Presently, about 75 percent by weight of materials in old automobiles (including most of the metals) are recovered and recycled. The remaining 25 percent of the shredder output, consisting of one-third plastics (typically around 220 pounds of twenty different types), one-third rubber and other elastomers, and one-third glass, fibers, and fluids, is generally landfilled. In the United States, this shredder "fluff" amounts to about 1 percent of total municipal solid waste. Sometimes, the fluff is contaminated with heavy metals and oils, or other hazardous materials. . . .
>
> In Germany, the landfilling of old automobile hulks and the shredder residues from automobile recycling operations is a growing problem. The German Government has proposed legislation that would require automakers to take back and recycle old automobiles at the end of their lifetime. This has stimulated German automakers to explore fundamental changes in automobile design that could result in more efficient materials management. . . .
>
> BMW recently built a pilot plant to study disassembly and recycling of recovered materials, and Volkswagen AG has constructed a similar facility. The goal of the BMW facility is to learn to make an automobile out of 100 percent reusable/recyclable parts by the year 2000. In 1991, BMW introduced a two-seat roadster model whose plastic body panels are designed for disassembly, and labeled as to resin type so they may be collected for recycling. . . .

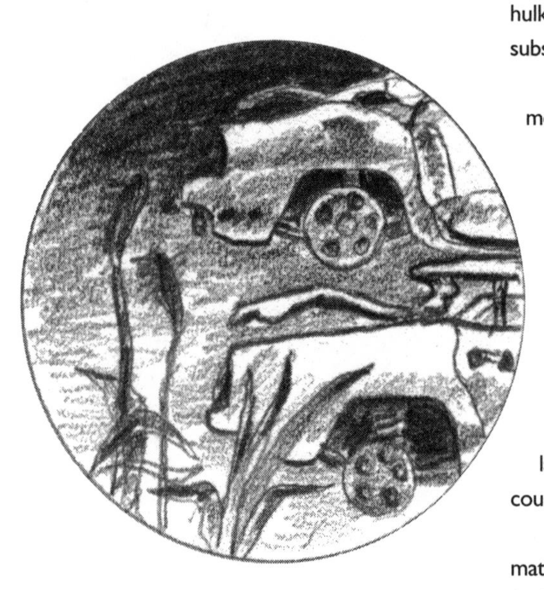

Some of us in urban areas take buses and ride trains. Many of us fly in airplanes for work and pleasure. Buses and trains can be more efficient than single-passenger cars when they are filled, but not when they are empty or partially occupied.

The costs of road transportation are enormous and include not only our direct personal expenses but also indirect burdens on ourselves and communities. If individuals had to pay for these costs, we'd probably have a much bigger outcry for more efficiency. Here is a chart for the U.S. and New York City.

TOTAL COSTS OF TRANSPORTATION

	USA	New York City
A. Direct Costs	billions of 1990 $	
auto	$510. 8	$6. 5
road construction & maintenance	$48. 1	$0. 9
school bus	$7. 5	$0. 1
truck	$272. 6	$13. 4
B. Other		
air pollution health & property	$1. 5	$0. 2
congestion	$142. 8	$4. 5
accident	$290. 4	$5. 2
noise	$1. 1	$0. 1

Let's take a look at the total energy demands of our modes of transportation. The following rounded-off figures are for trillion BTUs in 1989.

Intercity rail	18
Intercity bus	23
Motorcycles	26
Transit rail	43
School bus	62
Transit bus	77
Domestic air carriers	1600
Personal light trucks	4100
Automobiles	9000

Oil is our drug of choice: motor vehicles consume nearly half of the petroleum used in the United States. The background of the last fifty years has worried policymakers about our increased oil consumption and our vulnerability to decisions made by other governments, even aside from the threat of terrorist attacks or bombings. In 1960 the Organization of Petroleum Exporting Countries (OPEC) was formed to take control of pricing away from Western oil companies. The founding members were Iraq, Iran, Kuwait, Saudi Arabia, and Venezuela. Joining later were Algeria, Ecuador, Gabon, Indonesia, Libya, Nigeria, Qatar, and the United Arab Emirates. In 1973 the Arab countries embargoed sales to this country for five months. In response Congress created the Strategic Petroleum Reserve and established automobile economy standards. Alaska oil production began in 1977 because Congress assumes our supply of oil is running low and will become prohibitively expensive in the next fifty years. Our addiction to oil has caused destruction to the Earth (our body), toxic spills, and puts our future at risk.

SOLUTIONS

Opportunities are at hand to implement a new transportation vision. Granted, the financial stakes are high. But if environmentalists work with industry leaders to meet the challenge, both sides can win. If the goals of greater efficiency, alternative fuels, road containment, and bicycling can be made practical, we can establish a sustainable transport system.

ENERGY EFFICIENCY

Because emissions of carbon dioxide greatly disturb nature cycles, reducing them is one of the major goals of energy efficiency. Below is a chart that shows how much CO_2 is generated by gallon or mile and how you can figure the amount you produce through various means of transportation. Remember too that car air conditioners are the greatest sources of halocarbons, which eat the ozone layer. For every 3.3 pounds of CFCs, 17,500 pounds of CO_2 are released.

ANNUAL U.S. CO_2 EMISSIONS PER CAPITA TRANSPORTATION

Activity	Your use (units per yr)	CO_2 factor (lb CO_2/unit)	Annual emissions (lb CO_2 equivalent)
Automobile	gallons	22 lb/gal	
Air travel	miles	0. 9 lb/mile	
Bus, urban	miles	0. 7 lb/mile	
Bus, intercity	miles	0. 2 lb/mile	
Railway or subway	miles	0. 6 lb/mile	
Taxi or limousine	miles	1. 5 lb/mile	

A least-cost strategy to reduce 50% of CO_2 emissions over the next fifty years has been offered by a group of scientists affiliated with the American Council for an Energy Efficient Economy, the Alliance to Save Energy, the Natural Resources Defense Council, and the Union of Concerned Scientists. It calls for major investments in energy-efficient technology and energy conservation measures, a carbon tax of $25 per ton in addition to environmental "externality" costs, a shift to renewable fuels, and an end to sprawling development, with no further need for new highways after 1995. The authors point out that while some individuals may feel some burden, society as a whole benefits.

While airplanes, trains, and trucking are capable of considerable design improvement, a key factor is, of course, a more efficient car. Here is a description of what is possible:

The Green Car

John M. DeCicco and Deborah Gordon

What would the "green car" be? An oxymoron to some and an environmentally safe personal mobility machine to others. Think of the green car as an ideal toward which the nation must strive if it is to achieve an ecologically sustainable transportation system. Production, use, and disposal of such a car would consume no fossil fuels and generate no pollution. The greenness of a car depends not only on the machine, but also on how and when it is used. A car is greener with two people in it than it is with one, and it's greener still with three. A car is greenest if it's not used at all when there's a cleaner way to go: by foot, by bike, by transit, by wire ("telecommuting"). The green car's supporting infrastructure—roadways and fuel supply—must be built and maintained without net habitat degradation or greenhouse gas emissions. Finally, although the CO_2 emissions from a vehicle's use are now about ten times those associated with its manufacture, there must be parallel progress toward a greener industrial system focusing on reduced fossil fuel use and pollution, minimal waste, and the design of products for recycling or refurbishment.

While we cannot expect to quickly realize this vision, the industry does know how to make "greenish machines," vehicles which will greatly reduce the environmental impacts of each mile driven. Today's cars and light trucks average 20 MPG on the road, resulting in average CO_2 emissions of about 540 grams per mile. With automotive technologies now available and in development, light vehicle energy efficiency could be doubled. This would halve CO_2 emissions per mile. Population and economic activity will grow, driving transportation demand. But, by starting a switch to renewable fuels and pursuing a shift to more appropriate modes of urban travel, a generation from now the U. S. could achieve an absolute halving of CO_2 emissions from personal transportation.

What would be the "nuts and bolts" of a greenish machine? An electric drivetrain is a good bet. Electric motors have negligible direct emissions, operate at high efficiency over a range of loads, and draw no power at idle. When electric motors are used to brake the car during deceleration, they can act as generators to recover much of the energy that today's cars dissipate through friction. The significance of this *regenerative braking* must not be underestimated, since more and more driving is done under congested, stop-and-go conditions, in which most of the energy supplied by the engine is lost to braking. Electric motors are also quiet and durable and could be easily recycled or refurbished.

Zero tailpipe emissions would be a major boon in urban areas; this is why Los Angeles is leading the way to get electric vehicles on the road. However, the greenness of an electric vehicle depends on how clean and renewable the electricity generation system is.

To power an electric vehicle, we now have to rely on batteries—heavy, inefficient, and made with hazardous materials like lead and acid. The materials problems are not a fundamental limitation, since they could be dealt with through careful packaging and procedures for return and recycling by the battery supplier. The weight and performance limitations are, however, a much larger challenge. Major engineering breakthroughs are needed if batteries are ever to see widespread use as a sole source of on-board power.

More promising in the long run are fuel cells—devices that electrochemically convert a fuel into power. Hydrogen, supplied from a renewable resource such as biomass, is an ideal input for fuel cells. Although hydrogen storage is presently problematic, there are some promising options: metal hydrides, carbon, and an iron/water system. Hydrogen can also be carried in natural gas or methanol by using an on-board "reformer," a device to break the fuels into hydrogen and CO_2. Analysis by researchers at Princeton University suggests that such fuel cell systems look very promising as a long-run option for vehicles to be environmentally sustainable and have low lifecycle cost. A fuel cell electric vehicle would have high end-use efficiency, which is crucial for keeping any renewable fuel production to a manageable scale that avoids conflict with food production and habitat protection.

The first generation greenish machine could be a hybrid. The drivetrain would combine a small, efficient combustion engine with an electric motor and a medium-sized battery. The engine could be constrained to operate only under narrow conditions, maintaining optimum efficiency and minimal emissions. Battery range limitations would be eliminated, and the regenerative braking and efficiency benefits of an electric drivetrain would be realized. Use of hybrid vehicle technology could more than double the efficiency of light vehicles, pushing the on-road average of cars and

(Continued on page 128)

(Continued from page 127)

light trucks to 50 MPG without reducing size or compromising performance. Petroleum supplies would be stretched, and they would be used much more cleanly and efficiently. Hybrid vehicles could also operate on a diversity of fuels, with the choice dictated by regional energy resources and environmental constraints.

Best of all, many of the technologies needed to make an efficient hybrid vehicle are already on the shelf. Improved aerodynamics, low rolling resistance tires, high-efficiency mobile air conditioners, and other improved accessories already appear in new cars. A variety of refinements allows today's best engines to produce a given amount of power at less than half the size of older designs. Electronic control of ignition and intake/exhaust systems yields simultaneous lowering of emissions and improved torque—the rotational force needed to move the car. Further efficiency enhancements can follow from various "lean-burn" engine designs, including two-strokes and advanced diesels.

Nearly all automakers have prototypes of greener cars. No big breakthrough, just good engineering, is needed for practical hybrid vehicles. Steady research and development efforts could make fuel cell electric vehicles a reality for the next century. Ultimately, the challenge is much more a matter of political will than technical ability. With a national commitment to start heading in the right direction, we could soon be driving progressively greener machines down the road to an environmentally sustainable transportation future.

ALTERNATIVE FUELS

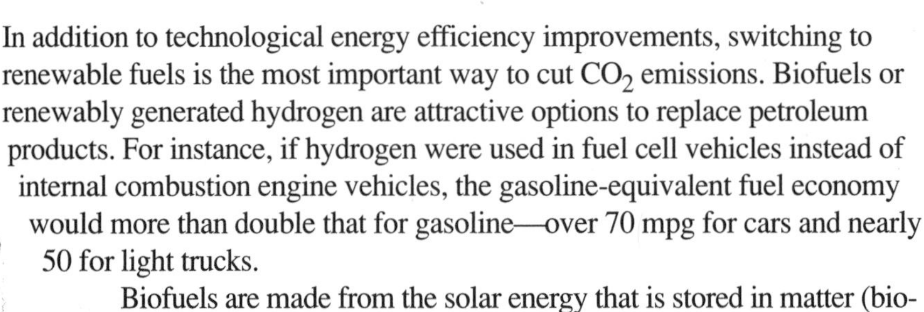

In addition to technological energy efficiency improvements, switching to renewable fuels is the most important way to cut CO_2 emissions. Biofuels or renewably generated hydrogen are attractive options to replace petroleum products. For instance, if hydrogen were used in fuel cell vehicles instead of internal combustion engine vehicles, the gasoline-equivalent fuel economy would more than double that for gasoline—over 70 mpg for cars and nearly 50 for light trucks.

Biofuels are made from the solar energy that is stored in matter (biomass), such as trees, crops, or human or animal waste. Currently, biomass technology has been used primarily to produce ethanol—used in gasohol, a clean-burning motor fuel, from corn and other agricultural products. In the future, ethanol will also be made from fast-growing tree crops. But these fuel crops are not sustainably grown at present because of chemical-intensive agriculture and use of fossil fuels in processing the ethanol. If the new technology is managed improperly, reliance on biofuels could produce its own major environmental problems, even if CO_2 emissions are reduced.

For instance, if ethanol fuels were substituted for gasoline and other transportation fuel, nearly 400 million acres of land would have to be cultivated for biomass. To meet all of our transportation and electrical energy needs, more than 900 million acres might have to be cultivated—nearly all of our arable land. Such intensive cultivation of biofuels on land could cause water pollution, erosion, loss of soil fertility, the spread of bioengineered organisms, and severe competition for other uses of land. Forests could be lost through conversion to "super tree farms."

Audubon would like to see coal, oil, biomass industry leaders, regulators, and environmentalists work together in planning the future of a sustainable biofuels program that would avoid any irreversible negative impacts. One idea is to identify indicator species that could warn of any dangers. Thus, a biodiversity protection policy could require a developer to demonstrate that a biomass energy project would strengthen local diversity and not diminish global biodiversity.

Another idea would be to use ecology in planning land use. For instance, if natural forests were replaced by monoculture plantations—i. e. , tree farms—the result could be a substantial loss of biodiversity. But if the biomass were grown on lands that were already degraded, biodiversity could be improved. Networks of mature patches of land could be connected by hedgerows, as in England, to ease the transition to biomass farms. Mature forests, including large trees, snags, and rotting logs, would act as reservoirs of organisms important to the health and productivity of ecosystems. Such buffer zones and wetlands would be part of the network.

Another way to maximize ecological diversity on intensively cultivated lands would be judicious use of cloning. Cloning is used to propagate genetically identical plants from cells or from stem and leaf cuttings from plants selected for a supposedly superior trait, such as fast growth or disease resistance. If cloned trees are planted as a monoculture, they have a detrimental effect on biodiversity, but a large menu of clones would provide more diversity. Another means is to interplant with nitrogen-fixing species, which reduces the need for fertilizer, thereby reducing groundwater contamination. In the long run, methods that promote more biodiversity ensure greater ecological health and productivity. Such are the considerations we must take into account when facing the future of biofuels.

Super Speed

One of the practical possibilities of superconductivity is the overcoming of friction and levitation that the manipulation of magnetic fields permits. In fifty years adaptation of this new technology could significantly alter our way of life. Imagine the possibilities. High-speed trains, even cars and trucks. Magnetic braking could be used to conserve energy that otherwise would be wasted in stop-and-go traffic.

The problem with rushing into a new technology is that we blunder into problems and then end up having to deal with the side effects, as happened with the advent of nuclear power. It may turn out that superconductivity will produce stray electromagnetic fields that cause cancer. The migratory signals of birds could be confused. It is possible to take precautions and insist that the industry shield these fields from the very start of commercialization. With any new technology on the horizon, meeting the risks with adequate measures in advance would go a long way to saving us headaches.

GLOBAL CLIMATE DISRUPTION OR TEMPERATURE MIGRATION

The volume and methods of transportation, as a result of industrial expansion and the population explosion of the last fifty years, have ended up being primary causes in human impact on the ecosphere. The gases spewed into the atmosphere have raised—or eventually will raise—the temperature, which in turn affects every element of ecosystems, biodiversity, and the major cycles of nature.

Over time the Earth has adjusted to the movements of continents, bombardment by debris from space, and shifts in relation to the sun that brought

THINKING LIKE AN

ECOSYSTEM

Climate and Temperature

The Earth's climate is a vast engine fueled by incoming solar radiation. This incoming energy is balanced by that which is re-radiated to space. It is not, however, equally balanced across the surface of the planet: net gains take place in the tropics while net losses occur at the poles. These temperature differences among latitudinal zones as well as analogous ones between land and sea are the basic driving force behind all meteorological phenomena. Winds and ocean currents redistribute heat from the tropics to temperate and polar latitudes, taking with them dust, nutrients, pollutants, particulate matter, gases, and even plants and animals. Solar radiation also drives the hydrologic cycle and creates the high and low pressure systems that determine regional climates.

Climate is not constant. It is influenced by a host of astrophysical, geophysical, atmospheric, and biological forces that together constitute the *climate system* and cause climatic conditions to change through time and across the face of the Earth. Any change, whether natural or anthropogenic, that affects the Earth's radiation balance can cause at least localized changes in climate.

Joel Carl Welty, zoologist and author of the extraordinary *Life of Birds*, said:

> Climate, more than any other physical factor, determines whether or not a given species will live in a given region. . . . Within vital limits, change in temperature works a greater hardship on an organism than a steady hot or a steady cold temperature.

A cyclical aspect of climate is light. Light changes in intensity, duration, wavelength (color), and direction on a daily basis between what we call night and day, and on a yearly basis in what we call seasons. These rhythms of light have a profound effect on creatures' anatomy, physiology, behavior, and distribution. Since the beginning of time climate has been a major factor in ecology.

on ice ages. None of these have destroyed our planet or extinguished all of life. In the past humans were a rather minor presence on the planet, but now, with our exponentially rising population, it is possible that we could destroy the conditions needed for our own survival. That is why we must plan ahead carefully.

Since the last ice age, the average temperature has risen nine degrees, and it is generally predicted that if we continue to pour greenhouse gases into the air at the present rate, the average surface temperature will rise by three to eight degrees by 2050 and every sixty years thereafter. We have to consider the risks of the situation. If we act as though the rise in temperature were consistent with the natural swings of climate variability, we can do nothing. If we perceive the rise as human-induced, through nonsustainable farming, road-building, and urbanization, we realize that instead of climate changes occurring naturally over thousands of years, the rate will take place over 100 to 200 years.

The search for what greenhouse theoreticians called the "warming signal" has been the subject of two enormous studies to plot the millions of temperature readings available, one led by researchers at East Anglia University and the other by James Hansen of NASA's Goddard Institute. Both concluded that global temperatures rose about one degree Fahrenheit during the last century. In 1988 Hansen testified before the Senate that he believed it had become apparent because in the last thirty years there had been three times the standard deviation. He said, "When we're talking about that degree of deviation, we have to feel it's pretty unlikely to be chance fluctuation. " Because that summer had been so hot, many people paid attention.

Armed with powerful computers, scientists use advanced mathematics to represent the physical workings of the atmosphere and to simulate the world's climate under varying conditions. But the models are as yet imprecise in making predictions.

Gazing into Our Greenhouse Future

Jon Luoma

One of NCAR's (National Center for Atmospheric Research)—and Steve Schneider's—key missions is to untangle the knot of knots:the inhalations and exhalations of Gaia, the workings of the planet's atmosphere. Schneider and his colleagues here at the Temple are among the world's leaders in trying to untangle the knot. And oh, it is tangled. Is the Earth really going to heat up? How much? How fast? To what effect?

NCAR is one of a handful of top purchasers of high-speed supercomputers in the world. . . . Supercomputer power is measured in "gigaflops," or billions of calculations per second, meaning that within about fifteen minutes a supercomputer can do as much arithmetic as a crack mathematician with a calculator, working full time, can do in—oh—roughly one ice age. . . .

While most scientists readily concede that there is no absolute proof that the world will become drastically hotter, many experts, including Schneider, point to two decades of evidence that, absolute proof or not, the risk appears to be extraordinarily high. If we wait to deal with the problem while scientists look for a conclusive answer, the hothouse century may already be upon us.

My own look into the global warming story suggests that we are careening into the future not with a sober climatologist in a Volvo, but at an extraordinary speed with a carbon dioxide-, methane-, and CFC-intoxicated teenager in a Yugo with loose steering. Probably on the wrong side of the expressway. . . .

Schneider is happy to disabuse anyone who will listen of a misconception or two. "The greenhouse effect," he says, "is not a 'controversial' theory. It is probably the best established principle in atmospheric science, and it has been well established for over one hundred years. "

The greenhouse effect, indeed, is responsible for warmth and life on Earth. It works like this: High in the atmosphere molecules of certain trace gases, principally carbon dioxide, allow heat radiation from the sun to filter through to the Earth, while also trapping near the Earth the infrared heat energy that is attempting to escape into space. Just like the panes of glass in a greenhouse, these gases keep the lower atmosphere of the Earth far warmer than it would otherwise be. Without them, in fact, temperatures on Earth would roughly approximate the refrigeration of Mars. . . .

No one can credibly dispute that levels of greenhouse gases in the atmosphere are steadily rising. The most compelling evidence of that comes from the CO_2 monitoring gadgets atop Mauna Loa, on the big island of Hawaii, far from any industrial or major urban pollution source. Continuous monitoring on the mountaintop since 1958 has shown a steady increase in atmospheric CO_2, rising from 315 parts per million to about 350 ppm today, an 11 percent increase in only three decades.

Further, no one can dispute that adding more of these greenhouse gases to the atmosphere should cause the Earth to warm up significantly, all other things being equal. But there's the rub. All things are not equal.

It is possible, just maybe, that the Earth has enough built-in self-regulating resistance to atmospheric alteration that the planet will not be devastated. That resistance comes from phenomena called "negative feedbacks." And the notion that negative feedbacks will greatly moderate the effects of greenhouse gases provides the entire basis for argument by those who deny that global warming is an actual threat.

Just what are these negative feedbacks? For one, trees and other green plants may respond to an abundance of carbon dioxide in the atmosphere by polysynthesizing at a

(Continued on page 132)

(Continued from page 131)

higher rate, and in the process use up much of the excess carbon dioxide. Similarly, exploding populations of algae phytoplankton in the warming oceans may take in much more carbon dioxide; or a warming may produce more bright, white clouds, which will reflect more solar heat back into space.

The hitch is that there are also "positive feedbacks." For instance, enormous amounts of methane are believed to be trapped in the Arctic permafrost. Warming the permafrost may release some methane, which will cause more warming and release more methane, and so on. Shrinking ice sheets may mean that the Earth reflects less heat back into space. And a warmer climate may induce more air conditioning, causing more fossil fuel burning, causing more CO_2 emissions, causing more heat. . . .

In any case a few scientists, pro-business policy-makers, chamber of commerce types, and other happy Pollyannas are relying almost entirely on negative feedbacks to balance, like magic, not only the straightforward warming itself but all the positive feedbacks as well. Although they can't come close to proving that they're right, maybe we'll get lucky. . . .

Mathematical modeling has its own host of problems. Essentially, a computer inhales all of the known data about climate, interactions between chemicals in the atmosphere, physical factors such as the location, size, and mass of land versus oceans, and thousands of other factors—the configuration of icefields, the exchange of heat between oceans and air, the metabolism of plankton in the sea and all the plants in a forest, projected human population growth, and industrialization. Such a Global Circulation Model then proceeds, through billions of calculations, to develop an atmosphere in a bottle—or, in this case, an atmosphere on a computer printout. . . .

It's a great idea. The problem is that it really doesn't work very well, just yet. . . .

Models "are marvels of mathematics and computer science," argues Reid Bryson of the University of Wisconsin, "but rather crude imitators of reality. ". . .

None are able to resolve their mathematical focus tightly enough, for example, to take fully into account the apparent negative feedback of the net cooling effects of the clouds. (This is complicated by the fact that scientists still aren't yet dead certain whether clouds actually cause cooling or warming.) Conversely, none have fully taken into account the positive feedback hypothesis forcefully advocated by scientist George Woodwell of the Woods Hole Research Center. According to Woodwell's analysis, increased soil temperatures will greatly increase the activity of decomposer bacteria in soils, which would release vast amounts of CO_2, a process that former U. S. Environmental Protection Agency scientist Daniel Lashof has called a "sleeping giant" in the greenhouse debate. . . .

"If we could 'only' cut emissions in half, we'd buy a great deal of time to study global warming and come up with better solutions. And if all our hypotheses are wrong—if it turns out to be an infrared herring—we will still have improved problems like acid rain, urban air pollution, and overconsumption of fossil fuels. "

Global warming is a subject that most of us are dependent on the specialists to figure out. But should we wait around for scientists to decide one way or another with absolute certainty? When chlorofluorocarbons (CFCs) were discovered to be ozone-eaters, the first alarms were sounded in the early 1970s, but it was not until 1987 that politicians of many nations gathered in Montreal and agreed to cut production by 50%. Meanwhile the damage was done. Common sense says act now before the destruction becomes too visible.

What are the predicted effects on what one scientist has called our "manscape"? (Keep in mind that when nature is assaulted, humans are not exempt.)

- The sea level will rise. When waters warm, they expand. Ice caps will melt. We will experience more flooding.
- There will be more intense and more frequent storms, driven by temperature differentials.
- There will be changes in biological productivity and the circulation patterns of oceans.
- The functions of ecosystems—community interactions, genetic phenomena and evolution, breeding and survivorship—will be altered. The capacity for change found in the systems of plants, animals, and microbes may be slower than climatic changes, making it difficult for them to adapt to changes. A decrease in population and in genetic diversity is predicted.

- Large-scale rearrangement of vegetation types will occur. Deserts move northward. Vegetation will increase in the north. Forests will be displaced and tree species killed. Species that live in coral reefs and mangrove swamps will be threatened.
- There will be impacts on various species. For instance, Arctic migratory shorebirds will decline because of loss of tundra habitats; because of drought, female elephant bands will fragment and increase copulation with non-dominant males; population sex ratios in crocodiles, alligators, and turtles will change because the sex of hatchlings depends on certain incubation temperatures. Migrations will be affected by the distribution of food sources. Volatilization of pheromones could disrupt social insects. Seasonality will affect fruits and flowers. Spring will come earlier and autumn later. Tropical wildlife will not be able to withstand soaring heat. The extinction of native species will increase.
- Wetlands will be flooded, inundated with saltwater, and eroded. Isolated sites, such as islands, will be threatened with obliteration.
- Wildlife remnants could be relegated to air-conditioned zoos.
- Soil moisture will decline, leading to an increase in fires as well as a change in species distribution. Pests will have new opportunities to spread.

We literally face the end of the natural world, as we know it today, and we don't know how to handle it. Mitigating the effects of temperature migrations may change nature conservators to emergency rescuers. It will require a great deal of foresight to protect species and their habitats as temperature zones move northward at a rate of ten miles per year. Furthermore, because we have so altered our landscape with roads and urban areas, it will be extremely difficult for flora and fauna to migrate. Establishing corridors that run north-south or extend inland from coastal marshes will be essential to prevent the genetic impoverishment of species. We will have to locate reserves in areas with heterogeneous soils, topography, and temperature and moisture regimes to permit species to migrate over short distances in order to appropriate areas in response to change.

Climate protection requires that we care about what happens to the planet in the next 400 years. As must be repeated until it sinks in, the gases that we spew out today will take a long time before they are absorbed into vegetation and oceans. Our children are the ones who will have to cope with the fossil wastes that we have produced in using energy for ourselves. Because of the long residence time of CO_2, we cannot treat it like the other pollutants. Our goal should be to reduce emissions to zero in the next 100 years. Thus, our energy policy should ensure that climate protection is part of the equation.

The good news is that this goal is attainable.

Bicycling Is Light on the Planet and Your Bank Account

Taken worldwide, bicycles outnumber cars two to one, and each year three times as many bicycles as cars are produced. Most of the world's 800 million bicycles are in Asia, with China accounting for about 300 million. While bicycles are essential in rural countries for people to get to their jobs, in Europe they are used more to supplement public transit systems. Denmark,

Germany, and the Netherlands are the most bicycle-friendly. Our country never got the bicycle habit. We know that bicycling keeps us physically fit. Isn't it strange that many of us ride stationary bikes indoors for exercise but don't bike outdoors? Nonetheless, today, possibly out of necessity, we have five times as many bicyclists as we did in 1960.

Increased bicycling would have an immediate effect on reducing air pollution and conserving oil instead of opening sensitive new areas to drilling. The EPA estimates that if the percentage of trips made by bicycle rose to 2.4% from approximately 0. 5%, the result would be a 5% reduction in carbon monoxide emissions. Before our cities get as smog-ridden as Mexico City, Athens, and Los Angeles, we'd be wise to institute more bicycling as a supplement to mass transit systems.

Small social innovations would help make bicycling easier for us to practice. Recently a pack was designed for bicycle commuters to wear on their backs. It helps keep clothes from wrinkling or getting dirty. Handy bike rental facilities would be convenient. Secure parking at railroad stations and office buildings would help. Other necessities include bike lanes, storage space, laws to stiffen prosecution for bike theft, encouragement by employers, inclusion in town planning, safety regulations, and state and federal support.

Too often bicycling is treated as a 'problem' by legislators and public officials. Instead, they should realize that [it] is part of the solution.

—League of American Wheelmen

While bikes have been popular in some college towns, more recently many communities are enthusiastically backing "greenway paths," which link paths through town and country. The transportation system is not an unchangeable monolith. Just as the advent of the automobile changed the face of the nation in two generations, our landscape can be altered again.

To deal with traffic congestion ecologically, we can add a lot more greenery instead of more concrete and asphalt roads and parking lots. Europeans construct islands of parks in order to "calm" traffic and pedestrians. To accommodate wildlife, we can build more underpasses for animals to crawl under traffic, and canopies of trees for birds to land on. We need overpasses for bicyclists and walkers, and noise barriers.

"With ecological roads," scientist Jan Beyea says, "wide-ranging animals would have a chance to survive, and short-ranging creatures would be able to recolonize opposite sides of the road after a disturbance. Humans would benefit in the short term from closer contact with nature and in the long run from reduced losses of biodiversity. More expensive? Yes, but perhaps it is time to pay the ecological damage costs that we have ignored too long. "

Ah, let us reduce stress on ourselves and the land through the benefits of walking.

Animal Transport

Nature has devised an array of schemes to get animals and seeds from place to place.

Wind is a prime agent. Hawks will look for clouds and thermal currents to give them a lift. Their broad wings help them catch big columns of air. Winds disperse seeds to new locations. After spiderlings hatch, some species crawl to the highest point they can find—up tall grasses or rocks or fences. Then they spin several long, thin lines of silk. When the wind comes along, they are carried along like a balloon for hundreds of feet or many miles until they are let down to live on their own.

Some organisms are cast out over water, eaten by birds, and defecated on land to be born again. Logs or tree branches fall into rivers and are carried out to sea, complete with organisms, insects, snakes, frogs, and occasional rodents and other small mammals living in them at the moment of departure. Frogs, crickets, and lice are designed to make broad leaps. Pollen, by sticking to the legs of insects, are carried to other plants. Burdock seeds hitch a ride on cygnets. Some vines and creeper plants, such as figs, anchor onto trees and climb them in order to reach light and air. Monkeys leap about tree canopies, entwining their long tails around branches for balance and security. Kangaroos have pouches to transport their young. Fish carry sediment and eggs in their mouths. Snakes slither.

No doubt our tractors copied caterpillar movement, and our airplane design the aerodynamics of birds, but what more is there to learn from the travel movements of creatures? We have yet to understand the awesome strength of seemingly fragile Monarch butterflies that fly thousands of miles a year, or the internal mechanisms that enable all migrating creatures to navigate them until they reach the appointed destination. In oceans exists a kind of transport that is unlike anything that takes place on land. Eggs drift for many miles before hatching. Transportation involves intricacies of space and time.

We ignore the subtle connections between nature cycles and living creatures to the peril of all. We have dammed rivers to improve irrigation for farmers, only to have the soil become too saturated with salt and the farms consequently abandoned. Moreover, dams have altered the seasonal fluctuations of rivers, upon which so much life depends. Countless fish are ripped to death by motor blades. These are examples of why we need to protect nature from our own foolhardiness.

I have met with but one or two persons in the course of my life who understood the art of Walking, that is, of taking walks,—who had a genius, so to speak, for sauntering. . . . The walking of which I speak has nothing in it akin to taking exercise, as it is called, as the sick take medicine at stated hours,—as the swinging of dumbbells or chairs; but is itself the enterprise and adventure of the day. . . . I come to my solitary woodland walk as the homesick go home. I thus dispose of the superfluous and see things as they really are, grand and beautiful.

—Henry David Thoreau

EcoQuiz

Q: Why do scientists disagree about global warming?

A: With very few exceptions, climatologists agree that the climate is warming, but their opinions vary on how fast it will change and what the results will be. The Greenhouse Effect is an established fact. A major influence is the oscillations between warm and cold currents in the tropical South Pacific, which could slow the pace of global warming. Another major question is how clouds will behave. As mostly water vapor, they could absorb radiation and amplify warming, but they also reflect solar radiation away from the Earth and thereby reduce warming.

Q: If temperatures do not rise steadily in the United States, doesn't that belie global warming? And why should the United States be concerned?

A: No, since the United States covers only a small fraction of the Earth's total surface. Regions vary. The globe as a whole could be warming while the United States shows no discernible trend in one direction or another. Because this country is responsible for more than 20% of the emissions that contribute to global warming, it would be wise for us to take the initiative in fighting it.

E C O S Y S T E M   P A T H W A Y S

Tourism and recreation can both help nature and harm it. Your actions can ensure that travel and recreation are fun-filled and healthy, both for us and the planet. Travel and recreation are dependent on clean and functioning ecosystems—and on our keeping them that way.

The negative impacts of recreation and travel are:

- Overvisitation exceeds the carrying capacity of the site.
- Visitors pollute, trample sensitive ecosystems, disturb wildlife, and exploit local inhabitants.
- Local people are given no participation in ecotourism development or financial benefit.

On the positive side:

- Ecotourism provides funds for protected-area purchase and management.
- Ecotourism provides needed income for local communities and governments.
- Ecotourists learn about unique environments and how to conduct themselves in them.
- Critical environments and wildlife are protected.

If we remember the following principles of ecology, we can be more responsible in our recreations:

- Everything is connected to everything else.
- The Earth is a delicate, closed life-support system that cannot tolerate unlimited development and its by-products.
- All environments have a carrying capacity, a ceiling which, like bank credit, cannot be exceeded without dire penalty.
- "Symbiosis" comes from two Greek words meaning "life together"; in nature are found an enormous variety of partnerships. One example is the way some animals depend on others to lead them to food, as egrets eat insects that buffalo flush. Let us not get in the way.
- Animals and vegetation have adapted to their particular ecosystems and biomes out of subtle relationships with the soil, water, and climate cycles. Moreover, at any given moment, all these factors are in a constant process of change.

Aldo Leopold once wrote that the capacity of the human species to show concern for the future of other species is a kind of superiority. May we use our greatness wisely!

10

RECREATION AND TOURISM

Guest author: Tensie Whelan

Garbage from ships pollutes water; debris drifts ashore

Animals contained in zoo

Fragile coastline

Condos and golf courses disrupt ecosystems

Tourist traffic around reefs breaks coral

Photographer too close to wildlife

Goose with neck caught in plastic ring

Golfer and backpacker having fun, but excess development means we face ugly vistas and eat polluted fish and water

CYCLES
WATER
SOIL

WHAT IS ECOTOURISM?

There are intense economic pressures on the people of the United States, Central America, and elsewhere to overexploit their natural resources. Many nations have established protected areas to guard against this. However, when the only way to obtain a meal is to mine the resources of a protected area, the protected area is going to lose. If we are to save any of our precious environment, we must provide people with alternatives.

Ecotourism—travel based on the ecology of the site, for example, bird-watching, hiking, swimming—done well can be a sustainable and relatively simple alternative. It promises employment and income to local communities and needed foreign exchange to national governments, while permitting the continued existence of the natural resource base. In fact, it cannot survive unless the resource on which it is based is protected. Ecotourism can empower local communities, giving them a sense of pride in their natural resources and control over their communities' development. It can educate travelers about the importance of the ecosystems they visit and actively

involve them in conservation efforts. In sum, it has the potential to maximize economic benefits and minimize environmental costs.

The potential of ecotourism for harm is not always realized, however. It can destroy both the environment and local communities. Travelers and the travel industry alike need to assess ecotourism's role in the sustainable development of natural areas and to answer the question: How can ecotourism be planned so that it is both ecologically sensitive and economically productive?

Tourism Is Big Business

According to the World Tourism Organization (WTO, a U.N. affiliate), tourism is the second-largest industry in the world, comprising 7% of the world trade in goods and services and producing $195 billion annually in domestic and international receipts. That's 390 million international tourists in 1988 (up 20 million from 1987), creating 74 million jobs in tourism (up from 65 million). In developing countries, tourism comprises one third of their trade in goods and services. WTO projects that tourism will become the world's largest industry by the year 2000. WTO also found that adventure travel (which includes ecotourism in the WTO definition) received almost 10% of the market in 1989 and is increasing at the rate of 30% a year.

The United States Department of Commerce estimates that by the year 2000, international tourism revenues will reach $30 billion. While no formal studies have been made, tour operators conjecture that ecotourism will comprise a significant portion of the total.

The most popular ecotourism activities are trekking/hiking, bird-watching, nature photography, wildlife safaris, camping, mountain climbing, fishing, river rafting/canoeing/kayaking, and botanical study. Nepal, Kenya, Tanzania, China, Mexico, Costa Rica, and Puerto Rico are the most popular ecotourism destinations.

Ecotourism is also popular in the United States; in 1989 there were 265 million recreational visits (both domestic and international) to the national parks system alone. Wyoming estimates total expenditures related to the consumptive (logging, mining, hunting) and nonconsumptive (recreational) use of its unique wildlife resource at nearly $1 billion annually.

Shortcomings of Ecotourism

LACK OF LOCAL PARTICIPATION

One of the most egregious shortcomings of most ecotourism projects is that the local people are not given any role in the planning process or implementation, and are forced off lands that were traditionally theirs to use. Not surprisingly, they become resentful of the "rich tourists" who supplant them, but more important, economic needs make it difficult for them not to overexploit the resources of areas that should be protected. Firewood, meat, agricultural land, sale of exotic wildlife—these means of subsistence have been removed, often with no viable alternative. And a high population rate means that the local population has an increasing number of mouths to feed.

Another problem is that income generated by tourism is very likely to bypass local communities. In Nepal, where local communities provide shelter and hospitality to trekkers, only 20 cents of the three dollars spent daily by the trekker stays in the villages.

Foreign tour operators are a large part of the problem in most countries. They frequently bring in their own supplies and staff, and hire very few locals to assist on their trips. A survey of 32 U.S.-based operators (41% of all U.S. ecotour operators) found that while 20 used local guides and interpreters, only 8 employed local managers or tour operators, 6 used local cooks, and 8 used local drivers. And, though 20 report they use local guides, it is likely that most are brought in from the larger cities and are not from the small communities where the tour takes place. The same survey found that while 40% of U.S. ecotour operators use rural and village accommodations, 21% use luxury hotels and 33% use other hotels (for example, 27% camp out).

The national economy of the host country is likely to do substantially better than the local economy. One study found that at least 50% of tourist expenditures in developing countries are likely to stay in the country. However, it is unusual to find those receipts, such as tourism taxes, channeled back to local communities or even to the management of the protected areas that generated the income. While steps to rectify this situation may have been taken recently by some more aware tour operators, we still have a long way to go.

LACK OF FUNDS FOR PROTECTED AREA MANAGEMENT

There are roughly 1000 national parks in the world today, mostly in the developed countries. Fewer than half of the developing countries contain them. While most countries do have some protected areas—there are about 7000 protected areas around the globe—the protection is often only on paper, because money and local support are lacking. Yet the success of ecotourism is dependent on the continued existence of these protected areas.

Over and over again, we find parks in crisis because not enough money is being spent on their management and protection. Governments often focus their attention on purchasing lands, but then fail to follow up with money for infrastructure and management. This is true in Costa Rica, where spending for parks (excluding acquisition) has remained at the same level for ten years; in Kenya, where until recently only $7 million of the $300 million generated by parks was returned to them; and in the United States, where park rangers have to supplement their salaries with food stamps, and parks such as the Adirondack State Park have become battlegrounds for developers. These financial problems are sometimes compounded by the fact that parks in developing countries charge woefully inadequate entry fees to foreign visitors, who can afford to pay a great deal more than the locals.

CARRYING CAPACITIES

Ironically, the survival of protected areas may be threatened by the very force that otherwise protects them—tourism. All protected areas have limited ecological and aesthetic carrying capacities. The ecological carrying capacity is reduced when the number of visitors and the characteristics of visitor use start to impact on the wildlife and degrade the ecosystem (e.g., disrupting mating habits and eroding soil). The aesthetic carrying capacity is reached when tourists encounter so many other tourists or see the impacts of other visitors (e.g., lack of watchable wildlife, litter, erosion, deforestation), that their enjoyment of the site is marred.

A survey of visitors to the Spanish Peaks Primitive Area in the United States, for example, found that they would be less willing to pay if the number of trail encounters were to increase from three to four, but enough would continue to come so that the payoff in terms of increased revenues would more than offset the loss due to persons being less willing to pay. However, when the number of expected trail encounters increases to five, the willingness to pay becomes so low that the aggregrate drops off sharply and the area begins to lose money.

While establishing the ecological capacity for a protected area seems essential, very few areas in the developing and developed worlds alike have identified carrying capacities. Nor have they determined how to avoid exceeding those carrying capacities.

The rapid increase in the number of ecotourists has overloaded fragile areas. Nepal has seen the number of its tourists increase about fivefold, from 45,000 in 1970 to 223,000 in 1986. Over the same period, the number of ecotourists (trekkers, mostly) tripled, from 12,600 to 33,600. This has resulted in the virtually overnight emergence of more than 200 mountain lodges and the clearing of large areas in order to supply fuel wood for lodges and trekkers. The visitor use of fuelwood for cooking, hot showers, and campfires is extravagant—a typical two-month climbing expedition may use as much as 8000 kilograms, while a traditional hearth burns 5000 kilograms in one year.

In the United States, many of the more accessible national and state parks are overwhelmed during the peak summer months. In Minnesota, where problems resemble those of other states, visits to the state's sixty-four parks increased from 6 million to 10 million in three years. Ten of the parks are subject to continual overcrowding. Increased visitation to parks nationwide has resulted in more roads, more parking lots, and more concessions built in the protected areas, often decreasing the aesthetic value of the park.

Often, park managers, conservationists, and governments determine to solve their carrying capacity problems by emphasizing quality rather than quantity. In other words, they target fewer people who can pay more. This may make sense from an environmental point of view, but it has elitist implications. In Rwanda, for example, visitors pay $170 a day to see Dian Fossey's gorillas in their mountain reserve. In order to keep the reserve accessible to Rwandans, the fee charged to locals is minimal. However, the reserve is no longer accessible to many foreign tourists. If this trend means that ecotourism becomes an industry only for the rich, average citizens will not be able to learn about other environments and wildlife and will be less inclined to fund or support protection efforts. (As an alternative, a lottery system could be introduced in which a *portion* of the visitors could be on a first-come, first-served basis.)

CONCLUSION

Ecotourism will not on its own save disappearing ecosystems. Nor will it alone liberate rural communities from the shackles of poverty. Unless it is planned to minimize environmental damage, maximize economic outcomes, and involve the local communities, it may actually harm the environment and local peoples.

But when ecotourism is planned as a tool for sustainable development, one that includes essential safeguards, it can indeed make an important contribution to the welfare of every person and every aspect of the environment. The challenge is to make sure that ecotourism doesn't occur willy-nilly wherever there is a demand for it, but that governments, tour operators, conservation groups, and local communities, among others, plan together where ecotourism sites should be established and how they should be managed. Then, fifty years from now, it will be possible for our grandchildren to enjoy the natural beauty and benefits associated with natural areas both near their homes and in other countries.

I have a deep feeling that in the beginning, we were much nearer to the animals than we are now. A state of communion existed between us, a kind of correspondence that we have lost, but can nonetheless regain. This was demonstrated to me once while I was making a film in the interior of Africa. As always in the bush, I had observed a rule of silence as if it were one of the most sacred commandments of the bush itself. When we ourselves had to speak, we learned to do so almost in whispers because there is no sound that carries further or grates so much on the nerves of the bush as the human voice. There was one conclusive moment, when I was able to walk, unarmed, to within a few yards of one of the most dangerous animals in Africa, the rhinoceros. He had made several dummy charges at me on other occasions, but on this early morning, when we had both barely come out of our sleep and the grass and the leaves were all pearl and silver with a heavy dew, I came upon him sunning himself in a little clearing which might have been tapestry for some legendary lady of the unicorn.

It was one of those rare timeless moments charged with a meaning uniquely its own. . . . An extraordinary feeling of harmony and of belonging was implicit and magnetic within it, and the rhinoceros was at the centre of it. I felt a foolhardy desire, as it seemed to my conscious self, to do something to express a strange inrush of gratitude for that privileged scene by going nearer the rhinoceros and somehow to make it clear that human beings were not just guns and violence, that somehow we too needed that sense of belonging which emanated from him and the bush. I whispered to my companions to stay where they were, stopped and told them what I intended to do. They tried hard to dissuade me. . . .

I walked slowly and evenly towards the rhinoceros. He turned slowly about to face me squarely, lowered his massive head so that his wide chin practically touched the ground and his scimitar horn was pointed straight at me. . . . I went steadily towards him until I was a mere three yards away. Then it was as if a signal passed from him to me that I had gone near enough and now had to observe what there was of distance left between the two of us; distance from there on was a matter of identity and dignity. I stood still and looked as steadily at him as he looked at me. He belonged, of course, to one of the oldest forms of mammal life. His species goes back to the age of the dinosaur and pterodactyl, and many hunters regard him as one of the most dangerous animals, one of the ugliest and most removed from any vestiges of animal reason. Yet, as I stood and looked at him there, I thought I saw through all that was considered inelegant and ugly in his appearance. I saw a strange first essay on the part of creation in the pattern of animal beauty. . . . Suddenly it was as if not only the gap of what we call time between him and me had been closed, but that a powerful feeling of emancipation was illuminating my war-darkened and industrialised senses.

—Sir Laurens Van Der Post, author and statesman

THINKING LIKE AN
ECOSYSTEM
Populations

Anyone who thinks about populations for a moment must be impressed by obvious differences in the numbers of individuals making up a population. There may be fifty trees of a particular species populating as large an area as an acre of forest—but there may be a million diatoms in a bucket of seawater. The density of the population in relation to the space that it occupies may exert considerable effect on the community. The abundance or scarcity of a population may fluctuate widely, but there are definite upper and lower limits to its density—determined by the energy flow in the community, restrictions of space and climate, the position of the organism in the fabric of the community, and the size and metabolism of the organism itself. Competition, predation, and of course human activity, such as recreation, hold the reins of a population in check.

In the North a predator rarely has an alternative prey if its regular source fails, unlike in southern tropical climes, where a predator can have any number of options. In the North the rise and fall of populations is more evident. The cycle of the lynx and its prey, the snowshoe hare, takes about ten years. Over the course of several years the hares grow to the point of overcrowding their habitat. This stress causes their livers to degenerate, creating a condition wherein if they get overly excited, they sink into a coma and die. The result is a crash in their numbers, followed by a crash in the numbers of lynx. Then a new cycle begins.

Populations vary in the dispersion of species as well as their capacity to grow. Environmental resistance will lower their potential. A species has four basic ways to deal with stress. It can decrease its birthrate or increase its death rate, as the snowshoe hare does; it can migrate to another area with a similar environment, but this can be difficult if the way is blocked by development; it can adapt and change through natural selection, which takes a long time and is unlikely if the species has little genetic diversity; or it can become extinct.

An example of the complexity of maintaining balance in populations was noted by John Sawhill in a *Nature Conservancy* magazine editorial. The Pacific sea otter was hunted to the brink of extinction; in 1900, only about 1000 otters survived. As the otter declined, so did the Pacific Coast populations of eagles, harbor seals, and fish. People were disturbed about what had happened to *their* food source. The problem was traced to the ocean's kelp beds. At the bottom of the food chain, kelp nourishes minute marine life that sustain the fish population that in turn supports eagles and seals. The kelp beds were being cut by sea urchins, who had lost their primary predator, the otter. When otter hunting was outlawed, the otter population rebounded, the urchin population dropped, the kelp flourished, fishing improved, and eagles and seals returned.

Sportspeople know full well that a duck marsh has a very low human carrying capacity. Too many hunters destroy the sport. The same is true with the most popular recreation of all: fishing. Is there anything that can be done? Clay Schoenfeld, in *Everybody's Ecology*, suggests that just as "when the number of basketball fans exceeds the capacity of the university fieldhouse, we ration the seats," so those who love to hunt and fish will have to cut down on their take. He also suggests getting outdoor "freeloaders" to kick in some resources. Freeloaders are all of us who hike, picnic, and sight-see, enjoying conservation programs that someone else pays for.

WHAT IS THE PERFECT VACATION?

According to magazine ads, the perfect vacation consists of: a tropical paradise, first-class hotel accommodations, a large service staff, five-star restaurants, "endless recreational options" (i.e., golf, tennis, volleyball, snorkeling), and no hassles. Most Americans set off in search of the perfect vacation in the summer, heading by car to national parks or seashores. Wild coastlines have all but disappeared due to widespread development. Vacations tend to have a manic quality about them, as people hurry through their agendas. In the 1930s and '40s people used to spend weeks or months at Yosemite National Park; now most arrive in the morning and are gone by dark.

The small percentage of our wilderness areas contained in national parks offer increasingly fewer opportunities for solitude and peace as they try to manage larger crowds. Roads, campgrounds, toilets, signs, fast-food stands, and clearly marked trails all dilute the possibility of anyone's being alone with nature. It becomes harder and harder "to get away from it all," which to many Americans would constitute the perfect vacation. As even the sacred sites of indigenous people are turned into tourist attractions, more vacationers refuse to divulge the whereabouts of pristine locations they have found in order to preserve them.

Americans say they want to meet natives. But only as long as they never have to leave their Hilton bubble.

—Skip Southworth, travel agent

DO'S AND DON'TS OF SUMMER CAMPING

DO have fun.
DO read up on the wildlife and ecosystem before your visit.
DO bring along a handy field guide.
DO pick out a site that has less visitation. Not only will you be helping to spread the load on the ecosystem, but you will also have a more "natural" experience with fewer people around.
DO use natural, biodegradable soaps and detergents if you are washing dishes or clothes.
DO use bark and branches from dead trees. Never strip living trees.
DON'T knock on birds' nests to make them come out. You will disturb their nesting and eating.
DON'T leave your litter behind. Pack out what you pack in.
DON'T "trailblaze." Keeping to the trails ensures that damage to the ecosystem stays limited to one spot.
DON'T feed the wildlife. Making them dependent on human largesse is not healthy for them.

Vacations are often packaged deals. In Texas and Florida you can go on hunting expeditions for imported big game in pens of 100 to 600 acres. Or you can pay handsomely to join a team of archeologists around the world and help with the research. Elderhostel offers a thick catalog of inexpensive educational/vacation choices to those in the autumn of life.

The traveller must be born again on the road, and earn a passport from the elements.

—Henry D. Thoreau

Whatever you decide is perfect for you, it would be wise to think of the energy costs involved. Those huge hotels and restaurants, for instance, are energy drains. Trading houses is an option. What's the best mode of transportation? Say, for example, you were heading toward Yellowstone National Park from the East or West coast. Taking a public bus and then camping out would consume the least energy resources. Flying in a full plane or riding in a full car would not be a bad choice. Bicycling once you are at your destination is not only very light in energy consumption, but tones your mind and body. You can figure the energy you would use with the chart on energy demand in terms of BTUs in chapter 9.

Here's a more in-depth look at the ecological impacts of some major travel and recreational activities, beginning with tips for photograph hunters.

National Wildlife Refuges: Shutterbug's Shangri-la

Gary Zahm

The golden-hued clouds, scattered across a fading blue-green New Mexico sky, had created the sunset that all photographers dream about. More important to me, however, was a blustery 30 mph northwest wind that funneled down the canyons of Chupadera Peak and across the marshlands of the Bosque del Apache National Wildlife Refuge. Silhouetted against the stunning backdrop, thousands of greater sandhill cranes fought the wind as they flew in search of the ideal roosting area.

Scanning the passing flocks, I picked out a small group and followed it through the viewfinder of my motordriven Canon F-1 and 400mm telephoto. With gunstock firmly nestled against my cheek and shoulder, I panned with the birds as they hung motionless against the wind. When I saw the first cranes drop their outstretched legs, I fired off 10 quick frames. Satisfied that I had captured a prize-winner, I headed back to my vehicle, parked on the refuge's public tour route just yards away.

Our national wildlife refuges provide superb opportunities for nature photography, usually without having to trudge far into the bushes. . . . Far in advance of visiting an unfamiliar refuge, you should write to the refuge manager and request copies of all appropriate leaflets, maps, and regulations. . . . Knowing about the presence of an unusual species—a concentration of snow geese on a newly cut field of corn, for instance—can guide the day's photography. . . .

As wildlife habitat continues to dwindle in the United States, the National Wildlife Refuge System and its diversity of readily accessible species may be viewed as the shutterbug's Shangri-la. For a rewarding experience, grab a map, locate your nearest refuge, bring your camera and lots of film, and take a low-budget photographic safari.

Cruise Lines Leave Trail of Trash

The waters in and around the island harbors of the Caribbean have become dumping grounds for the cruise line industry that plies the region. The assorted trash tossed overboard—foam cups, plastic bags, cans, and other refuse—threatens both sea life and the region's large tourism industry.

Eyewitness accounts from fishermen, pleasure boaters, and disgruntled crewmembers reveal that the ships dump bags of trash overboard at night, often as they are leaving harbor or are close enough to shore so that the trash washes in on the tides. In some instances, the dumping violates international law.

The impact on wildlife can be devastating. Endangered sea turtles eat floating plastic bags, mistaking them for jellyfish, and the result can be fatal. Seabirds, too, choke on small pieces of plastic.

Islanders fear that the multimillion-dollar tourist economy will suffer from the tons of garbage that litter the water and beaches—the two prime attractions for vacationers to the region. "Plastic is greatly destroying our economic potential," Bahamian Prime Minister Lynden O. Pindling has said. Meanwhile, the cruise liners provide little economic benefit to the Caribbean islands—the passengers are encouraged to spend their money on board.

Some of the cruise lines say they have started to reduce trash flow and lessen the environmental impact of their ships. "We know there are problems. Nothing is foolproof," says David Whitten, director of quality assurance for Royal Caribbean Cruise Lines. "But we have a sewage system and compactor. Our ships have two incinerators, one on standby, that burn plastics and anything burnable. I wouldn't want to breathe what comes out of the smokestack, but we have a two-stage burner. It's better than it was. We can discharge the ash because it's classified as 'other garbage.' "

According to conservation groups, the industry has created other problems for the region as well. Eyewitnesses give accounts of ocean liners dragging anchors, leaving large swatches of devastated reefs. Dredging along Great Guana Cay in the Bahamas to accommodate cruise ships has caused extensive damage to the rich coral reef along the island.

Without stricter regulation, the environmental impact of cruise ships will likely increase in the coming decade.

When Echoes and Exhaust Fade, How Much Damage Remains?

For many backcountry hikers, bird-watchers, and skiers, aggravation is spelled "ORV. " Off-road vehicles shatter the wilderness silence, crush vegetation, terrify wildlife, leave rutted trails, and cause erosion.

From the beaches of Long Island to the deserts of Southern California, increasing numbers of high-octane thrill seekers are blazing their own trails in four-wheel-drive cars and trucks, dune and swamp buggies, motorcycles, snowmobiles, and the latest craze, fat-tired tricycles called all-terrain cycles or ATCs. Sales of ATCs have risen dramatically.

Environmentalists claim that ORVs are as damaging as they are disruptive and should be restricted to private lands and limited areas of public property. As the popularity of ORVs has exploded, stewards of our public lands have wrestled with the ways to manage the motorized hordes. How much environmental damage do ORVs cause? And can their use be made compatible with nonmotorized uses of public lands?

In a study spanning eight years, Audubon scientists have gathered data on Big Cypress National Preserve in southern Florida to help answer these questions.

Big Cypress, part of the Everglades ecosystem, is made up of flat, wet, seasonally flooded savannas and dense swamps. During the rainy season it can be negotiated only with a vehicle that can run in shin-deep water. Off-road conveyances are not just a sporty means of transportation for area residents, but are a way of life. Deer hunters, for example, roll out an imaginative array of home-built machines riding on huge aircraft tires or tank treads. Others use airboats (which can operate in just a few inches of water or, as one enthusiast put it, "on a good heavy dew") and the increasingly popular balloon-tired ATCs.

Much of the preserve is crisscrossed with ORV tracks, and deeply rutted roads mark jump-off points from the main highway. In 1978 the National Park Service asked Audubon's Ecological Research Unit to assess the damage and to determine how quickly nature can obliterate the tracks.

The research unit was a team of scientists specializing in improving management of natural areas. It was headed by Mike Duever, a Ph.D. ecologist, and included his wife, Jean McCollom, also an ecologist, and Larry Riopelle, a biologist/pilot.

The research team selected the most common vehicle types for study: five kinds of swamp buggy, airboat, all-terrain cycles, and tracked vehicle (sort of a homemade tank). Then they determined the predominant habitats in the Big Cypress preserve: woodlands of small, scattered cypress; pinelands; and marl, sand, and peat marshes. After carefully mapping out study sites in each terrain, they began making tracks with the ORVs.

On one plot in each study site, they made a single pass with each ORV; in another they made enough passes to damage the vegetation; and in the third they kept driving, with painstaking destructiveness, until the lane was rutted. Sometimes that took dozens of passes. "It was enough to make you never want to climb on an ORV again," said chief driver Riopelle.

Just after the tests were completed, and again after one growing season, the team measured visual impacts, soil compaction, rut depth, and, in great detail, the effects on vegetation.

Many of the test tracks were clearly visible a year later. The researchers found that the water level was the single most important factor influencing severity of the damage and recovery time. Damage was the heaviest when the ground was very wet or covered with shallow water. Vehicle weight relative to tire size also proved important. The heavy, tracked vehicles and swamp buggies did the most damage, while airboats, which glide over the surface, left few visible scars.

In 1985, seven years after the initial studies were done, the park service asked the Ecological Research Unit to reassess its study plots. Surprisingly, many of the tracks were still apparent. In one area of small cypress trees, a track made by a single pass of a wheeled buggy was still visible. An ATC track also remained obvious. In general, the pinelands had the best recovery and the marl marshes the least.

The general conclusions of the study have greatly increased our understanding of what happens to habitat when the ORVs leave and the dust clears. Duever emphasizes that the specific results are not necessarily transferable to other parts of the country and that more research in other areas is needed.

"Public land managers must have hard data to support their policies," Duever said, "or they'll never be able to balance competing demands for the resources."

Duever believes that ORVs can have a place on public lands if their use is managed to keep long-term damage to a minimum, and if the off-road enthusiasts cooperate with regulatory plans.

GOLFING
Hazards of the Game
Jolee Edmondson

Feast your weary, smog-sore eyes on them: flawless, ultragreen havens surrounded by suburbs and country and even wilderness. So pristine, so halcyon, so pastoral, so in tune with nature, such a tonic to look at. Who would ever imagine that golf courses could be hazardous to the environment?

In the last twenty years, since the dawn of ecological awareness, a case has been building against those seemingly innocuous places where polite people in pastel outfits chase little white balls. Some environmentalists go so far as to label golf courses dangerous toxic-waste sites, while others conservatively suggest that they are a potential problem. Either way, it is clear that fairways are no longer just playgrounds but hot issues as well. Golf courses are encountering the same hard scrutiny as freeways and factory yards.

"Today, when a developer builds a golf course, he has to go through at least two years of bureaucratic delays—one environmental hearing after another—before he gets approval to start," says Roger Rulewich, senior associate at Robert Trent Jones, Inc. of Montclair, New Jersey, arguably the world's most prominent golf course architectural firm. "Sometimes it gets to the point where they throw up their hands and give up on the idea. And it's going to get even tougher."

What a breeze it was a couple of decades ago to carpet the land with fairways. No environmental impact statements to file. No environmental consultants to call in. No public inquests. It was merely a matter of finding the right spot and bulldozing. Everybody perceived golf courses as pleasant recreational facilities or even greenbelts. Or they didn't think about them at all. Veteran golf course architect Lawrence Packard recalls that in 1970, when he designed the Island Course at Innisbrook—a resort in South Florida—the first six and a half holes were made from a marsh his crew filled in with sand. An abominable act by modern precepts. "I know that we wouldn't be allowed to do that today," concedes Packard. "But back then there were no laws."

There are more than twelve thousand golf courses in the United States. An average of 110 courses are built annually, each consuming approximately 150 acres. This steady gobbling up of land is bound to have an effect on the environment, the most obvious being the further depletion of rapidly shrinking wildlife habitats. As with any kind of development in rural or semirural regions, the construction of a golf course means a major upheaval for native creatures. But not total destruction of habitat. Fairways are not made of chrome and glass, but grass, which can cause confusion among wildlife species. Here are these enticing, meadowlike expanses fringed by myriad types of vegetation. Instead of fleeing the area, some animals are drawn back to what appears to be lush grazing grounds, only to be met with flying balls, electric carts, and a relentless parade of humans.

Most golfers relish the sight of wildlife—to a point. When huge gaggles of geese defecate on their fairways and root up their turf, and deer gouge their greens, enchantment turns to exasperation. At a number of golf courses the conflict between man and beast has become intense. . . .

It is in the sphere of pesticides and fertilizers that golf courses come under the heaviest fire from environmentalists. Fairways and greens are regularly doused with volumes of chemical turf-care products. Hence the burning question: How polluting is a golf course to the community or woodlands in which it nestles? One side claims that there is no potential for runoff of contaminated water from golf courses, while the other side insists that runoff is a frequent occurrence. And the battle goes on. Two things are certain: (1) There is little documentation to support either argument. (2) Golf course developers and superintendents are facing mounting restrictions in terms of water management.

In the lawless past, the builders of golf courses paid no heed to erosion, letting soil and water flow where they might while their layouts were under construction. Nowadays, stringent preventive measures are mandatory. Fences, hay bales, and gravel dams are common sights on evolving fairways. At a newly opened country club in Orlando, Florida, named

(Continued on page 148)

(Continued from page 147)

MetroWest, the developer was forced by local environmental agencies to contain every drop of water on his course by building retention basins throughout the property, a tedious and costly procedure. . . .

Not to be ignored is the threat posed by surface contamination.

There are countless bird mortalities on golf courses every year, many of which are traced directly to turf that has been saturated with chemicals. A dramatic case in point: On the afternoon of May 7, 1984, a prevalently used pesticide called Diazinon was applied to three fairways at Seawane Country Club in Hewlett Harbor, New York. A flock of brant came to feed on one of the treated fairways. By nine in the morning several hundred of the fowl were reported dead on the course and in the adjacent harbor. Over the next three days the fatality count climbed to an estimated seven hundred. A large number of the dead birds were submitted (along with sod samples) to the pathology unit of the New York State Department for Environmental Conservation. Diagnosis: Death caused by ingestion of poison, i.e., Diazinon.

Birds are not the only living things that have fallen victim to chemically treated fairways. In August 1982 Navy Lieutenant George Prior, thirty, left his home in Arlington, Virginia, to play golf at the Army Navy Country Club. The turf was in superb condition, having been recently sprayed with Daconil—a compound particularly effective in eradicating brown spots. After his game, Prior complained of nausea and a headache. By the end of his third day on the course, he was seriously ill. When a fast-spreading rash appeared on his stomach, he checked himself into Bethesda's National Naval Medical Center. The rash developed into huge blisters, and his internal organs began to fail. Less than two weeks after entering the hospital, Prior died of a heart attack. Dr. Jonathon Lord, an expert Navy forensic pathologist, found Prior's shoes, clubs, and golf balls coated with Daconil and concluded that the cause of death was severe allergic reaction.

Regardless of all the accusations and negative implications, it would be a gross distortion to portray golf course superintendents as the Bad Guys, scoundrels who run around fiendishly poisoning fairways. Far from it. Most of these men, particularly the seven thousand who are members of the Golf Course Superintendents Association of America (GCSAA), are conscientious (and) intent on learning more about the interchange between golf courses and the environment. . . .

Perhaps the real culprit in all this is the American golfer. He is spoiled rotten. He wants his fairways verdant and velvety, his greens looking like emerald oases. The slightest blemish, and he gets ornery. Golf courses in this country are traditionally overmanicured. Thus, the ritual use of pesticides, fertilizers, fungicides, herbicides, and other chemicals.

American golfers would do the environment a favor by imitating the Scots, who originated the sport. Their links were designed by Mother Nature. They just sort of happened five centuries ago. The grasses were fertilized by the droppings of indigenous birds and were kept cut by rabbits. Bunkers were formed by sheep and other animals that burrowed into the turf as protection against the elements. Years and years of receding oceanic tides rendered the sandy soil on which they're laid out. The enduring result: shaggy, wide-open playing fields with random clumps of gorse and crater-like "pot bunkers." The greens consist mainly of native fescues. Maintenance on these links is minimal. Indeed, golf in Scotland is a different game from that which flourishes in this country. And the environment is the benefactor.

Zoos Respond to the Call of the Wild

Because of the precarious position of wildlife today, many zoos have changed their practice from exhibiting wild animals that could be replaced from the wild when they died to propagating as many of their own wild animals as they can. About one sixth of all species of mammals and one twelfth of all of the world's species of birds have been bred in zoos within the last few years. A growing number of zoos also sponsor research and support conservation in the field for a variety of endangered species, from penguins to mountain gorillas, because without effective measures for preserving native habitat and population strains, captive propagation programs cannot ensure species survival.

Zoos have launched the following programs to enhance their effectiveness in preserving wild species:

1. The Species Survival Plan (SSP) is a coordinated effort of world-class zoos to breed "world herds" of species endangered in the wild in order to preserve genetic variability. It is administered by the American Association of Zoological Parks and Aquariums (AAZPA). The participating institutions have developed species survival plans for sixty-seven species, not all listed as endangered. To avoid the loss of an entire captive population to disease or disaster, the goal is to maintain populations of each species in a number of different zoos.
2. To aid their breeding programs, the International Species Inventory System (ISIS) has been developed. It is a computerized inventory of more than 60,000 individual zoo animals.
3. The dream of dedicated zoo and aquarium professionals is to see a time when, through captive propagation, stable populations of animals that are dwindling or gone from the wild will be reintroduced to protected natural habitats. In recent years, this dream for zoo-born animals has become a reality, most notably with the return of captive-born Arabian oryx to Oman and golden lion tamarins to Brazil.

Nature's Tourists

Migrants are the travelers in nature. Bobolinks travel 7000 miles one way. The longest journey among mammals is made by gray whales. In late autumn they leave their feeding grounds in the Bering Sea and travel 6000 miles south along the west coast of the United States to Mexico. In the spring they return by the same route. The journey takes about ninety days. Caribou travel about 1400 miles in their migration, and Alaskan fur seals about 3000.

E c o Q u i z

Rate Yourself: Are You an Ecotourist?

Check a, b, or c.

1. How can tour operators best help local people to benefit from ecotourism?
a) build large hotels in the community___
b) train local guides___
c) ask local people to plan the ecotourism project with them___

2. Which of the following activities is not ecotourism:
a) bird-watching in Brazil __
b) suntanning in the Bahamas__
c) hiking in the Himalayas__

3. What kind of human food can you give to wildlife when you are on a camping trip?
a) bread __
b) leftover meat__
c) neither a nor b

4. How is citizen science related to ecotourism?
a) scientists teaching citizens about environmental impacts of ecotourism __
b) citizens undertaking scientific projects as part of a travel program__
c) scientists taking ecotours__

5. What is one of the most negative impacts of golf courses on the environment?
a) deforestation__
b) loss of wildlife habitat__
c) neither a nor b

6. How long does it take for the tracks made by all-terrain vehicles to disappear?
a) by the following season__
b) six months__
c) more than five years__

CORRECT ANSWERS: 1-c, 2-b, 3-c, 4-b, 5-b, 6-c

PART IV

Land and Ocean Ecology

E C O S Y S T E M   P A T H W A Y S

America's public lands get hard use. Even abuse, many say. The national forests illustrate the challenges of managing a resource from which everybody wants to take a slice. They serve the innumerable—and often conflicting—public needs for recreation, wildlife preservation, and the utilization of natural resources.

- Timber companies use forests' wood products.
- Ranchers secure grazing permits for sheep and cattle.
- Miners extract oil, gas, coal, and minerals.
- Hunters and fishermen harvest the abundant wildlife.
- Boaters enjoy a rich variety of lakes and streams.
- Hikers explore remote back country.
- Bird-watchers discover regional "specialties."
- Game managers carry out programs for endangered species.

Controversy erupts when a segment of the public detects other activities on forest lands that conflict with their own. Roads, for instance, make forest resources easily accessible for consumptive users such as loggers, miners, hunters, and squatters. But they are seen as destructive intrusions by nonconsumptive users such as hikers and campers.

The doctrine of multiple use on public lands ought to reflect democracy in action. A tug of war arises when one group sees them as a commodity, to be exploited for profit, and the other group sees them as a backdrop for recreational activities. Perhaps both sides will, in the end, be best served if these lands are seen as complex but fragile ecosystems that can be destroyed by abuse.

> ***In my own field, forestry, group A is quite content to grow trees like cabbages, with cellulose as the basic forest commodity. It feels no inhibition against violence; its ideology is agronomic. Group B, on the other hand, sees forestry as fundamentally different from agronomy because it employs natural species, and manages a natural environment rather than creating an artificial one. Group B prefers natural reproduction on principle. It worries on biotic as well as economic grounds about the loss of species like chestnut, and the threatened loss of the white pines. It worries about a whole series of secondary forest functions: wildlife, recreation, watersheds, wilderness areas. To my mind, Group B feels the stirrings of an ecological conscience.***
>
> **—Aldo Leopold,**
> ***A Sand County Almanac***

11

OUR PUBLIC LANDS

Guest author: Frank Graham, Jr.

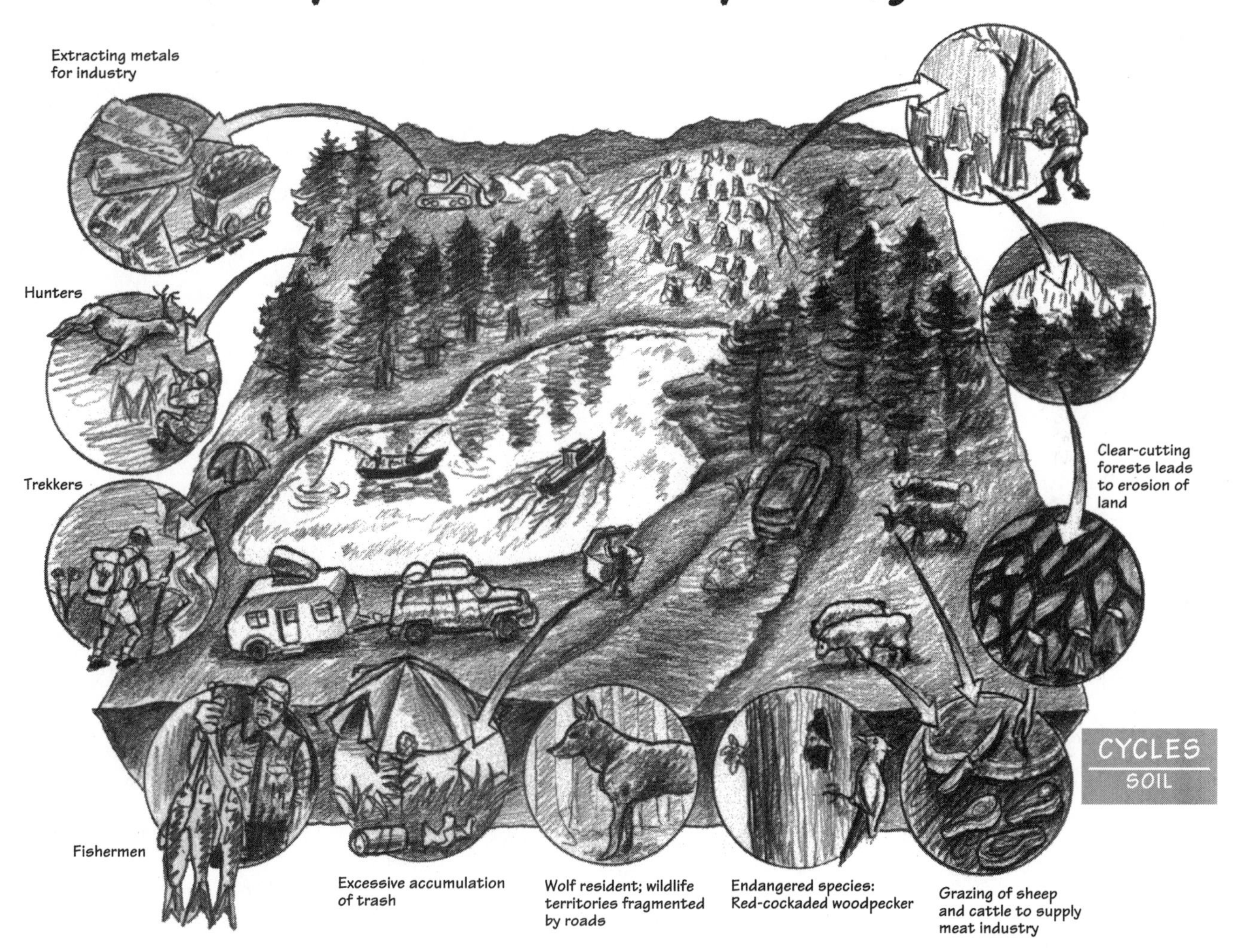

CYCLES
SOIL

PUBLIC LANDS: WHOSE ARE THEY?

That's an easy question. The public lands are *your* lands. This point is often missed, even by environmentalists. Intimidation works too often in these matters, and the general public is shunted aside when the time comes to make decisions about how the land is to be used. Legislators or bureaucrats give away public resources, and exploiters see themselves as the sole decision-makers on what they can take for profit. When that happens, the *common wealth* of the nation is diminished.

Example: the national forests. In theory, those forests, which are the property of all the people of the United States, are to be managed on a "multiple use" basis. This means that the agency that manages the national forests, the Forest Service, must keep in mind that a variety of uses receive full and equal protection, among them fish and wildlife preservation, watershed integrity, outdoor recreation, range and timber development, and wilderness values. But if someone comes in and cuts down all the trees, of what good is the forest to all the rest of the "users"?

"The forest will grow back," is the common reply of the exploiter. But soils and ecosystems are not rebuilt in a human lifetime. The hastily regrown "plantation" will lack the wildlife and wilderness values of a mature wild forest.

The exploiters of our public lands don't always want to see it that way. A local developer who wants to clear the land around a scenic lake within the boundary of New York State's Adirondack Park rails against an "outsider" from Buffalo or Brooklyn who campaigns to stop the project. A rancher in Wyoming cannot understand why a public lands advocate in Ohio raises a protest about the overgrazing of cattle on government rangeland. Yet the protesters have as much right to call for controls on the exploitation of *their* land as commercial interests have to request permits to carry out their projects. The developers have their way so often because they wield superior political clout.

On the other hand, environmentalists have found that in many cases they can gain as much clout, perhaps even more, by banding together and proving the justice of their cause. It takes time and money. But more and more people believe that it is worth the time and money to save a beautiful lake or a majestic forest.

Remember, *this* land is *your* land!

Each town should have a park or rather a primitive forest of five hundred acres, where a stick should never be cut for fuel.

—Henry D. Thoreau

PUBLIC LANDS: WHAT ARE THEY?

The federal government—you and I—owns one third of the land in the United States. These lands are managed for a variety of purposes by a variety of agencies. They include:

National Parks

The National Park Service, an agency of the Department of the Interior, manages more than 350 national parks, monuments, seashores, recreation areas, monuments, historical parks, memorials, and other properties comprising some 80 million acres.

National Wildlife Refuges

The Fish and Wildlife Service, also a part of the Department of the Interior, manages a system of 478 refuges in every state and five U.S. territories, comprising nearly 90 million acres. About 77 million acres of that total lie in Alaska, where both the Yukon Delta and Alaska National Wildlife Refuge cover over 19 million acres.

National Forests

The Forest Service, an agency of the Department of Agriculture, manages more than 191 million acres of forests and grasslands. Most of these lands were originally set aside for watershed protection and timber management. But Congress also directed they provide such multiple use benefits as wildlife protection, livestock grazing, mineral development, and outdoor recreation.

Federal Lands

The Bureau of Land Management, created by Congress in 1946 to manage assorted lands and assigned to the Department of the Interior, is the granddaddy of all landowning agencies. It holds 269 million acres, or about 48% of federal lands, and administers the mineral rights on additional millions of acres.

Department of Defense

The Army, Navy, Air Force, and Corps of Engineers hold nearly 30 million acres of bases, training areas, and flood control areas, most of them in Western states. Because of their large size and tight security, many of the department's holdings remain important if uncertain reservoirs for wildlife.

Department of Energy

These lands, which include such nightmare memorials to the nuclear arms race as the Nevada Test Site and the Hanford, Washington, Nuclear Reservation, comprise more than 1.5 million acres.

Another large segment of the nation's lands is controlled by state and local agencies. They range from a village green or recreation site to such vital urban refuges as New York City's Central Park and San Francisco's Golden Gate Park to New York State's Adirondack Park.

The Adirondack Park is the largest of any kind in the country—at about 6 million acres far larger than the two giants among our national parks, Yellowstone and Mount McKinley (each around 2 million acres). Because of the complexities of its creation, 60% of the Adirondack Park is privately owned, though the private and public elements are welded together by regulations promulgated through the Adirondack Park Agency and often fiercely contested by local residents and developers.

Anne LaBastille, wildlife ecologist and author of *Woodswoman; Assignment: Wildlife; Women and Wilderness; Beyond Black Bear Lake; Mama Poc;* and *The Wilderness World of Anne LaBastille,* has been an inhabitant of and guide in the Adirondacks for years. She reflects:

> **Why do I continue to bumble through the woods at night on mushy snow? Carry impossible loads by backpack and canoe? Go for backcountry saunters rather than shopping mall sprees? Cut and split firewood instead of turning up a thermostat? Build a little cabin to write at instead of buying a condo to relax in?**
>
> **Perhaps it's because the world around me seems to be so complex and materialistic. It's my small rebellion to keep myself in pioneerlike fitness, to promote creativity, and to maintain a sense of adventure in my life. It's also my desire to exist in tune with sound ecological and ethical principles—that is, "small is beautiful," and "simplicity is best."**
>
> **Much, much has changed since I bought a piece of wild forest beside Black Bear Lake, built my one-room cabin, and moved in with the idyllic notion of writing and living frugally and tranquilly. . . . There is that harassment of professional demands and public curiosity. There are the intrusions of mail, jets, and boats. There is the fracturing of silence. Also, the slow corrosion of my metal roof, pollution of my drinking water, and insidious poisoning of trees, fish, frogs, and maybe me by acid rain.**
>
> **Recognizing this duality, I realize more strongly than ever that the only way to handle this ambivalence is to fight the dark side—whatever it**

WHY WILDERNESS?

The least anthropocentric wilderness benefit derives from the very recent idea that nonhuman life and even wild ecosystems themselves have *intrinsic value* and the right to exist. From this perspective wilderness is not *for* humans at all, and wilderness preservation testifies to the human capacity for restraint. A designated wilderness, in this sense, is a gesture of planetary modesty and a way of demonstrating that humans are members, not masters, of the community of life. . . . Wilderness is the best place to learn humility, dependency and reverence for all life.

—Roderick Frazier Nash

is—with sharp, intelligent skirmishes. Then you retreat, rest, and restore yourself in quiet, beautiful places. Thus you can gain strength and inspiration for the next battle.

The clash of differing interests has not always served the public lands well. But the public does have a voice, and how that voice is used will determine the quality of these national treasures that are passed on to future generations.

THE LIVING SOIL

Soil is the offspring of rocks. Not surprisingly, because it must be broken down into loose material and then enriched before it is able to sustain life, the process of building good soil takes thousands of years. Carefully husbanded, soil is a basic natural resource, sharing with water the capacity to serve as a nurturing medium in which both plants and animals can grow.

The soil cycle begins with the weathering of rocks. Over long periods of time, wind and water, ice and gases, relentlessly erode the rock surface of our planet, breaking it down into fine grains which, like rocks, are essentially sterile. The proper mix has yet to appear. But now this disassembled material is capable of being penetrated by other elements. Air and water droplets invade the spaces between the grains of rock. Then tiny living organisms arrive. Some like worms, insects, or mites under their own power, others like bits of plant material and even tiny spiders, are carried in by the wind. As those fragments of life go about the business of living, they convert minerals such as nitrogen and phosphorus that are already in the rock to a form that can be used by other plants and animals. Dying and decomposing, they add their own rich organic substance to the mix, slowly building a soil environment that nourishes the plants that in turn nourish the teeming variety of animal life, including humans, in a healthy ecosystem.

In soils of varying richness all over our continent, seedlings find a substratum into which to sink strong roots and draw up nourishing minerals and organic matter. Some of the seedlings grow into mighty trees, in their mass providing food and shelter for all the other plants and animals of a flourishing forest. Human ingenuity unfortunately has the power, through carelessness or ignorance (as, for instance, in poor farming or forestry practices), to destroy the soil and bring the age-old cycle to a premature end. The primary goal of our network of public lands ought to be the preservation of that fragmented bedrock, with all the plants and animals that make up the forest ecosystem.

EVERGLADES IN TROUBLE

Two facts stick out above all others when we begin to talk about Everglades National Park, at the southern tip of Florida. The impetus for the dedication in 1947 of this jewel in our National Park System was the year-round presence in the Glades of the most glorious assemblage of wading birds on the North American continent. Since then the number of waders—including herons, egrets, ibises, and storks—nesting in the park has declined by nearly 90%.

The complex ecosystem of the Everglades has been called "a river of

BUREAU OF LAND MANAGEMENT A.K.A. BUREAU OF LIVESTOCK AND MINING

Its turf is what is left after all the disposal programs and the transfers by executive order or by acts of Congress to other federal agencies. Not until 1976—two centuries after Congress took dominion over the nation—did it recognize these sprawling, diverse, and beautiful lands as worthy of retention for national purposes. Not until then did Congress give the secretary of the interior and his Bureau of Land Management coherent policy direction for taking care of them with the Federal Land Policy and Management Act. In critical ways the act's conservationist intent has been more honored in the breach than the observance.

Ninety-nine percent of the bureau's 269 million acres lie within the contiguous western states and in Alaska. These are not mostly flat and uninteresting desert, as casual observers might conclude when driving through the Great Basin on highways routed and designed for speed and not for scenic enjoyment. They are remarkably diverse in physical and biological characteristics: to the more sensitive observer, enchanting in their beauty.

The bureau also handles the leasing or selling of mineral rights on about sixty-eight million acres where the U.S. government disposed of the surface but kept the mineral estate. Unfortunately, for most of its history, the bureau has been the servant of private mining interests, as well as private livestock interests who want a free hand in grazing—and overgrazing—the land.

—Charles H. Callison

grass." From the air, one sees compact islands of subtropical trees rising from a vast prairie of tall, sharp-bladed plants aptly named sawgrass. Blanched streaks mark the paths gouged across the prairie's underlying muck by speeding airboats. At times the sweep of sawgrass masks the shallow river that is the Everglades' substance, but inevitably the flow emerges again in glistening ribbons winding among mangrove jungles where the land grades imperceptibly into Florida Bay.

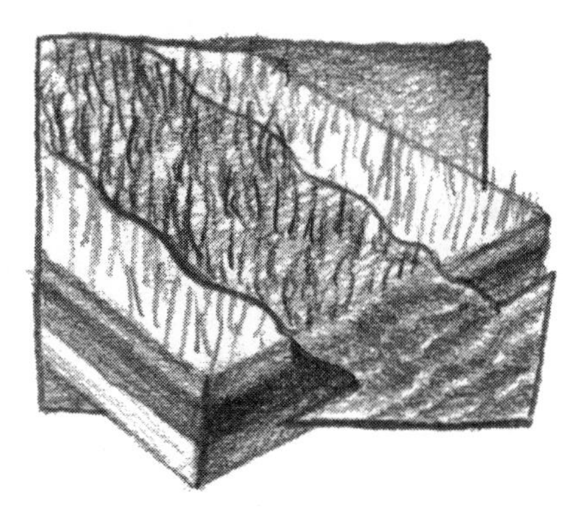

But only on penetrating the pattern at ground level does one confront the most incredibly various array of life on this continent. Plants of temperate and subtropic zones meet and mingle. Graceful wading birds in their elegant plumage share the watery places with drabber ducks and sandpipers. Bald eagles perch in dying trees. In varying numbers there are deer, otters, manatees (sea cows), black bears (that staple of our national parks), tarpon, butterflies, bobcats, alligators, crocodiles, rattlesnakes, and even an occasional Florida panther, which prowls the glades on its apparent path to extinction.

But a century of drainage schemes began to alter the original fragile water pattern of the Everglades, diverting water into the sea to create dry land for farms and towns. The situation worsened during the 1950s and '60s as a human tide of tourists and new residents began to overwhelm southern Florida. At first the waders, adapted down through the ages to shuttling among the Glades, mangrove islands, and tidal flats, were not seriously disturbed. But this invasion prompted more projects for both drainage and water supplies, expanding the system to 1400 miles of canals and levees, lowering water levels and decreasing sheet flow (which is the movement of a shallow expanse of water across the landscape) from Lake Okeechobee. The area of wetlands in the Glades shrank by more than a third.

As the national park was cut off from its natural flow of water and left to maintain itself largely on rainfall, disastrous fires began to sweep the Everglades. This was a crisis situation, and state and federal managers met to work out a plan to save the ecosystem. Floodgates were opened, permitting water to flow into the park from impoundments to the north. But this system, too, was badly flawed. Although the delivery system was meant to mimic natural sheet flow from the north, the plan permitted additional releases of water as part of the flood control system. What an Audubon biologist has called "unnatural pulses of water" caused levels to rise at inopportune times. The elaborate patterns of flooding and drying to which wading birds like the wood storks had adapted now broke down.

"Even park officials thought the management plan was a good idea," a park biologist recalls. "Everybody thought it was just a problem of water volume. They didn't realize that the timing and distribution of water flow are just as important. The nesting colonies began to fade away."

Citizen action had saved the Everglades twice before. Untold numbers of wading birds, mostly herons and egrets, were shot during the late nineteenth and early twentieth centuries, supplying plumes that fashion then decreed for women's hats. Audubon societies were established in response to the killing, and an aroused public fought for bird-protection laws. By the 1930s, Audubon wardens were protecting huge rookeries that made the Everglades once more a "paradise for birds." After World War II, Audubon joined Florida conservationists, including Marjory Stoneman Douglas (the author known as champion of the Everglades), in lobbying for the legislation that created the national park.

In recent years, Audubon has strengthened the alternative public voice that counteracts development pressures in and around the national park.

Working closely with the conservation-oriented Everglades Coalition, Audubon President Peter A. A. Berle expressed the environmental priorities for the region:

- Lobby hard for the acquisition by federal and state governments of key areas that are outside the park but are critical to the Everglades' survival;
- Stop the discharge of polluted water to public lands from agricultural areas south of Lake Okeechobee;
- Develop a new federal wetlands policy and manual that mandates the protection of wetlands on public lands.

Ugly wildlife manager?

The seasonal cycle of rising and falling water is the dominant fact in the lives of all the animals of the Everglades. In this ecological drama the alligator often plays a role as the natural hydraulic engineer, much as the beaver does on mountain streams. Big alligators frequently dig ponds at low places in the Glades and return to these private retreats each dry season, deepening and enlarging the pond as the years pass. Such ponds, called 'gator holes, provide places where marsh snails, frogs, and fish can survive the dry months to repopulate the Glades when summer rains come and the water rises again.

— William B. Robertson, *Everglades—The Park Story*

Heartbeat of the Everglades

Nearly a century of experience by Audubon wardens, wildlife managers, and hydrologists in south Florida lies behind a new "Vision Document" prepared by Audubon's Everglades Team. The team recognizes that the region must be seen by water managers as a whole—as a complete ecosystem. The great pool of water that lay at the heart of the Glades functioned like a heartbeat, "with water expanding and contracting over the marsh in an annual rhythm."

Now the system has been ravaged. The watershed around the Kissimmee Prairie and Lake Okeechobee to the north has been nearly severed from the Everglades, vast acreage of wetlands drained or paved over, and huge volumes of water that once pulsed through the system diverted to agriculture or development or even dumped into the sea. But the team describes its vision of a different scenario if state and federal authorities grow determined to restore a semblance of the natural water cycles to this priceless land: "A restored and healthy environment is critical for the economy, water supply, and quality of life for south Floridians. A successful restoration campaign will provide an opportunity to demonstrate our leadership in the protection and restoration of wetlands nationally and internationally, and ensure that we pass to our children and grandchildren a healthy environment. Populations of shrimp, lobsters, and numerous marine fish are linked with freshwater flows out of the Everglades."

Restoration of flows would lead to improvements in the commercial fisheries that depend on these populations. Historically, Lake Okeechobee supported commercial fisheries and could again support one. Furthermore, millions of tourists visit the Everglades system every year to experience the enjoyment of hiking, camping, canoeing, and fishing in the natural environment of south Florida. The wetlands of south Florida are critical to the water supply for human use in south Florida. These wetlands function to recharge well-fields and to store water. A healthy Everglades has large recreational potential. The Everglades Ecosystem is an international treasure.

CRY WOLF!

Does your voice count in wildlife protection? It does if you believe that environmental activism is an act of faith in democracy.

The wolf almost rivals the snake in the antipathy it has stirred among humans for centuries. Perhaps the story of Little Red Riding Hood somehow enters the human nervous system and afflicts it with a lifelong case of fear and loathing. Whatever it is, the history of the extirpation of the gray wolf from state after state during the march of European settlement to the Pacific shore stands as a mark of shame to many Americans. This large predator, once the most widespread of all modern wild mammals throughout the northern hemisphere, but thought to be a menace to livestock, was shot, poisoned, and trapped until only remnants of the population were left in the lower forty-eight states.

A mark of shame, perhaps, in the story of man's inhumanity to other species, but even most admirers of the wolf felt that the battle was lost forever. Recently, however, there have been attempts to reintroduce the gray wolf to parts of its former habitat in the Western states. Proponents consider the most promising site, where a small population might survive until it has had an opportunity to establish itself, is Yellowstone National Park.

There was a predicted outcry from industries and individuals who do not believe a place remains in the United States for a large predator. The specter of ravaged ranches, diminished hunting for sportsmen, and even wolf attacks on women and children was raised.

Conservationists responded with an attempt to put the issue in perspective. The Fish and Wildlife Service, which manages problems affecting endangered species such as the wolf, called for massive input by the public, and groups decided to fight myth with a roster of facts.

"From numerous wolf researchers we know that 'wolf predation alone has not caused rapid declines in ungulates in any mainland ecosystems,' " Audubon's wolf specialist wrote to chapter activists. "From Minnesota, where 1550–1750 wolves share habitat with over 300,000 livestock, we know that confirmed highest losses have been .25 per thousand for cattle and 1.44 per thousand for sheep. Finally, we know that there are no documented cases of healthy, wild wolves killing humans in North America. The old myths die hard."

The following points constitute a reasoned checklist that anyone can adapt for wildlife issues in other parts of the country:

1. Wolf recovery, to be successful, must be based on sound biological principles and the Endangered Species Act.
2. Wolf recovery must be accomplished in the larger Greater Yellowstone Area, not just in Yellowstone National Park.
3. Any introduction effort must not compromise or conflict with naturally recolonizing wolves in Idaho, Montana, or Wyoming.
4. As the major predator on North American ungulates, the wolf is crucial to the natural functioning of the ecosystem and the regulation of prey populations.

5. Every national survey, and surveys of all states bordering Yellowstone, have shown strong majority support for wolf recovery.
6. Any control of depredating wolves must be confined to confirmed cases only, and must be administered by federal, or federal-state, teams. Private take (i.e., killing) of wolves should be severely limited until full recovery is achieved.
7. Control efforts must be prompt, aimed only at offending wolves, and adequately staffed and funded at all times. A fund to compensate livestock producers with confirmed losses to wolves is the key to gaining local support for recovery efforts.

Speak out!

> **The main trouble in dealing with ecological issues is that we underestimate them. The ecosystems that support all living things are the most complex entities we know anything about in the universe. . . . Today, armed with all the devices of mechanical ingenuity, we are redoing the Earth. We add confusion to complexity. Our 'developments' are a welter of discordant conditions that we seldom dignify as a system. The biosphere has become a patchwork of disturbed communities maintained as unsightly back alleys of the human citadel. The occupation forces do not know what was there before or what is likely to come after. A look at what man hath wrought suggests that we got into the construction business without benefit of blueprinting.**
>
> **—Durward L. Allen, in his address to the 1985 North American Wildlife and Natural Resource Conference**

Deadly Refuges

Government wildlife refuges sometimes turn into death traps for birds and other animals.

One recent winter, thousands of ducks, geese, and sandhill cranes died of avian cholera on two National Wildlife Refuges at 7000 feet above sea level in the San Luis Valley of southern Colorado. The climate can be very harsh on the high valley, but streams and rivers (including the Rio Grande) flow out of the surrounding mountains and provide abundant water and marshes that have attracted migratory waterfowl for untold centuries.

As European settlers converted the valley floor to farms and ranches, their wells tapped the groundwater and their ditches carried the Rio Grande's flow to distant parts of the valley for irrigation. The once-abundant wildlife habitat began to dry up. However, the great flocks of waterfowl—prized by hunters—were considered worthy of salvation, and after World War II the federal government began to buy valley land, along with deep wells, from local farmers. The Fish and Wildlife Service created the Monte Vista National Wildlife Refuge in 1953 and the Alamosa National Wildlife Refuge in 1962, a total of 25,000 protected acres, and the valley once more became a haven for migrating and wintering waterfowl.

The lure for birds at Monte Vista and Alamosa during all seasons is the abundance of both open water and food. The deepest artesian wells, with their powerful flow, stay comparatively free of ice in the coldest weather. Waste grain in the surrounding fields and a standing crop grown especially for birds on the refuges create dense populations of them in winter and spring.

In an earlier time wild birds had the freedom to spread out all across the rich feeding and breeding areas that a pristine continent could offer. Constantly on the move, they were unlikely to encounter situations in which massive epidemics could flourish, and indeed none of the early explorers in the American West reported seeing the huge concentrated die-offs that have become commonplace in our day.

But now, with habitat shrinking, waterfowl are often forced into avian ghettos, where their numbers may cause habitat deterioration. Their wastes, and perhaps the high number of rotting carcasses that tend to accumulate in overcrowded wildlife communities, become the vehicles by which disease organisms are passed around.

"Outside pressures from society tend to encourage poor management practices on public lands," says a wildlife biologist. "We have many groups that *want* to see birds concentrated for their own convenience—hunters who want to shoot them, birders and photographers who want to see them, and business people who want to cater to those groups."

Moving between modern refuges are "Typhoid Marys," birds that have not succumbed to the disease yet carry the deadly organism. And when the carrier arrives on a refuge where wild birds are crammed like chickens in a poultry house, the pathogen finds ideal conditions for reproduction and spread. Though hard figures are difficult to come by, some biologists guess that diseases kill at least 5 million waterfowl every year. Avian cholera, avian tuberculosis, duck plague, botulism, and avian pox have become lamented additions to the ornithologist's vocabulary.

When I brought you into the garden land to eat its goodly fruits, you entered and defiled my land, you made my heritage loathsome.

—Jeremiah 2:7

"There's good evidence that tinkering with water levels for irrigation or other diversion projects can touch off big die-offs from botulism," a veterinarian said after a series of major wildfowl disasters at Utah's Bear Lake Delta. "When water levels are drawn down, huge numbers of invertebrate animals die along with the aquatic vegetation, and bacteria can thrive in the decaying animal protein."

The ultimate "fix" for poisoning is to provide more room for birds and wildlife in our increasingly crowded human world. That step will not be made until citizen calls for it become a lot louder than they are now. Government agencies have tried to make breeding and wintering grounds at each end of the line secure for migratory birds, yet for many species the weak links are now the feeding and resting places along their migration routes.

Displaced Species

One of the major problems on public lands today is the population explosion of exotic plants and animals. These are, if you will, weed species—brought here by humans in the past and turned loose through accident or ignorance. In their own way, these species are as much of a headache to wildlife managers as imported insect pests are to farmers.

Most of the "exotics" have few predators and parasites in their new homeland, so their populations grow unchecked. They tend to destroy habitats, often out-competing native species for scarce food and shelter. They can even change normal ecological processes, interfering with predator-prey or plant-animal relationships built up over long periods of time.

Exotic species problems are everywhere in our national parks:

- Wild boars roam Great Smoky Mountain National Park, as well as Haleakala and Hawaiian Volcanoes national parks in Hawaii. Brought to a private hunting preserve in Tennessee in 1912, the boars escaped and now occupy most of the park, destroying native plants, eating small animals, even competing with black bears for fruits and nuts in the fall. Two endemic species—a salamander and a snail—have been put at risk.

THINKING LIKE AN

ECOSYSTEM

Succession

A forest doesn't just appear on the face of the Earth. A forest is a process, often going through a cycle of hundreds of years as the soil is prepared, predecessors rise and vanish, and the big trees finally take hold. In the beginning there may be only a stretch of poor soil, spotted here and there with primitive plants. But one day, a few seeds of small plants drift in on the wind and take hold in crevices. Year after year, as these small plants grow, flower and make new seeds, and then die, their remains decay to enrich the soil. New seeds drift in, or are dropped there by passing birds and other animals, and a variety of plants flourish. The story of plant succession is still new and a subject of some controversy. But the outlines of it can be suggested here. The first shrubs and trees sprout up on this plot of ground. Alders perhaps, adding badly needed nutrients to the soil. Birches and poplars, which are usually small, short-lived trees, march in, perhaps from stands that have grown nearby for years. The succession of plants is well underway. Finally, the seeds of pines take root. They grow slowly at first, but flourish in the shade and shelter provided by the mature birches. The young pines shoulder up through the smaller trees, taking away their needed sunlight and hastening their end. At last the pines become the dominant species. Ecologists refer to them as comprising a mature forest. They reign until a catastrophe—perhaps a fire, or a man-made disturbance, destroys the forest. And then the process begins all over again.

Along with plant succession goes animal succession. When the ground is bare, not many animals can find food or shelter. Those creatures that live in the soil or burrow in it will be present. In the early weed stage, seed-eating birds and mice and rabbits will find food. When the vegetation gets taller and bushier, more birds come; they are able to nest there. In the tree stage mammals, such as chipmunks, deer, squirrels, bear, raccoons, and opossums also will be able to find food and shelter.

Many forests or prairies grow where there once was a lake or a pond. In the beginning the water is lined with plants. After many years these die and fall to the bottom, slowly building up sediment. On the shallow edges plants such as sedges, bulrushes, or cattails form a mat over what was once water. Over time more plants will grow, and the surrounding soil will become firmer and firmer. Eventually a tree will get started, then another, and so on until a forest has taken over.

Along with these topside changes go alterations in the water life. At first the lake would have supported fish that could live in cold water. The shallower, warmer edges would have supported fish that required warmer water. As the bottom fills in, the water becomes warmer. More insects can live on the submerged plants. Bass would be fed by the insects and smaller fish. As the area becomes reedier, more birds would make nests; muskrats can thrive. Finally, as the prairie or forest takes over, land animals would replace ducks, fish, turtles, and other aquatic life.

These successive changes are in process constantly and are interfered with by thoughtless human activities.

- Feral burros caused extensive destruction in Grand Canyon National Park and Death Valley National Monument. Abandoned by miners in those areas in the late 1800s, these animals destroyed scarce desert vegetation, caused severe erosion, damaged prehistoric sites, and competed for food and water with native species such as bighorn sheep. Park officials kept burros under control for a while by shooting, but when some animal lovers protested, costly live trapping was required to remove most of the animals.
- Exotic plants have disrupted many park ecosystems, driving out native plants and thus interrupting the food chain to the detriment of native animals. Tamarisk, native to the Mediterranean, has invaded springs and waterways in several desert parks, its deep roots tapping water that is needed by local animals. Hawaii Volcanoes National Park, like so much of the Hawaiian Islands, has lost many native species to the invasion by exotics. "In Everglades National Park, Australian pines (*Casuarina*), introduced as a windbreak and now common in Florida, change the deposition of sand on beaches, thus reducing nesting sites for the American crocodile and for sea turtles," according to the *Audubon Wildlife Report* (1985). "Another troublesome exotic, *Melaleuca*, was aerially seeded in 1936 to help 'dry up' the Everglades. A third exotic, Brazilian pepper, also is widespread and prevents native species from occupying certain areas. Management consists of pulling small plants out or using herbicides on larger specimens, but efforts are small compared to the size of the problem. Further, because the park is surrounded by other land where exotics are common, exotic control will always be at least a maintenance problem for the park."

Bringing in plants and animals from abroad often seems like a good idea, but they generally end up as somebody else's problem.

Shredding the New Frontier

In 1980 President Jimmy Carter signed into law the Alaska National Interest Lands Conservation Act. It was the most dramatic step taken in our time to preserve land for the public.

The act placed over 100 million acres in Alaska under various forms of protection. It established or substantially enlarged sixteen national wildlife refuges and fifteen parks, monuments, and preserves. The act also placed over 56 million acres of national parks, refuges, and forests in the National Wilderness Preservation System—designed to preserve their resources permanently.

One unit, recommended for wilderness designation but rejected by the Nixon White House because of the possibility of oil on the site, was the Alaska National Wildlife Refuge.

"Bordering the Arctic Ocean, the refuge is the summer calving ground of the Porcupine caribou herd, which numbers 170,000 animals," David Rains Wallace wrote in *Life in the Balance*. "The herd's spring migration from wintering grounds in spruce forests southeast of the refuge is one of the greatest wildlife spectacles in North America. Dozens of shorebird, waterfowl, and songbird species also breed on the coastal tundra,

and several hundred thousand snow geese visit the area during fall and spring migrations. Musk oxen, wolves, Dall's sheep, grizzly and black bears, moose, wolverines, arctic foxes, and other species inhabit the refuge year round. A few polar bear females den on the coast."

Given an opening, oil and gas producers applied to Congress for entry into the refuge to begin exploration. Activities of this kind in 1984 and 1985 caused some damage to tundra vegetation and disturbed wildlife. Despite common knowledge that whatever oil is there would provide only a few months' supply for the nation, congressional authorization was making headway until the enormous *Exxon Valdez* oil spill in Prince William Sound in 1989.

The battle continues, and promises to be a long one. Bills have been introduced by conservationists to designate the refuge as a wilderness, off-limits to resource extraction, while the development side has introduced bills to open the refuge for leasing to the oil companies.

Ultimately, the people of the United States, acting through their elected officials and legislators, will decide.

> There will always be proposals to use environmentally valuable land such as wilderness and wildlife refuges for energy and mineral development. The pressure of proposals for development in these areas cannot be relieved by granting one-time access. Each time the price of energy or minerals jumps, or whenever a new technology allowing recovery of formerly inaccessible resources is developed, engineers have an incentive to return to an area for a closer look. Plans for development in environmentally sensitive areas, shelved due to economic or technological constraints, may be revived decades later. Only statutory protection, such as that granted by inclusion in the wilderness system, can provide long-term protection. A nation, like ours, with a 200-year history should look at the wilderness and wildlife preservation issue in a time frame that spans hundreds of years, not mere decades. Only with such a perspective can we pass on to succeeding generations these living laboratories of natural history that are still intact.
>
> —Alex Stege and Jan Beyea, in the *Annual Review of Energy*, 1986

E c o Q u i z

In the debate about drilling for oil in the Arctic National Wildlife Refuge, here are some questions and the environmentalist response.

Q: Doesn't the country need the oil in the Arctic National Wildlife Refuge if we are to have a secure energy future?

A: According to the Department of the Interior, there is an 80% chance that no commercially recoverable oil will be found in the coastal plain of the refuge. At best, there is an 18% chance that there exists about one month's supply of oil in the coastal plain and a less than 1% chance of an eighteenth-month supply. Any oil that is found won't be available for at least ten years.

Q: But won't whatever oil is there be put to good use? Won't it help solve our energy problems?

A: Hardly, because even if oil is found, it will represent at most only 3% of current imports. We can save almost 300,000 barrels of oil a day (the best-case estimate for the refuge) just by increasing the efficiency of our cars by two miles per gallon.

Q: What's the harm in drilling?

A: One harm is the effect drilling would have on wildlife. According to the Fish and Wildlife Service, more than 450,000 acres of caribou calving range and migratory stopover areas for snow geese in the refuge would be affected by development. At least 12,000 acres would be lost to construction, while a much larger area would be crisscrossed by pipeline roads, base camps, gravel pits, and sewage treatment plants, and contaminated by pollutants. Thousands of people would be needed at such an oil installation, which would resemble a major industrial complex and could bring about declines approaching 50% in regional musk ox and caribou populations and in snow geese using the area.

Q: But doesn't the oil industry say it has a good environmental record at Prudhoe Bay and other Alaskan oil fields?

A: Large-scale oil exploration and development always has a lasting impact on the environment. Other Alaskan oil operations generate huge quantities of toxic drilling-waste fluids. Oil spills are a common occurrence, even aside from the *Exxon Valdez* disaster. Oil operations have spilled at least 4 million gallons of oil and other hazardous chemicals onto the tundra since 1972. The largest clean-up of PCBs (a known cancer-causing substance) in the nation took place at Alaska's Kenai National Wildlife Refuge, where ARCO incinerated 87,000 tons of PCB-contaminated materials.

Q: But when you take a close look at it, isn't the Arctic National Wildlife Refuge a barren desert?

A: This 19-million-acre refuge is the only pristine area in North America that protects a complete range of arctic ecosystems. The most biologically productive portion of the refuge is the coastal plain, which is where the oil industry wants to drill. More than 160 species of animals, including musk oxen, arctic foxes, grizzly bears, and wolves live there. More than 130 species of migrant birds use the area for breeding and feeding. Its rivers and coastal waters contain critical habitat for some sixty species of fish.

Q: There's plenty of wilderness for everybody. Why do we need this little piece of it?

A: There is very little of our wilderness left: only about 8% of U.S. land, mostly in Alaska. Half of that is legally protected, half has yet to be. The Arctic Refuge's coastal plain is the only stretch of the arctic Alaskan coastline not already open to development.

E C O S Y S T E M   P A T H W A Y S

Genesis granted humans "dominion over the fish of the sea, and over the fowl of the air, and over every living thing that moveth upon the Earth." Until very recently, that injunction was taken mainly as a license to exploit wildlife to the fullest extent possible, which in the worst cases led to extinction.

The consequences for wild America have forced enlightened humans to view "dominion's" other side, which is not exploitation but responsibility.

Although it saddens many lovers of the natural world to admit it, much of the wild has been taken out of wildlife. Paradoxically, in most cases today, wildlife preservation implies wildlife management.

Every concerned environmentalist and biologist has become an ecologist who lists among his or her aims to

- Manage forests and wetlands for biodiversity;
- Look for "indicator species"—animals or plants especially vulnerable to pollution or excessive disturbance—as aids to detecting breakdowns in ecosystems;
- Focus on prevention rather than cure, trying to alleviate human impacts on fragile environments *before* the threat of extinction appears within them;
- Preserve habitat for species throughout their life cycles, as in creating refuges for migratory birds along their flyways as well as on their breeding and wintering grounds.

Enlightened stewardship is now wildlife's only hope.

12

OUR WILDLIFE

Guest author: Frank Graham, Jr.

The visitor to an old-growth forest in the American Northwest finds a wild community that is unique. This kind of forest, which is fast vanishing in the face of excessive logging, is dominated, as its name implies, by ancient trees. But it can be understood only if seen, not as a commodity to be exploited, but as an organism nourished and sustained by the climate, topography, and other living things with which it evolved.

Thriving in the region's heavy rains and fertile soils, these forests consist of many kinds of cone-bearing trees. Sitka spruce, western hemlock, Douglas fir, silver fir, and mountain hemlock are each dominant in areas best suited to them, and form the forest canopy.

"The canopies also support lichens of the genus *Lobaria*, which fix nitrogen from the atmosphere and replenish the soil when they fall to Earth," writes Andy Feeney in the *Audubon Wildlife Report* (1989–1990). He adds that the canopies of ancient forests also provide old, moss-covered branches and high nesting sites that may be critical to the nesting success of threatened birds such as marbled murrelets. Other species, red crossbills and pine siskins, come to feed on cones that grow in the upper branches.

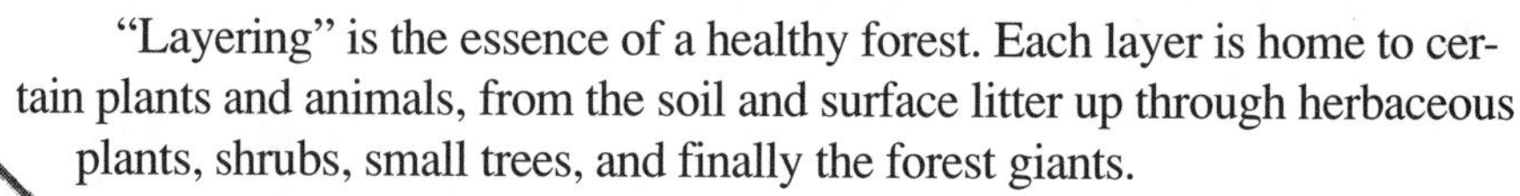

"Layering" is the essence of a healthy forest. Each layer is home to certain plants and animals, from the soil and surface litter up through herbaceous plants, shrubs, small trees, and finally the forest giants.

All elements play their roles. Stringlike fungi serve as an extension of the tree's root system, snaking out through the soil to extract water and nutrients, which are then absorbed by the tree. The fungi also contain nitrogen-fixing bacteria, which further enrich the soil.

Squirrels, voles, and shrews eat the fruiting bodies of these fungi. The fungi are spread through the forest in animal feces, making nitrogen-fixing organisms available to the next generation of trees.

Death plays a major role. As an occasional big tree dies and falls, sunshafts pour through openings in the canopy, nourishing new plant life below. Herbs and shrubs, then smaller trees, spring up. A teeming understory develops around the fallen giant. Yet the decaying tree also radiates life. As its fibers break down, they give nourishment and moisture to the soil.

"In our forests of the Northwest, a 400-year-old Douglas-fir usually lasts between 200 and 250 years as a fallen tree," writes ecologist Chris Maser. "An 800-year-old Douglas-fir takes about 400 years to decompose and recycle into the system. After death, it serves an entirely different suite of functions that are necessary to keep the forest going."

The lungless Oregon salamander lives in the moisture around the rotting tree. If the tree happened to fall across a stream, it may serve as a dam, backing up the water and creating a pond for beavers. Beetles and other insects burrow into the trunk, where they are preyed on by small mammals and woodpeckers.

Then, as the inner wood softens, it is easily hollowed out for shelters by foxes and various birds. The foxes, owls, and other predators feed on the smaller animals and, themselves eventually dying, enrich the soil with their decaying bodies.

That old phrase "the web of nature" takes on new meaning as scientists study the interplay of species in the ancient forest.

On Biodiversity

Organisms are all the more remarkable in combination. Pull out the flower from its crannied retreat, shake the soil from the roots into the cupped hand, magnify it for close examination. The black earth is alive with a riot of algae, fungi, nematodes, mites, springtails, enchytraeid worms, thousands of species of bacteria. The handful may be only a tiny fragment of one ecosystem, but because of the genetic codes of its residents it holds more order than can be found on the surfaces of all the planets combined. It is a sample of the living force that runs the Earth—and will continue to do so with or without us.

—E. O. Wilson

Extinction Is Forever

The history of planet Earth is a record of change—volcanoes erupting, glaciers spreading and retreating, species coming into being and passing away.

We know from finding their remains in rocks that dinosaurs and saber-toothed tigers once lived in North America. Mammoths roamed where New York's Borough of the Bronx stands today. All those animals have vanished. They are extinct.

Those extinctions took place over periods that are short in the geological timescale but long in terms of human lifetimes. The animals have been replaced by other creatures that are truly spectacular in their numbers and diversity. The world is filled with thousands of forms that, from childhood on, we have learned to treasure—lilies and elephants, oaks and tigers, apes and butterflies.

Now many plants and animals are leaving our planet much faster than earlier species did. We humans are changing the world so extensively that life forms with which we evolved over vast stretches of time can no longer live here. Extinction is speeding up. It took millennia for dinosaurs to vanish, but a number of species that were plentiful when the European settlers came to these shores are now gone or on the way out.

- The gray wolf was shot and poisoned by ranchers (and even government agents) because it sometimes attacked sheep and cattle. It was almost extirpated from this country. Attempts to reintroduce the wolf to the wild in Yellowstone National Park are still bitterly resisted by many nearby landholders.
- The passenger pigeon, which once darkened the sky over our forests, was hunted to extinction for sport and food—often food for pigs and dogs. The last of its kind died in the Cincinnati Zoo in 1914.
- The bald eagle, our national bird, nearly vanished from the United States after World War II because of DDT and other long-pesticide residues in the environment. When the eagles ate fish contaminated by pesticide runoff in lakes and rivers, they laid eggs that wouldn't hatch. A ban on those "hard" pesticides went into effect just in time.
- The black-footed ferret was given up for lost when ranchers destroyed the prairie dog towns where the ferret found prey and shelter. Only an aggressive captive-breeding program by state and federal biologists saved the species from extinction, and it has been reintroduced into the wild in Wyoming.
- The red-cockaded woodpecker digs its nest in the soft wood of aging pines and other trees in our Southern forests. Those trees, however, are generally the first ones cut for timber and so the bird's habitat is almost gone. Its plight is hauntingly similar to that of the ivory-billed woodpecker, which became extinct in this country when its old nesting and feeding trees disappeared.

If humanity survives to look back on the 20th century, future historians will regard the mass extinctions of plant and animal species, now under way, as the development with the most lasting consequences for us.

—Jared M. Diamond, Professor of Physiology, University of California

The Endangered Species Act remains the last barrrier to extinction for those species that have already plummeted to alarmingly low numbers. Under the Act's provisions, federal officials are obliged to protect habitat vital to the survival of the affected species, while scientists are directed to devise a plan to help it on the road to recovery.

There have been huge die-offs in the past, when many species disappeared, discarded by evolution in a doodling with life forms that is heartless, mindless, merciless, but also unmalicious, intentionless, random. The high extinction rate at the moment is unique within our span of recorded time, so it surprises us; but mass extinctions are not unique. What should unnerve us is that large waves of extinction have always wiped out the culprits. When organisms were too abundant, dominating the

Earth and ruining the environment, they went extinct with countless other animals. Then a new form of ooze or mouse started evolution all over again. So it's not that large numbers of animals haven't gone extinct before, or that nature can't take care of herself. It's that when nature does, things start off from scratch in a new line of evolution that may not include us. . . . It is possible that we may also become extinct and, if we do, we will not be the only species that sabotaged itself. Just the only one that could have prevented it. . . . If we'd like our species to hang around a while, we must be vigilant about what we're doing to other species, because evolution is a set of handshakes, not a list of winners.

—Diane Ackerman, author of *Whale by Moonlight, Natural History of the Senses,* and five volumes of poetry

Save That Plant!

Today, about one in every ten forms of life on Earth is in trouble. That includes insects, fish, clams—and plants.

Green plants support the chain of life. Without them, other forms of life could not exist. Aside from a few creatures living underwater near hot ocean springs, they are the only living things that can manufacture food, turning complex chemicals into nourishment through the process of photosynthesis. Plants, including fungi, herbs, shrubs, and trees, form the base of the food chain: plant-eating animals nourish themselves on leaves, seeds, nuts, fruits, pollen, nectar, bark, and roots; while predators feed on the plant-eaters (and sometimes on each other).

Trees and other plants provide homes for a great many animals. A hummingbird, a mouse, a beetle, an inchworm, a bee, and perhaps a dozen or more animals may depend on a single kind of plant for food and shelter. The survival of fungi or bromeliads may be tied to that plant. If the plant becomes extinct, its "dependents" may likewise disappear forever.

The irony here is that protection of plants under the Endangered Species Act is incomplete. An endangered plant is protected from "taking" only on federal lands. It receives no protection at all under the act on private lands.

The Passenger Pigeon

The pigeon was a biological storm. He was the lightning that played between two opposing potentials of intolerable intensity: the fat of the land and the oxygen of the air. Yearly the feathered tempest roared up, down, and across the continent, sucking up the laden fruits of forest and prairie, burning them in a traveling blast of life. Like any other chain reaction, the pigeon could survive no diminution of his own furious intensity. When the pigeoners subtracted from his numbers, and the pioneers chopped gaps in the continuity of his fuel, his flame guttered out with hardly a sputter or even a wisp of smoke.

—Aldo Leopold, in *A Sand County Almanac*

On Saving the Diversity of Species

It is as artist and scientist that [man] is unique, the only being able, in considerable measure, to understand and appreciate the world of which he is a part. And that world is fast slipping away from him—the real world, that is—to be replaced by an artificial world, built of concrete, steel, and chrome. Our eyes are upon the vastness of space, our dreams of other worlds, unexplored, unimaginable. Once again the thrill of the primeval, the challenge of unwarped nature! Yet our own tired planet still has its frontiers: my own back yard is full of them, full of creatures that put to shame the science-fiction writer's men of Mars. We would do well to spend less time reaching for stars, to value some things above comfort and the expansion of our economy. We would do well, now and then, to stretch out on the good Earth with a notebook, camera, or sketch pad and chronicle the lives of some of our less-important neighbors. —Howard Ensign Evans, in *Wasp Farm*

The Activist

The first line of defense for American wildlife is the citizen activist. The activist may have no background in biology or wildlife management, bringing to the cause little more than a love for wild things and a determination to protect them. He or she is the true "amateur," for that word can be traced back to the Latin *amare*: to love.

A century ago America's wildlife was in crisis. Protective laws were few and mostly unenforceable. A hunter, in a single day, might kill hundreds of wild birds and sell their flesh or feathers on the open market. Many species seemed to be slipping away. Indeed, several once-abundant kinds of birds became extinct before anyone noticed they were gone.

By the 1890s the buffalo had been hunted to the point of extinction on the Western plains. The passenger pigeon and the Carolina parakeet were already over the brink, while many kinds of gulls, terns, herons, and egrets were killed for the plumes or feathers used to trim women's hats and dresses. Their doom was apparently sealed.

At this point a great public outcry arose against the slaughter. An unlikely mix of people—sportsmen, women of fashion, teachers, poets, painters, and businessmen—came together and founded the first Audubon Societies (named for the great painter of birds, John James Audubon). In 1905, a number of those state wildlife conservation societies created a federation that eventually became the National Audubon Society.

"The object of this organization," wrote its first president, businessman William Dutcher, "is to be a barrier between wild birds and animals and a very large unthinking class, and a smaller but more harmful class of selfish people. The unthinking, or, in plain English, the ignorant class, we hope to reach through educational channels, while the selfish people we shall control through the enforcement of wise laws, reservations or bird refuges, and the warden system."

It was a far-reaching vision, but those early conservationists put it into practice. They lobbied in legislatures and government agencies for protective laws and regulations. A succession of successes, culminating in the Migratory

Bird Treaty Act of 1918, gave protection to most kinds of wild birds and set bag limits for those species that could be legally hunted.

They also acquired land and created Audubon sanctuaries, guarded by dedicated wardens. When Guy Bradley, an Audubon warden, was murdered in 1905 by plume hunters in Florida Bay, President Theodore Roosevelt publicly lamented the tragedy from his office in the White House. Education programs helped to alter public opinion and showed generations of Americans the economic and esthetic value of predatory birds such as hawks and owls.

Other wildlife conservation organizations followed Audubon into the field. Some battles were won, but today's citizen activists face a battery of environmental threats to wildlife: the clearing of vital habitat for roads and developments, the draining of wetlands for agriculture, the poisoning of lakes and streams by toxic chemicals, and the destruction of old-growth forests. Global warming looms on the horizon.

ECO HELPER

Every State Should Have a Leo Drey

Donald Jackson

A disaffected junior executive for a St. Louis shoe company, Leo Drey decided to use his inheritance to become a conservation-minded land baron. He bought his first tract of badly abused Ozark woodland in 1951. Today his domain, "Pioneer Forest," embraces 160,000 acres, with an emphasis on selective logging that preserves the hills' scenic integrity. . . .

The mill site is one of nine parks and natural areas that he purchased and leases to state land agencies through the nonprofit L-A-D (for Leo Albert Drey) Foundation. "The idea of the foundation is to preserve outstanding remnants of the landscape that were here when the settlers came," he says. Leo holds veto power over the management plans for the sites. . . .

The L-A-D Foundation has saved some of the most sublime venues in the Ozarks from development. "Leo's things are caves and springs, canyons and natural bridges, and history," David Bedan says. Drey's preservationist gorge rises when his commitment to retaining an area's natural character clashes with the desire of state officials to ease public access. At one site the state wanted to improve a trail to a canyon rim while Leo wanted the existing trail blocked off. Leo won. "He's death on trails," says John Wylie. "He seems to think visitation means destruction.". . .

"I don't have the same obligations as the state," Drey replies. "I don't think access for the public should be too easy. There's a distance filter effect. A visitor should have to struggle a bit." The last word, as usual, belonged to Leo, who is one of the few private citizens in the country who can defy a government land agency on conservation grounds and get away with it.

A Man and His Refuge

Laura Riley

Tom Atkeson was a junior biologist when Wheeler National Wildlife Refuge on the Tennessee River was new in the late 1930s, and over the decades he has overseen its fields, forests, waterways, and natural inhabitants in a special way no one could have imagined. For during World War II, a mine blew up in his face, leaving him sightless and without hands. . . .

For the past thirty years, Atkeson has written a weekly outdoor column which is printed in half a dozen area newspapers and aired over radio and TV stations. Through it he gets across some of his ideas about outdoor safety and wildlife, "sugar-coating it" with anecdotes and lore: Deer are attracted to music, he reports ("European hunters sometimes hire a violinist to accompany them and hide in a thicket while the musician plays seductive strains from Bach or Beethoven"); the world's best insect and tick repellent is the Avon lady's Skin-So-Soft ("Use it full strength on neck, wrists, and lower legs"). He even gives recipes for sassafras tea, wild blackberry dumplings, catfish stew, or striped bass with brown mustard and Tabasco sauce. . . .

There have been sheaves of commendations, awards, and citations, which he ships to his children to keep. There has been a Tom Atkeson Day in Birmingham, with the keys to the city, and the Alabama Governor's Award. In 1984 the state legislature passed an act officially recognizing his outstanding work in conservation, with the scroll entered in the Alabama archives.

Wildlife Destruction = Human Misery

We talked of the country where we were, of the beings best fitted to live and prosper here, not only of our species, but of all species, and also of the enormous destruction of everything here, except the rocks; the aborigines themselves melting away before the encroachments of the white man, who looks without pity upon the decrease of the devoted Indian, from whom he rifles home, food, clothing, and life. For as the Deer, the Caribou, and all other game is killed for the dollar which its skin brings in, the Indian must search in vain over the deserted country for that on which he is accustomed to feed, till, worn out by sorrow, despair, and want, he either goes far from his early haunts to others, which in time will be similarly invaded, or he lies on the rocky seashore and dies. We are often told rum kills the Indian; I think not; it is oftener the want of food, the loss of hope as he loses sight of all that was once abundant, before the white man intruded on his land and killed off the wild quadrupeds and birds with which he has fed and clothed himself since his creation. Nature herself seems perishing.

—John James Audubon, in *The Labrador Journal,* 1833

Mother Goose Syndrome

Most humans have an ambivalent attitude toward wild animals: they divide the animal kingdom into good guys and bad guys. Good animals are those that look handsome or peaceful or cuddly, while bad animals don't fit the viewer's notions of the body beautiful or acceptable behavior. Biologists call this attitude the Mother Goose Syndrome.

A man once chided an Audubon warden for his efforts on behalf of the California condor, a huge vulture that feeds on carrion.

"What good do you see in a creature that eats dead things?" the man asked.

"Tell me, sir," the warden interrupted. "Do you eat live things?"

To the biologist or wildlife conservationist, there is no such thing as repulsive form or behavior in the animal kingdom. Each organism has evolved to fit neatly into its niche in an ecosystem. Its physical and behavioral features are appropriate to its existence—the elephant's trunk, the fly's predisposition to breed in filth, the crocodile's protective scales, the wolf's fangs, or the slug's slimy skin.

To understand nature is to understand the arachnologist who enjoys looking through his microscope at the faces of spiders, which he finds "beautiful." Or the entomologist who retired and bought an old farm on the most mosquito-ridden marsh he could find, the better to study the creatures that fascinated him.

The form that functions is elegant enough.

Humans and Wildlife: A Complex Relationship

We need another and a wiser and perhaps a more mystical concept of animals. Remote from universal nature, and living by complicated artifice, man in civilization surveys the creature through the glass of his knowledge and sees thereby a feather magnified and the whole image in distortion. We patronize them for their incompleteness, for their tragic fate of having taken form so far below ourselves. And therein we err, and greatly err. For the animal shall not be measured by man. In a world older and more complete than ours they move finished and complete, gifted with extensions of the senses we have lost or never attained, living by voices we shall never hear. They are not brethren, they are not underlings; they are other nations caught with ourselves in the net of life and time, fellow prisoners of the splendour and travail of the Earth. —Henry Beston, in *The Outermost House*

WETLANDS

Wetlands are areas where water is the primary factor controlling the environment and its associated plants and animal life for at least part of the year. These are the transitional habitats between rivers, bays, lakes, and other aquatic places and dry upland areas.

The soil, often dry, is, at least seasonally, saturated or covered by water. Water levels on the surface may vary during the year from zero to six feet or more. In the coterminous forty-eight United States, nearly 94% of the wetlands are located inland and another 6% are saltwater coastal wetlands.

Wetlands generally are rich in plant life, which attracts large numbers of animals. Some 5000 species of plants, 190 species of amphibians, and one-third of all bird species in the United States are found in wetlands. More than half the marine sport fish caught in this country depend on wetland estuaries for survival, especially during the early stages of their lives.

People need these areas, too. Wetlands store and convey floodwaters, thereby reducing flooding. They improve water quality by trapping pollutants in their sediments, by converting toxic pollutants through biochemical processes to less harmful substances, and by the uptake of pollutants in wetland vegetation. They improve groundwater recharge and stabilize shorelines. And they are vital to the nation's commercial fisheries: about two thirds of our principal commercial fish depend on estuaries and salt marshes for their spawning or nursery grounds.

Yet, traditionally, wetlands have always been waste places, or worse: remember the Great Grimpen Mire, a focus of fear, where Sherlock Holmes pursued the killer in "The Hound of the Baskervilles." The human instinct has been, over the centuries, to obliterate wetlands by draining, dredging, filling, leveling, and flooding. The region that now covers our forty-eight coterminous states contained an estimated 221 million wetland acres in Colonial times, but lost more than half of them during the intervening two centuries. Ten states have lost 70% or more, while California has lost over 90%.

Eco Koan

A seeker, who notices similarities between animal and human behavior, asks the Eco Guru, "What creatures are humans most like?"

The Eco Guru answers, "Hogs. Humans take all the land and cover it over. Humans take all the water and pollute it."

WETLANDS SANCTUARY

The 2200-acre Lillian Annette Rowe Sanctuary, which belongs to the National Audubon Society, demonstrates the tension between the protection of wetland habitat for wildlife and competing human demands in the modern world.

The sanctuary lies in the shallow Platte River Valley, just east of Kearney, Nebraska. It protects river channels and wet meadows that are critical to the lives of several species of migratory birds, including the whooping crane, sandhill crane, bald eagle, least tern, and piping plover. But the river's annual flow is only 30% of what it was 100 years ago. The water has not simply disappeared; it has been diverted by a variety of engineering projects, including dams and irrigation schemes on the river's North and South forks.

The once "mile-wide and foot-deep" Platte is now often only a few hundred feet wide, heavily forested islands standing where water once flowed. The forceful flow of spring meltwater from the mountains, that used to flood the channels and rip the plants from the riverbed, is now a trickle that has escaped the control projects upstream.

The Rowe Sanctuary's manager, his staff, and volunteer helpers spend a great deal of time each year doing the job that the river once did—clearing the sandbars of vegetation. Because of their efforts, thousands of sandhill cranes and several endangered whooping cranes are able to feed and roost in and along the Platte for a critical few weeks during each spring migration.

In addition, threatened piping plovers and least terns can now find some of the bare sandbars that are essential for their nesting. Bald eagles have open water for winter foraging. Thousands of ducks and geese locate safe and healthy spring migration stopover sites. And, for about six weeks each spring, visitors to Rowe Sanctuary can see one of the most stirring wildlife spectacles in North America, as tens of thousands of sandhill cranes roost and feed along what is now the remnants of a great river.

In recent years, the National Audubon Society and other activist groups throughout the country successfully battled the proposed Two Forks Dam, an enormous water project whose completion would have signaled the end of the river and the age-old flights of the cranes that depend upon it.

Six Steps to Save a Wetland

To protect one wetland in your lifetime is an enormous contribution. Below is a guide for an individual or a group.

1—Identify Wetlands in Your Area
Review available maps such as the Fish and Wildlife Service National Wetlands Inventory maps, Geological Survey topographic maps, state resource maps, and local and regional planning maps. For a National Wetlands Inventory map, call 1-800-USA-MAPS.

2—Assess the Condition of One or More Wetlands
Make an inventory of the plants and animals and if possible document their presence with photos. What type of wetland is it—swamp, bog, freshwater marsh, or riparian area? Is it wet year-round or only in certain seasons? Describe the wetlands and surrounding land-use, such as subdivisions, highways, and agriculture. Identify actual or potential threats such as dumping or construction. Continue to monitor them regularly.

3—Identify the Owners
Contact them and discuss their plans for the site. Try to find out if it has already been altered in any way.

4—Develop a Strategy for Protection
Choices include focusing on saving a specific threatened wetland, developing a master plan for the community, organizing a "wetlands watch" to monitor existing threats, lobbying for local ordinances to protect wetlands, educating the community, and acquiring key wetlands.

5—Identify Protective Mechanisms
Determine the zoning for the area. Find out which protective laws might apply: the Clean Water Act, the Rivers and Harbors Act, the Endangered Species Act, the National Environmental Policy Act, the Coastal Zone Management Act, or state and local laws affecting planning, zoning, and wetlands. Contact local regulatory agencies that could help.

6—Be Proactive!
Help set up local policies, legislation, and action plans before a threat arises. Attend local planning, zoning, and wetlands meetings and public hearings. Make known your concern for wetland protection.

The trouble with both [terms] "rainforests" and "wetlands" is that they appeal to the ego and the superego. Swamps appeal to something arising from the reptile brain itself (obviously a fan of swamps)—the id. In fact, I would go so far as to say not only that swamps appeal to the id but that if we were to psychoanalyze the landscape we would find the broad, sunlit uplands of the ego; the craggy peaks of the superego, to be savored only after punishing climbs; and the powerfully aromatic swamp of the id—fecund, moist, mysterious, crawling with phallic serpents and alligators, warm, inviting penetration but always threatening to swallow us up, to eat us alive. —James Gorman

THINKING LIKE AN

ECOSYSTEM

Predation

The subject is deceptively simple: Who eats whom?

But humans have always seen it from a variety of viewpoints: Little Red Riding Hood's grandmother undoubtedly took a different view of predation than a modern wildlife biologist. And the child at a picnic who expresses disgust at the sight of a robin gobbling a plump, writhing worm is all-oblivious to the part some unfortunate steer played in supplying the picnic's hamburgers.

Predation is the act of one animal preying upon, or eating, another one. The winner is the predator, the loser is the prey. An animal can spend its life as the predator, then abruptly become the prey: a robin that lives by preying on worms may one day fall victim to that efficient predator—a house cat.

Although humans mourn the fall of the robin to a cat, or the loss of the cat to an opportunistic coyote, predation shapes all life on Earth. A bird that consumes large numbers of leaf-eating bugs may help preserve a forest or a field of wheat. Moreover, predator and prey evolve in concert. As the bug becomes better at hiding, the bird must fine-tune its hunting skills to keep up. As the antelope increases in speed and wariness, the lion develops more responsive muscles and more effective hunting techniques.

One of the common defenses against predation is cryptic coloration. For some creatures camouflage is used as a strategy to disguise attack. Spiders and praying mantises sit in flowers that they match in color in order to deceive passing insects. Coloration does not help at night. A most remarkable demonstration of nocturnal attack—and counterattack—is told by Paul Ehrlich in *The Machinery of Nature*.

"The moths' most important nocturnal predators, bats, hunt insects by generating ultrasonic pulses—sounds with frequencies much higher than can be detected by the human ear. The returning echoes of the pulses indicate the location of flying prey, much in the manner that returning sound pulses from sonar equipment are used by surface vessels to locate submarines. Using these echoes, . . .bats have an astonishing ability to assemble a picture of their environment . . . [and] to allow them to identify the species of prey. . . . Some moths have evolved ears that detect the sonar. . . . Some go even further—they attempt to 'jam' the bat's sonar with their own electronic sounds. The bats, in their turn, have evolved countermeasures to the moths' defenses, such as shifting their sonar frequencies out of the frequency range of maximum sensitivity of moth ears. (Complicating the situation for moths are parasitic ear mites that destroy the moths' ability to hear.)"

Who eats whom is both high drama and a complex ecological story.

Parasites and Predators

Some years ago, when the Department of Agriculture opened a laboratory in Niles, Michigan, and word got around that it was to be a rearing station for "parasites," most of the neighbors became uneasy. It took some explaining to make them see that their uneasiness was directed at the wrong critters.

The real threat to the region were cereal leaf beetles, which had somehow entered the Midwest from Europe and by the 1960s were a major pest in grain fields. When pesticides failed to halt their spread, Department of

Agriculture scientists turned to biological control. Field biologists under contract to the department scoured Europe for the parasites and predators that kept the beetles under natural control in their homeland.

When researchers found a number of likely candidates, they shipped them to the United States and eventually on for further study at the new laboratory. The candidates were gnat-sized members of the wasp family known as parasitic Hymenoptera.

"The news that we were raising parasites and spreading them around the countryside scared the hell out of some people," the lab's director recalled. "Of course, those people were thinking of tapeworms or cattle ticks or something like that. It's a good thing they didn't know they were *wasps* besides!"

These tiny parasitic wasps neither build nests nor use their "stingers" directly as weapons. The stinger functions as an egg-laying organ, which the wasp uses to deposit one or more eggs on or inside its victim. After it hatches, the infant wasp feeds remorselessly, growing rapidly in its natural shelter until the victim dies. Then it pupates and eventually emerges as an adult to mate and begin the cycle all over again.

The value of these insects in pest control is that, unlike predators that prey on many different kinds of animals, they generally search out a single species, or several closely related species, to attack. They are programmed by nature to attack only a specific host. The use of parasites is far cheaper and safer than applying insecticides to the open environment.

But what's to prevent an insect, imported from abroad to kill pests, from eventually turning into a pest itself? The answer is that these natural enemies are highly adapted to the insects they have always preyed on. They have evolved with them through the ages, adapting to overcome the defenses of their particular victim. In their absence, the parasitoids will starve. No parasitoid is released in biological control until specialists have determined its binding ties to the pest it was imported to fight. In a sense, these natural enemies are the sum of the adaptations they have made to deal with their victim—the length and strength of the ovipositor needed to implant an egg in its body, the ability to detect the prey's distinctive chemical cues, the inherited characteristics that time their reproductive cycles to the developing stages of the pest.

A century of practice has demonstrated that after their food habits have been carefully studied and confirmed under laboratory conditions, they can be

Controlling Pests

In the end, cliché or not, the solution lies with an educated and concerned public. Probably the most fundamental difference between the insect and ourselves is that the insect, with little capacity for learning, survives by his extraordinary adaptability to all environments and all changes that man engineers, while man survives because of his superior learning ability. Clearly we have the intellectual ability to triumph over the insects' superior adaptive ability. We may also already possess the knowledge to answer the hard questions and solve some of the pressing problems. But do we have the wisdom and, above all, the unselfishness to use this knowledge effectively for the benefit of mankind?

—Vincent Dethier, *Man's Plague: Insects and Agriculture*

The Balance of Nature

In some quarters nowadays it is fashionable to dismiss the balance of nature as a state of affairs that prevailed in an earlier, simpler world—a state that has now been so thoroughly upset that we might as well forget it. Some find this a convenient assumption, but as a chart for a course of action it is highly dangerous. The balance of nature is not the same today as in Pleistocene times, but it is still there: a complex, precise, and highly integrated system of relationships between living things which cannot safely be ignored any more than the law of gravity can be defied with impunity by a man perched on the edge of a cliff. The balance of nature is not a status quo; it is fluid, ever shifting, in a constant state of adjustment. Man, too, is part of this balance. Sometimes the balance is in his favor; sometimes—and all too often through his own activities—it is shifted to his disadvantage.

—Rachel Carson, in *Silent Spring*

safely released into the open environment in the knowledge that they will not attack beneficial species and become pests themselves.

Technically, these insects aren't parasites. A true parasite, such as a flea, lives in or on its host for some part of its life cycle, and although it often weakens its victim, seldom kills it. But the parasitic insects valuable in biological control are intermediate between parasites and predators. They are, in a sense, specialized predators, and so entomologists prefer to call them "parasitoids," meaning parasitelike.

When studies at the Michigan lab had demonstrated that several species of the tiny wasps could survive in the Midwestern climate and diligently search out cereal leaf beetles, four species were released into farmlands. They succeeded beyond the fondest hopes of farmers and biologists, spreading through the region and helping to bring the pest beetles down to tolerable levels without the excessive costs of widespread chemical use.

As a wise scientist has said, "The object of man's game with nature is not to win, but to keep on playing."

Migration's Mysteries

Birds, the most mobile of terrestrial animals, are constantly on the move. About 80% of the 650 birds that breed regularly north of the Mexican border migrate to some extent, some for only a couple of hundred miles, others all the way to the tropics or deep into the Southern Hemisphere.

They leave the northern forests in fall before snow and ice shut down the supply of insects, amphibians, and other small animal life on which many of them depend for food. Townsend's warblers, retreating from breeding sites in the conifers of Alaska and British Columbia, may venture no farther than oak and laurel stands in northern California or the environs of Santa Barbara. Others of the same species fly on, over the scrub and desert lands of southeastern California and northwestern Mexico to winter in the mountain forests of the Sierra Madre in Mexico and Central America.

"Altitudinal migration" sometimes takes place in the Far West. Jays, juncos, and chickadees abandon their summer homes in the mountain forests and fly to lower elevations as cold weather comes on. Mountain quail, at the same time, *walk* down the slopes in small flocks.

On longer migrations, the birds often come together in flocks of several species, perhaps gaining some protection from predators by the intimidating element of numbers. Young birds, on their first long flight, may benefit from tagging along with the more experienced adults on that species' traditional migration route.

Whooping cranes, for instance, remain solicitous of their young, one or more families banding together on their fall migration from the nesting grounds in northwestern Canada to the Gulf Coast of Texas. The white-and-rust-colored offspring of that year fly and feed alongside their parents en route, and the family ties may persist on the wintering grounds and on the whoopers' flight north again in the spring.

What triggers migration? For a long time, scientists believed that birds timed their departure for the South with the onset of cold weather and the consequent decline of their food supply. But in most cases the migrants set off *before* the weather turns nasty. Swallows often depart while the tiny insects still fill the air of late summer.

So there is a mix of answers, external forces mingling with internal physiological changes, that send a bird on its way. In spring, increasing daylight causes internal reactions that stimulate the swelling of the sex glands, leading to restlessness and then to departure. There may be changes in rainfall or wind direction, or perhaps some internal clock that is set in motion by the species' evolutionary history and prompts migration.

Another mystery that still challenges scientists is the mechanism by which these tiny bundles of feathers, some flying by day, others by night, find their way across thousands of hazardous, wind-tossed miles to their destination. The blackpoll warbler, which migrates between the tropical forests of Brazil and the boreal woodlands of northwestern Canada, weighs only an ounce, and thus could be mailed anywhere in the United States for the price of a first-class postage stamp. Yet, twice a year, these mites launch themselves out over the ocean with almost as much punctuality as the tides themselves.

Some birds may rely on natural landmarks, and indeed they often hug the coastlines or mountain ridges. Others undoubtedly use the sun or stars. But what do they use for a compass on overcast nights? Some studies suggest that birds may sometimes make use of the Earth's magnetic field: pigeons are known to have iron-rich tissue in their heads.

Anyone with the capacity to be thrilled by such mysteries must take note of the Kirtland's warbler. Each spring, virtually the entire world population of that species (only a little more than 500 birds) fly from their winter home in the Caribbean to a small area of northern Michigan, where they nest only under jack pines.

That is precision and consistency, and a lasting cause for wonder.

When I used to rise in the morning last autumn, and see the swallows and martins clustering on the chimneys and thatch of the neighboring cottages, I could not help being touched with a secret delight, mixed with some degree of mortification; with delight, to observe with how much ardor and punctuality those poor little birds obeyed the strong impulse towards migration, or hiding, imprinted on their minds by their great Creator; and with some degree of mortification, when I reflected that, after all our pains and inquiries, we are yet not quite certain to what regions they do migrate; and are still farther embarrassed to find that some do not actually migrate at all. —Gilbert White, in *The Natural History of Selborne* (1789)

EcoQuiz

In the debate over protecting wildlife, here are the key questions and the ecologist's responses:

Q: Does the protection of endangered species mean that the nation's economy will suffer?

A: Not at all. In fact, not protecting our diversity of plants and animals—and their habitats—will cripple our economy in the long run. One example is Florida's Everglades: the degradation of this vast wetland ecosystem has damaged the state's multibillion-dollar commercial and sport fishing industries.

Q: But doesn't the Endangered Species Act stifle economic development?

A: Survey after survey shows that environmental laws have had very little adverse impact on economic viability, either locally or nationally. For example, the Fish and Wildlife Service reviewed 120,000 projects between 1979 and 1991 to determine whether they would hamper the recovery of threatened or endangered species. All but thirty-four of the projects were approved.

Q: Aren't endangered plants and animals being protected at the expense of people?

A: Remember, humans are a part of the web of all life on Earth. By reducing the diversity of species, we put at risk the well-being of ourselves and our children. Half of our medicines come from plants and animals, and a rich diversity is essential for maintaining vigorous agricultural crops and fish stocks.

Q: When you get right down to it, hasn't the Endangered Species Act been a failure?

A: On the contrary, this law has made the difference between existence and extinction for dozens of North American species, including the bald eagle, the peregrine falcon, the California condor, and the black-footed ferret. Populations of nearly half of all the species so far listed as threatened or endangered are now stable or increasing.

Q: So we've saved the "crown jewels" of our wildlife. Can we relax now?

A: Not at all. The pace of rescue is slowing down of late. Complicating amendments to the Act in recent years, combined with intense pressure on the Fish and Wildlife Service, often distort the original intent to provide fast and efficient relief to plants and animals on the brink. The Service broke and ran at least a couple of times, delaying the listing of beleaguered species until the completion of the projects that threatened them. Four species of endemic freshwater mussels, for instance, disappeared during the building of the Tennessee-Tombigbee Waterway. Too often, developers get what developers want. Everybody wants to save the bald eagle. But species without charisma become expendable when they stand in the way of a bright new mega-dam. The highest use of the Endangered Species Act would be to save not simply the glamour critters, but the splendid diversity that makes up an ecosystem. As Aldo Leopold, the most famous of American wildlife managers, put it, "The first requisite of intelligent tinkering is to save all the pieces."

E C O S Y S T E M   P A T H W A Y S

Water is our lifeblood. It covers more than 70% of the planet. Our bodies are made mostly of water. Water is the great carrier and dissolver. It also protects life from rapid changes in temperature. Because of the way it absorbs heat when it converts into vapor and releases heat when vapor condenses back to water, it is a major factor in the distribution of heat throughout the world. Thus, the water cycle and climate cycles are closely connected. Water plays a large role in the carbon cycle as well. Oceans are believed to absorb half the carbon dioxide produced by humans. But if the ocean warms as the result of climate changes, it will have a lower chemical capacity to retain carbon dioxide.

Plankton—single-celled and multicelled organisms—are the foundation of the sea's food web. Phytoplankton (plant species) trap solar energy and by photosynthesis convert that energy into food for themselves. Small zooplankton (animal species) like copepods eat phytoplankton and are in turn food for larger zooplankton, like krill, which are in turn food for fish and larger sea creatures such as whales. Phytoplankton make 80% of the Earth's breathable oxygen. Plankton multiply especially in spring when thaws bring about an upwelling of nutrients from the depths. They devour remnants of plants, fish skeletons, industrial wastes, sewage, and some of the junk we put into the water. While plankton communities are essential to the survival of the ecosphere, in recent years they have been signaling that they cannot continue to bear the levels of contamination or organic matter we have been inflicting upon them. If the demand for oxygen exceeds the supply, or the number of phytoplankton is reduced during die-offs, the result is less oxygen in the water and eventually death for all life. Fish, floating belly-up as a result of not being able to breathe, wave a bright red flag to us.

In thinking about our oceans, it is also important to recall the food web and the transfer of energy. For a human at the top of the food pyramid to gain a pound, the sea must produce half a ton of living matter. With each transfer of material, there is about a 90% loss in energy. Thus, 1000 pounds of plant plankton produce 100 pounds of animal plankton, which in turn yield 10 pounds of fish—the amount needed by a human to gain a pound.

Our oceans are home to millions of species of marine plants and animals as well as being sources of iron, sand, gravel, phosphates, magnesium, oil, natural gas, and many other valuable resources. But they are being raided by people who are more concerned with extracting quick profits than with stewarding fish and other marine animals. Consider the following wasteful practices:

- Every year, 12 to 20 billion pounds of unwanted sea life, called "bycatch," are taken during efforts to catch other species, then dumped overboard. An estimated 20% of all fish catch never reaches the consumer.

- 10 pounds of juvenile fish and other nonmarketed fish are dumped overboard dead for every pound of shrimp caught in the southern United States.

  Because of the popularity of barbecued red snapper, they are so overfished and the spawning stock so reduced that they are in or near a state of collapse. Many baby snappers are also killed by shrimp trawlers.

- The National Marine Fisheries Service says 65 of the 153 species it assesses are overfished in U.S. waters, meaning they are being caught faster than they can replace themselves. Some of these are Atlantic cod, salmon, and halibut; haddock; flounder; sharks; hard clams; swordfish; bluefin tuna; king mackerel; Caribbean reef fish; grouper; lobsters; albacore tuna; Pacific striped bass; blue marlin; rockfish; perch; abalone; and all salmon except Alaskan. This is a large-scale SOS. (Save Our Seas)!

- Drift nets, up to forty miles long, hang as curtains of death. Not only do they decimate populations of fish, but they also catch, entangle, and drown marine mammals, turtles, and migratory seabirds. Drift nets have been banned by the U.N., but there has been some compliance and some defiance in the North Pacific. A study shows that 33,000 to 44,000 endangered sea turtles drown in shrimp nets each year. The Steller's sea lion recently became a threatened species because of losses from fishing gear.

- Seabirds get caught in fishing gear too. Killed annually are an estimated one million seabirds of at least twenty-one species, more than 6000 seals and sea lions, more than 10,000 whales and dolphins, and 44,000 albatross.

- The ocean has been used as a dump site for sewage sludge, industrial waste, hazardous chemicals, soil dredged from contaminated harbors, and radioactive wastes generated by medical, research, and mining activities. Contamination also occurs in the form of daily oil spills, groundwater runoff from streets and farms, sewer releases, and acid rain. Many of the distressing happenings, such as fatal epidemics among dolphins, giant tumors on turtles, PCBs in the flesh of fish, and shellfish poisoned by algae blooms (which in turn poison people) have occurred in the last thirty years.

(Continued on page 184)

13

OUR OCEANS AND MARINE LIFE

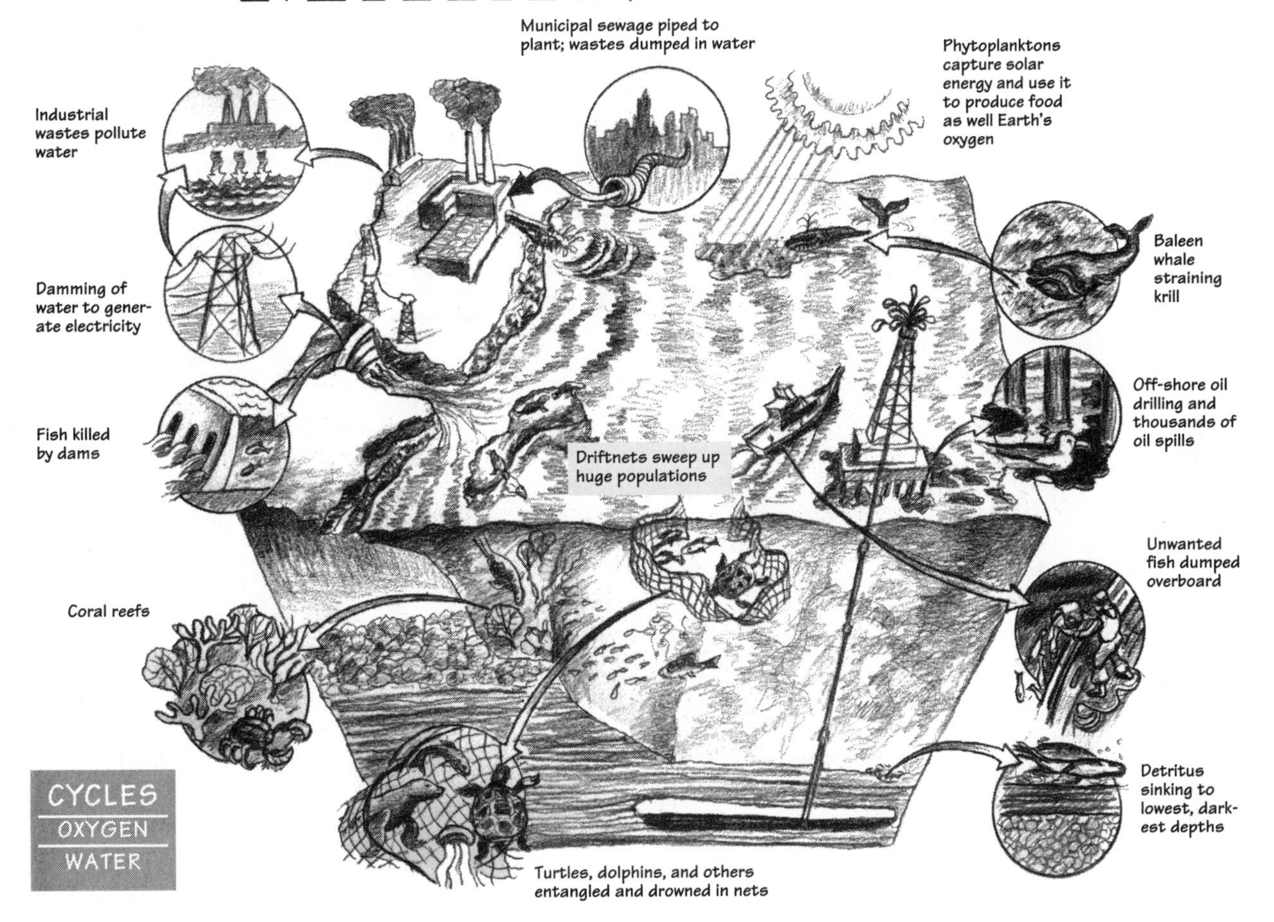

CYCLES
OXYGEN
WATER

FISH ARE WILDLIFE TOO

Carl Safina became a marine ecologist who in his work for Audubon has devoted himself and aroused others to protect our troubled oceans and marine life, just as we protect land wildlife. He writes:

> Traveling on mysterious migrations, shimmering schools and shadowy predators inhabit the oceans. Among them are some of the most beautiful and awesome of creatures. From our vantage point at the shore, though, they are well hidden beneath the waves—literally overlooked.
>
> Fish evade our view and remain unfamiliar. While we can get to know other wildlife—like birds—through our kitchen window, no one can hang a tuna feeder in their backyard. Yet fish are wildlife, too. Imagine if our main experience with birds was the poultry section in the supermarket or a bowl of chicken broth. Unfortunately, our experience with ocean fish is confined largely to opening cans and to glaze-eyed carcasses whose odor is appreciated mostly by our cats. Because fish are largely out of sight and out of mind until we kill them, we tend not to shed sympathy on their predicament. . . .

(Continued from page 182)

- Dams block the access of fish to spawning streams, and by altering water flow, they interfere with fishes' migrational rhythms. Since the late 1960s, more than 100 dams have eliminated 80% to100% of the migration and spawning areas of important fish, such as salmon, in certain regions.
- Oil and gas operations. Most are now located in the Gulf of Mexico, Texas, Mississippi, Louisiana, and Alabama, but pressure has been mounted to open areas off Florida and Alaska. Some areas are more sensitive than others. In Louisiana thirty-six square miles of wetlands are lost each year as the land sinks after being pumped.
- Coral reef bleaching and dying is increasing everywhere.
- Population growth and consequent development along the coasts have decimated wetlands. About 110 million people, almost half of the U.S. population, live in coastal areas. We build more and more homes by the water each decade. The support services and infrastructure needed—sewage treatment, roads, factories, shopping centers, parking lots, airports—all destroy the natural environment that attracted people there in the first place.

Waterfronts should be made safe for all to live, work, and dream by. Many people have seen beloved coastal shores permanently altered, and can identify with Carl Safina's reminiscence about the Long Island, New York, he knew as a boy:

On warm spring evenings before dusk, at the time of year when the breeze smells of moist earth and the redwings' song textures the air with the exuberance of living, my father and I would head for the pebbly shore of the Sound to hunt striped bass. On the way through town we would as often as not be stopped at the railroad tracks as the businessmen solemnly descended from the platform, gray men in gray suits with all the hope and happiness gone out of them. On the shores of the Sound, with the sparkle of water, the coursing of terns, the iridescence and intricacy of a fresh-caught bass, the world seemed unspeakably beautiful, and—I remember this clearly. . . so real. It seemed so real. The pollution I had heard of seemed remote and the bulldozers an unfathomable menace, and I resolved to devote myself to protecting such beauty.

That beach where my father and I hunted for striped bass, and where the world struck me as so excruciatingly beautiful, that beach is now a block of houses. The area is now off-limits; a sign warns that the profoundly enriching experiences I was fortunate to have known are now illegal. But of course the sign is unnecessary. The area is not worth visiting. The place no longer exists.

A very limited survey of particulars conveys a sense of the unprecedented pressures that are focused on ocean animals. Consider sharks. Compared to the relatively high reproductive potential of most other fish, sharks reproduce at a snail's pace. They are slow to mature, and they bear few young annually. This aspect of their biology reflects their niche as top predators. For millions of years sharks gave more grief than they got; their reproductive biology evolved in an environment of very low mortality. Their life history entails investing little energy in reproduction because, in an evolutionary sense, sharks simply don't expect much trouble. Now, for these quintessential predatory masters, the party's over. Shark mortality has exceeded their reproductive capacity by nearly 40 percent annually. Eleven percent of the commercial catch is landed while 89 percent was discarded dead as bycatch in the swordfish, tuna, shrimp, and squid fisheries. The U.S. National Marine Fisheries Service states that "continued overfishing will result in a collapse" of shark populations. A federal "emergency plan" intended to extend a hand to sharks was gummed to death by political red tape and its implementation delayed repeatedly for three years, while the attack on sharks continued apace, driving some species to such low levels that it would take decades for them to recover even if fishing ended today.

Calling it "the last buffalo hunt," one of Safina's targets has been the overfishing of bluefin tuna, whose adult population has been reduced 90%

since 1975. Audubon petitioned to have international trade in bluefish suspended so the species could recover, by getting the bluefin listed as an endangered species under the Convention on International Trade in Endangered Species. This petition was intended to put pressure on the Atlantic tuna commission to reduce allowed catches, but so far the involved countries have been unable to reach a consensus to allow the bluefin to replenish, so Audubon and others are continuing to pursue a listing for the beleagured bluefin.

Writer Ted Williams describes the history of these remarkable creatures in relation to humans:

THE LAST BLUEFIN HUNT

Ted Williams

It seemed to me impossible any fish could be so big, or so beautiful. Her silver, almost slimeless flanks were washed with blues and greens, and the little finlets behind her scythelike ventral and rear dorsal fins glowed bright yellow.

Among bony fishes only marlin are bigger and faster, and not so much on either count. Bluefin tuna can hit 55 miles per hour, at which speed pelvic, pectoral, and front dorsal fins all retract into slots. Even in relation to body weight, their hearts are immense. The gills of bluefins are immense, too; and instead of pumping water through them like lesser fish, they reverse the process, pushing their gills through the water. Always they swim with mouths agape, supercharging their blood-rich muscles with oxygen after the fashion of a ramjet engine; and like aircraft, they have horizontal stabilizers protruding from the base of their hard, sickle tails. Restrained from swimming, they suffocate and drown. They can live for more than 30 years and attain weights of at least 1,500 pounds. These elk-size pelagic wanderers, among the most advanced of fishes, are warmblooded, at times maintaining body temperatures 38 degrees above that of surrounding seawater. . . .

I remember when they weren't even thought of as a commodity. By 1980 the Cat Cay tournament was releasing all fish, but only a few years earlier giant bluefins taken at major tournaments had been hoisted by the tail, written upon, photographed, then dumped off the dock. Hemingway, revealing as much about the mindset of his age as about his own personality, was known to get drunk and use them for punching bags.

Then in the early 1960s four West Coast purse seiners steamed east and opened a major commercial fishery on Atlantic bluefins. Targeting juvenile fish for the canneries, they killed 5,000 metric tons' worth in a single season. Concurrently the Japanese arrived to mine the giants (defined in the trade as fish over 310 pounds) everywhere they could find them, including the Gulf of Mexico—the only known spawning area of the genetically distinct western Atlantic population. In 1962 they set 12 million nautical miles of longlines festooned with baited hooks, enough to girdle the globe 500 times. By decade's end it was clear to anyone paying attention that the resource was in trouble.

In 1976 the California-based tuna industry—whose vessels were regularly seized by foreign governments for violating exclusive fishery zones and which regularly relied on U.S.taxpayers to bail them out—lobbied all tunas out of the purview of the Magnuson Fishery Conservation and Management Act, which had been passed that year to control foreign fishing within 200 miles of the United States.

As the 1970s ended, New England fish dealers pioneered a giant-bluefin export market in Japan, where the fish had long been relished raw as sushi (with rice) and sashimi (plain), but where western stock had been largely unavailable because no one had figured out how to get it there fast, fresh, and cost-effectively. . . . When the Japanese decide something is beautiful or delicious, no price is too high. By 1986 giants were fetching $12 a pound. Last winter a single fish sold in Tokyo for $68,503—$95.65 a pound. Raw bluefin, arranged in beautiful flower patterns, can retail for as much as $350 a pound. The Japanese claim they depend on it, that it's vital to their culture and diet.

Fact: Next to the insects, mollusks are the most numerous group of animals on Earth, with an estimated 130,000 different species. They are also probably the most widely distributed group of animals, and are found on land and in fresh and salt water from the polar regions to the tropics.

THINKING LIKE AN

ECOSYSTEM

Ocean and Earth Zones

Oceans have two principal life zones: coastal and open sea. The coastal zone is the relatively warm, nutrient-rich, shallow water that extends from the high-tide mark on land to the gently sloping edge of the continental shelf—the submerged part of the continents. The coastal zone, including wetlands and estuaries, represents less than 10% of the world's ocean area but contains 90% of all ocean species. Nutrients wash from the land and are deposited by rivers into shallow coastal waters. Surface winds and ocean currents stir up the resulting deposits of nutrient-rich sediments from the ocean bottom. Much of the ocean's voluminous activity goes unseen by human eyes.

Oceans in the temperate latitudes have seasonal gradations that are dependent on solar radiation, temperature, and the cycling of nutrients. In winter, when the surface waters are cold and poorly illuminated, plankton growth is inhibited. Minerals and decaying organisms accumulate on the bottom. Later, storms stir vertical currents, which carry them to the surface. In spring, as the top layer of water warms, the plankton make use of the abundant nutrients that accumulated over the winter, increasing from 1000-fold to as much as 60,000-fold. The animals that feed on them burst in numbers too. By summer the supply is used up. In summer's calm, the colder, heavier, richer bottom waters cannot cross the temperature barrier to the surface. But autumn's falling temperatures and rising gales once more rouse the ocean depths and permit mineral nutrients to mix with the upper layers. This growth of plankton is not as great as spring's.

Climate on Earth is greatly influenced by the "southern oscillation," in which the surface waters around the Equator in the eastern Pacific go from one temperature extreme to another every three to six years. On the one hand is the massive strip of abnormally warm water that from time to time stretches westward along the Equator from South America (often referred to as El Niño), and on the other is an abnormally cold stretch of water.

Another mystery of the ocean involves the way the bottom is constantly being replaced. Tectonic plates move throughout the ocean floor, bearing the continents. Dr. Sylvia Earle, who logs many an hour plumbing the depths of the ocean, has been prompted to say that people who have climbed the mountains and seen the forests on lands believe that the planet has been fully explored. "But in fact many of the forests have yet to be seen for the first time. They just happen to be underwater. We're still explorers. Perhaps the greatest era is just beginning."

> Amidst the din of rushing waters, the noise from the stones as they rattled one over the other, . . . night and day, may be heard along the whole course of the torrent. The sound spoke eloquently to the geologist; the thousands and thousands of stones, which, striking against each other, made the one dull uniform sound, were all hurrying in one direction. It was like thinking on time, where the minute that now glides past is irrecoverable. So was it with these stones; the ocean is their eternity, and each note of that wild music told of one more step towards their destiny. It is not possible for the mind to comprehend, except by a slow process, any effect which is produced by a cause repeated so often. Calling to mind that whole races of animals have passed away from the face of the Earth, and that during this whole period stones have gone rattling onwards in their course, I have thought to myself, can any mountains, any continent, withstand such waste.
>
> — Charles Darwin, *Journal of Researches . . . During the Voyage of H.M.S. Beagle Round the World,* 1835

INHABITANTS OF THE DEEP

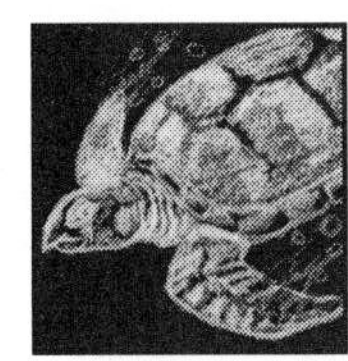

PONDS

Henry D. Thoreau was the first American naturalist to study a pond ecosystem. He observed Walden Pond's wildlife, tracked the rise and fall of its water level, measured its temperature at various depths, and calculated how the temperature differences affected the biology of the pond. He contemplated its color and vegetation even under ice.

Teeming with interdependent communities, life underwater has many niches, partnerships, and movements that are often difficult to see. As on land, there is an ecological web of producers, consumers, and decomposers.

Many microscopic plants float in the pond, not only near the top of the water, but also as far down as light penetrates into the pond. This floating mass of plants, together with some equally tiny animals, makes up plankton.

Of course, there are also rooted plants in and around a pond. Some emerge from the water (cattails, rushes, pickerelweed, arrowhead); some have parts floating on the water (water lilies, watershield); and others grow completely submerged (most pond-weeds, water milfoil). These rooted plants also make food. The main functions of rooted plants to animals are to provide a way for them to get in and out of the water and to provide shelter for aquatic animals and a place where they can lay their eggs.

A pond has many herbivores—animals that feed directly on plants. The herbivores are in turn consumed by carnivorous pond animals. Not all of the herbivores meet their end in the jaws of a carnivore. Many die for other reasons, and their decomposing bodies add nutrients to the water and pond bottom. The most abundant herbivores are the planktonic animals, such as tiny crustaceans, especially water fleas (cladocera) and copepods, which graze on the plant plankton. Other small animals that feed directly on plants are tadpoles of frogs and toads, mayfly nymphs, water scavenger beetles, some mos-

quito larvae, snails, and clams. Some small animals—which also are food for larger carnivores—feed on a wide variety of things, including plants, organic debris, and planktonic animals. These omnivorous feeders include protozoa, hydra, planaria, rotifers, roundworms, waterboatmen, caddis fly larvae, and some mosquito larvae.

The carnivorous animals of the pond prey on the much more numerous herbivores. Some aquatic carnivores are small, such as diving beetles, back swimmers, whirligig beetles, dragonfly nymphs, damselfly nymphs, water striders, and most minnows. We are most familiar with the larger pond animals, such as sunfish, bass, pike, turtles, frogs, and water snakes. These larger consumers do not restrict their diet to herbivores, but also feed on the smaller carnivores—or sometimes on each other.

A pond's decomposers consume dead plant and animal matter. In breaking down this material, they return to the pond nutrients used by the plants. In the depths of the pond, bacteria act on organic material that drifts down to them. In shallow water some flatworms, roundworms, crayfish, and turtles feed on the dead material and aid in the recycling job. Throughout the pond, aquatic fungi join in breaking down organic matter.

In any community, there are some forms that live as parasites. In the pond, parasitic worms and protozoa live, mostly unseen, in or on the bodies of animals. Many of the brightly colored water mites spend their lives as animal parasites.

Fish Alliances

Some fish, like birds, build nests, and some males play important father roles. Sticklebacks, for instance. Less than six inches long, with sharp spines on their backs, they have powerful jaws that crush shrimp and other small animals for food. In the spring a male's belly turns deep red. He finds a territory for his nest and threatens any other male who comes near. He gathers bits of plants in the pond, using sticky threads from his body to glue them to the nest, which he shapes like an open tube. To attract a female, he will do a dance, pointing to his nest. The female enters, leaves her eggs, and swims away. The male fertilizes the eggs. He also moves his fins like fans, pulsing water with plenty of oxygen over the eggs so they won't die. Tiny fish hatch a week later and venture from the nest. If they swim away, their father catches them in his mouth and spits them back into the nest. He guards them until they are old enough to care for themselves.

Fish are very creative in their symbiotic alliances. For instance, the shrimp and the goby. As Ann McGovern and Eugenie Clark write in *The Desert Beneath the Sea*:

> A goby, less than six inches long, guards the entrance to a burrow built in the sand. It shares its home with a bulldozer shrimp that is nearly blind. . . . The bulldozer shrimp uses its big claws to bulldoze out the sand that keeps falling into their home. As the shrimp works, it keeps one of its feelers (antenna) on the goby's body. The goby usually stands guard, as still as a statue. But if the goby senses danger, it vibrates its body. The shrimp's antenna picks up the vibration. That's the signal for the shrimp to disappear down the hole, with the goby following quickly after.

Sea Turtles

These ancient reptiles have been plying oceanic waters for 90 million years but it is only recently that we have begun to understand their breeding and migratory needs.

Worldwide, there are seven species of sea turtles: the green, flatback, loggerhead, hawksbill, Kemp's ridley, olive ridley, and leatherback. Unlike most of the terrestrial and freshwater turtles, the marine turtles cannot withdraw their heads and legs into the sanctuary of their shells. Consequently, these animals have very thick skin on exposed surfaces and heavily fortified skulls as some measure of protection against predators such as sharks. The seven species avoid competing with each other for food and nesting habitat by specializing in their diet and behavior.

Sea turtles are famous for their long migrations between nesting and feeding grounds. Biologists have been able to learn about sea turtle wanderings by affixing tags to their flippers, much in the same way that information is gathered by banding birds. Green turtles travel up to 1500 miles one way from their nesting sites to very specific feeding areas; apparently they are very purposeful in these pilgrimages. Some species scatter widely; the ridleys and the leatherback disperse throughout tropical and semitropical waters when they are no longer concerned with domestic activities. Leatherbacks have been recovered 3000 miles away from where they were originally tagged. We know something about where these turtles go, but how they are able to find their way remains a puzzle. Although many research projects have tried to unravel the mysteries of their navigational ability, thus far these critters have kept secret the key to their success.

And we may never know. The survival of all seven species of sea turtle is precarious throughout the globe. The reasons are many. Sea turtles are hunted for meat, leather, and shells; their eggs are collected; they are caught and drowned in shrimping nets and ensnared in drift nets; their nesting beaches are taken over by resorts and beach houses; and they ingest plastic trash that pollutes the ocean. All seven species are protected by CITES, the Convention on International Trade in Endangered Species. On the U.S. endangered species list, the hawksbill, Kemp's ridley, and leatherback are listed as endangered; the loggerhead, green, and olive ridley are considered threatened.

Marine Mammals

Marine mammals—whales and dolphins, seals, sea otters, and manatees—are a fraternity of evolutionary engineers and emigrants. Over the vast course of geologic time, they have taken the most dramatic developmental step by returning to the sea, abandoning terrestrial ties made when they first left their briny origins. In doing so, they became the architects of some of the most profound anatomical changes that have occurred in mammalian evolution. This metamorphosis, carried out over millennia, resulted in highly specialized, intelligent creatures with few or no natural enemies. The evolution of the marine mammals is an epic chapter in Earth's history, an incredible journey that is

changing course once again, but in a tragic direction. This time marine mammals may face an insurmountable challenge: the exploitation of their kind by Earth's other great evolutionary success—humans.

The smallest of the marine mammals, the seventy-to-hundred-pound sea otter, was pushed to the threshold of extinction during the eighteenth and nineteenth centuries by hunters who prized its luxurious coat. Now protected, its population is gradually increasing again along the northern Pacific coast.

Unlike the other marine mammals, sea otters do not have blubber. Instead, they rely on a rich, dense fur that traps countless microscopic air pockets, preventing water from reaching and chilling the skin. Otters keep their coats in top condition by fastidiously grooming throughout the day, and are especially vulnerable if their fur is soiled. Oil spills bring certain death to these appealing creatures.

Coral Reef Fantasia

Coral reefs are the largest structures assembled by life on Earth. They are to the oceans what rain forests are to land. The magic about coral reefs is this: from the surface the ocean looks opaque, but underneath is an immensely spacious, transparent habitat of fish and coral in astonishing shapes, colors, and dimensions. Here are brain and star corals, polyps like stone flowers, graceful anemone bouquets, sea fans, sponges like giant cactus, green and gold sea whips, fish in dazzling designs—black angelfish with purple stripes, red coneys with blue polkadots, blue-and-yellow-striped grunts, a giant turtle gliding like a spaceship. The only way to see the wonders of underwater life is to put on a face mask and dive or snorkel.

Dee Scarr was an English and debating teacher in Florida for six years when her love for scuba diving led to a dive-instructing job on San Salvador, a Bahamian Island. Now she does an entirely different kind of teaching: taking people underwater to show them how to make friends with creatures often thought to be dangerous, like moray eels, spiny sea urchins, sharks, and barracuda. Her aim, she states, is "to show that the sea is not populated with animals hoping to bite, stab, sting or injure people, but that the sea is a gentle place with animals willing to return affection."

The trouble is too many people are not content to just leave the beautiful fish and coral in their habitat. If they would only witness and not touch, things would be fine, but a boom in sales of coral and tropical fish to aquarium hobbyists has led to collectors raiding the reefs. In two years, for instance, collectors carried off between 500 and 600 tons of Florida Reef Tract.

The State of the Reef

Kenneth Brower

A report on the health of the world's reefs—a State of the Reef Message—is not easy to formulate. We know little about how coral communities are supposed to be, and it is hard to be sure when something is wrong with them.

Western science has come very late to tropical reefs. The first great reef biologist was Charles Darwin. In the century and a half since Darwin deduced, correctly, the manner in which coral atolls form, we have only begun to catalog the reef's inhabitants, to unravel the reef's webs of interdependency, to measure its cycles, to guess at its workings. The coral reef has been called "The Ultimate Ecosystem." There is no more complex natural community on the face of the planet, unless it is tropical forest. If the eye of the diver suffers its moments of vertigo in contemplating the reef, then so does the eye of science. . . .

We are in the middle of a revolution in our way of looking at the reef. Ten years ago it was commonplace to speak of coral reefs as "fragile." Tropical marine organisms live at temperatures closer to their upper thermal limits than do temperate organisms; they live closer to their lower oxygen limits; they require clear water and are sensitive to the smallest increases in turbidity. The dead reefs in the wake of the crown-of-thorns, the demise of reefs in sedimental places like Hawaii's Kaneohe Bay, did indeed make it seem that reefs are delicate systems. It now appears that they are tougher and more resilient than we thought. . . .

How are we to know which of these fluctuations are natural and which are caused by man? How, in light of our ignorance of the reefs, are we to manage and conserve them?

Despite the millions of Australian and American dollars that have gone into research on Acanthaster planci, the cause of outbreaks of the crown-of-thorns is still unknown. The starfish remains nearly as mysterious as some alien thing from the stars.

No one who writes on Acanthaster planci can avoid the science-fiction metaphor. The body plan is from a grade-B sci-fi movie—a spiny, many-armed disk, slow-moving but inexorable. There is something low-budget even in the color scheme—vague tans and purples applied without much skill. The spines are poisonous. The starfish can regenerate from parts of itself. (The natives of several archipelagoes overlooked this trait in early control efforts, when they tried to kill starfish by cutting them up. They succeeded only in seeding their reefs with dragon's teeth.) The starfish has manners as horrible as any movie beast's, everting its stomach through its mouth to envelop coral with its digestive system. Feeding, crowns-of-thorns release chemicals that attract more crowns-of-thorns. They attack the reef en masse, forming fronts, mats, moving in a sort of glacial feeding frenzy. . . .

The great dither of human scientists over the menace of Acanthaster planci is amusing, in a way. As a threat to coral reefs, the crown-of-thorns, for all its movie-monster spines, has nothing on Homo sapiens.

The plague on the reef is rich men. In the Caribbean, the anchors and mooring lines of yachts do considerable damage to corals. Drifting lobster traps—lost by Caribbean fishermen who could never afford to eat their own catch—go ghost-fishing among the corals. Monofilament fishing line garrotes coral heads, and scuba divers vacationing from high latitudes step on them. In the Caribbean especially, but in other seas as well, the construction of resort hotels near the reef produces siltation, and the operation of those hotels generates effluents that are harmful to coral. . . .

The plague on the reef is poor men. Most of the planet's reefs are in the Third World; reef destruction is largely a problem of undeveloped nations.

The reefs of the Philippines are the most diverse in the world, and the most threatened. Philippine reefs support more than 2,000 species of fish (the Great Barrier Reef, for comparison, supports around 1,500), yet a recent study of 619 Philippine reefs showed that 70 percent were dead or dying. . . .

Dredging, dynamite fishing, coral-collecting, shell-collecting, and the aquarium-fish trade are all destroying coral communities throughout the archipelago. The fish-dynamiters are poor men destroying their children's future in order to feed them today. The aquarium-fish collectors are poor men who dive in fins cut from plywood, breathing from hookah hoses stuck in their mouths, and carrying small detergent bottles refilled with sodium cyanide. The divers call the chemical "magic."

A squirt into the coral, and all the fish hiding there come out spinning and jerking. Any angelfish, triggerfish, squirrelfish, or blue tangs the diver catches in his hand. Most of the rest are unmarketable and left spiraling and fluttering amid poisoned corals. For every fish captured alive by cyanide, nine die. . . .

(Continued on page 192)

(Continued from page 191)

The plague on the reef is Chinese. Giant clams, genus Tridacna, are just a memory on the reefs of China, a consequence not so much of human poverty and numbers—though China can certainly boast those—as of passionate, half-scrutable gastronomic craving. . . . Their own reef larder looted, Southeast Asian clammers moved outward into the western Indo-Pacific. As the big clams disappeared from those waters too, the clammers pushed east. In Fiji the giant clams are on their way to extinction, largely because of the depredations of the Taiwanese. Stocks of big clams of several species have declined precipitously on reefs near inhabited Fijian islands, where they are collected and eaten by Fijians, but they are plummeting also on uninhabited reefs, where they are falling prey to Taiwanese poachers. . . .

The plague on the reef is Japanese. One of the last undisturbed coral reefs of the Okinawa region, Shiraho Reef of Ishigaki Island, in the Ryukyu chain, is threatened by airport construction. . . .

The plague on the reef is Africans. Of the seven Tanzanian reefs recently recommended for World Heritage status, five are now dead, blasted by Tanzanian fish-dynamiters. The plague is Indonesians and Polynesians and Melanesians. Island peoples were once fine stewards of their reefs, but with Westernization the old management skills and sensibilities are departing. The plague is any of us with a spear gun and scuba gear in his closet, as I have in mine, or a cowrie on his desk, like the cowrie staring me in the face right now. . . .

The prospects for reefs are bleak, but not entirely so.

"A reef will regrow again, if you leave it alone," Sue Wells told me. . . .

"Pacific Islanders invented all the basic fisheries conservation measures," [Robert] Johannes told me recently. "The measures that we invented about ninety years ago, they invented hundreds of years ago—limited entry, closed areas, closed seasons, restrictions on fishing gear." . . .

The Great Barrier Reef Marine Park, established by act of the Australian Commonwealth in 1975, is, many think, the best thing to happen to reefs since zooxanthellae. Marine-park planners all across the tropics are looking toward that great park for example. At present, 2000 reef complexes and most of the 1250-mile length of the Great Barrier Reef are given one form of protection or another. The basic tenet of the park's planners is that they will manage people, not ecosystems. This is a wise and appropriately humble first principle. The reef ecosystem is too complex for human tinkering and will be left to take care of itself.

Management is through an intricate zoning system: General Use "A" zones (where prawn trawling is permitted), General Use "B" (where trawling is not), Marine National Park "A" and "B" zones, scientific-research zones, preservation zones, replenishment areas, seasonal-closure areas. The system is overseen by a somewhat bewildering assortment of agencies: the Great Barrier Reef Marine Park Authority (GBRMPA), the Queensland National Parks and Wildlife Service, and the Queensland Fisheries Service among them. In Australia it seems to work.

How well it will work in the Third World, where most reefs lie, and where nations are far less wealthy and educated, is a question asked by many reef scientists.

MARINE CONSERVATION RULES OF THUMB

Getting consensus about quotas among commercial fishermen, sport fishermen, charter boat owners, scientists, and environmentalists is difficult. Here are guidelines for ecologically sound actions.

1. "Thou shalt protect juveniles." Regulations need to be adopted that will protect individual fish until they spawn, and keep fishing mortality rates low enough to maintain high spawning populations.
2. Eliminate by-catch waste of nontargeted species and indiscriminate fishing gear, such as driftnets and longlines up to thirty miles long strewn with hooks.
3. Seriously reduce pollution.

Chesapeake Bay is a tragic example of the problems caused by water turbidity and massive algae growth that deprive an ecosystem of the oxygen needed for plants and animals to flourish. Many fish fail to reproduce, causing massive declines in striped bass, among others. The death of eggs and larvae is also attributed to pesticides, acid rain, dams, and diversion of freshwater, which causes degradation of the bay's prime low-salinity nurseries.

Our favorite rivers and lakes have often been closed to swimming. If the water looks dirty and uninviting to us, imagine what it must be to fish who have to breathe in it. Like many rivers, the mighty Mississippi is lined with industries that dump into it. One hundred sixty-four toxic waste disposal sites line a three-mile strip of the Niagara River. Change is possible. The Great Lakes, which were seriously contaminated in the 1960s, have improved since 1972, when Canada and the United States agreed on a joint pollution-control project. Now they are threatened by attempts to erect expensive structures to regulate their water levels so that property owners won't have to worry about flooding. But such tinkering could cause massive loss of habitat and pollution.

Ecological destruction from oil spills is out of control. By some estimates, 240 million gallons of used crankcase oil get into our waterways every year—42 times more oil than the *Exxon Valdez* spilled in Alaska. A single quart of oil is enough to pollute 250,000 gallons of water. Oil in a big lake kills plankton, fish eggs, and other sensitive organisms. It can stay in sediments for years, perhaps for a century.

Fourth rule of thumb: preserve habitats and the integrity of ecosystems. The loss of habitat is the worst it has ever been, and is the biggest long-term threat to the future viability of fisheries.

Wetland loss and degradation has been staggering. By the mid-1970s over half our salt marshes and mangrove swamps, some of the most productive lands anywhere, were destroyed. California has lost 90% of its coastal wetlands. The Gulf of Mexico shoreline is expected to retreat inland thirty-three miles in some areas over the next fifty years. Global warming trends will cause inundation of coastal habitats.

Many of us pass by without noticing that estuarine nursery areas are dying of thirst due to excessive freshwater diversions from incoming rivers. Freshwater inflow controls the biological productivity of estuaries. Diversion of more than about 30% of normal freshwater flows into estuaries results in excess salinity, decreased nutrients, increased pollutant exposure due to reduced flushing, destruction of migration routes and spawning areas, and contamination of freshwater sources for humans. Many spawning areas are wiped out.

Seagrass meadows are also vital to save. They grow very fast and produce high amounts of food. They provide anchorage for attachment of other food-organisms and protective cover from predators. They stabilize the area by preventing erosion and filtering water. Some fish rely on their leaves for laying their eggs. Larvae and juveniles depend on them for food. Because the 4 million acres of seagrass meadows in the South are not included in wetland inventories or recognized formally, they are terribly vulnerable to development.

The preciousness of water requires a clear national policy for conserving habitats and strengthening such programs as the Clean Water Act. Laws need to be tougher and funding greater to restore wetlands, reefs, and species. Let us act out of love for the future of the ecosphere and put muscle into our laws.

Sport Fishing

Avid fisherman Carl Safina has learned from his experiences and changed his ways. After a huge struggle catching a mako, he reflects:

> One thing was obvious: The mako is one of the most beautiful and perfectly evolved predators on the planet. The writing of Henry Beston came to mind: "In a world older and more complete than ours, they move finished and complete." Why had I killed this thing? Was this a good thing to do to such a magnificent, such a powerful, such a mysterious creature? I wasn't sure, and for the first time I felt a surge of real remorse. The fisherman in me was thrilled; the hunter was excited; the naturalist was awed; but the philosopher was ill at ease. In the time it took to catch this fish, the fisherman, the hunter, and the naturalist had taken their turns. Now, the philosopher was on deck. I was profoundly humbled by this fish.
>
> I looked at all the equipment it had taken to catch this fish. Fiberglass, stainless steel, engines, fuel, electronics, etc. How many people's efforts, really, had gone into getting to, finding, subduing and killing this fish, this one fish that an hour ago had been simply swimming the sea just as its kind had done for millions of years? And here I had come into its ancestral home, a dangerous newcomer, part of the first generation in the history of mankind to have these toys, to amuse ourselves with this vestige of our past called "sport."

About the special responsibilities of recreational fishermen in fishery management, he declares:

> Now, the point of sport fishing is to enjoy oneself. And nothing is more enjoyable while fishing than to be catching lots of fish. For that reason, when the fishing is good, anglers have three choices: (1) stop fishing as soon as you get enough fish to eat, (2) keep fishing and load a cooler with fish that you will later throw away or spend all night cleaning and two days trying to give to neighbors (neighbors will usually retaliate by dumping so many wheelbarrows of zucchini and tomatoes in your driveway that it becomes impossible to move your beach buggy or boat trailer), or (3) start releasing fish and keep on enjoying yourself.
>
> People who have fished with me since I was a kid know that my natural tendency was toward being a fish hog, and that it took some discipline to limit myself to keeping only what I could easily use. I went through all the stages; I used to like to pose for photos with a pile of dead fish that seemed to prove what a good fisherman I was, and later on I thought I was a hot-shot if I sold some bass or tuna. It was an evolution for me to really get into catch-and-release, but now releasing fish feels like a major part of the fun. And fishing seems a kinder, gentler sport.
>
> One thing the Striped Bass Plan, with its increasing size limit, has taught me is that it is just as satisfying to come back from a nighttime striper blitz with one really nice fish on the stringer or in the cooler as it is to come back with a cooler full of dead fish. That's because the sport in fishing ends when the fish is fought to a standstill, whether it is killed or not. And it's simply beautiful to watch a big bass scoot away, a shark slink out of sight, or a power-

ful tuna course into the depths. Plus, after a long day or night of fishing, who wants to be bothered cleaning a huge pile of dead fish, or, worst of all, demeaning oneself and one's sport by running around trying to unload some tuna before they spoil? Releasing a big fish makes me feel like a fat cat, like I can afford not to be greedy.

He says that many "recreational" fishermen become part-time commercial fishermen when they sell rather than release what they don't want, which is hypocritical of them. A conservation ethic restores the integrity of the sport and the health of marine ecosystems.

EcoQuiz

Q: Why are amphibians disappearing?

A: Because of acid rain in ponds, pesticides, habitat destruction, and the thinning of Earth's ozone layer, which lets in radiation of the sun, and viruses.

Q: Why do we think toads cause warts?

A: Perhaps because they have small bumps on their heads and backs that produce poisons. These poisons damage the mouths of predators and can kill them, although they are usually harmless to humans.

Q: What have people hunted turtles for?

A: Their eggs and meat; their shells for jewelry and ornaments; their skins for shoes, bags, and belts.

Q: Why is Alaska so important to marine conservation?

A: It is surrounded on three sides by two oceans and three seas. It constitutes 57% of the U.S. coastline and supports some of the most spectacular assemblages of fish, marine birds, and mammals in the world. Many of these species would suffer if the temperature heats up because they would have nowhere to go. Many marine creatures now suffer from overfishing.

Q: What are the most important ecological support systems to preserve?

A: Plankton because of its importance in producing food and maintaining atmospheric stability; estuaries and wetlands because of their importance as breeding and nursery grounds; breeding areas and seasons; migratory staging areas; coral reefs; kelp forests; and seagrass beds.

PART V

Cultural Ecology

ECOSYSTEM PATHWAYS

Having learned how thoroughly imbedded we are in the matrix of cycles, we must recognize our common political responsibility to protect them. We need policies and laws that ensure well-being for seven generations to come.

More and more our newspapers report on conflicts that arise in various parts of the country over environmental issues. In Washington, congressional leaders listen to testimony from "experts"; lobbyists pressure for bills; officials seek election by catering to their voters. Often decisions are made by people who have not received an ecological education and who may be blind to the ecological implications of their acts.

As stated in the *Household Ecoteam Workbook*, "Everything that is normally described as an 'environmental' problem could be more accurately called an environmental symptom of a human problem." Because it is generally agreed that the larger public must be protected for the benefit of all, we have numerous treaties, laws, and regulations that attempt to do this. But we Americans, who pride ourselves on individualism and free enterprise, often find trouble lies in enforcing regulations against private interests. Ultimately, enforcement is not as effective as individuals who care in their hearts about nature. Audubon conservationist Jane Lyons has said, "For better or for worse, we humans have unlimited capacity and power to destroy other species. It is our collective knowledge, wisdom, and compassion that prevent us from doing so."

The dominant environmental challenges, on the national as well as local level, are global warming, ancient forests and wildernesses, wetlands, sustainable development, population growth, energy efficiency, clean air and water, and endangered species. It is also important to:

- Attend to problems before they become crises.

- Focus not only on wildlife management but on conservation biology, which means being concerned with broad areas of habitat rather than individual species.

- Develop business opportunities and apply technology to conservation goals, thus saving ourselves the huge costs of cleaning up environmental damage before it happens by putting business and technology through a "green" filter at the outset.

- Combat subversive strategies, such as attempts to weaken proposed laws through delays and loopholes; emasculate the budget for the EPA and other agencies so they can't enforce regulations; threaten to withdraw advertising funds from pro-environmental media; showcase environmental projects while continuing to do business as usual; give the appearance of protecting the public by adopting environmental slogans; brand environmentalists as radical and anti-American; and sue individuals and environmental groups in order to intimidate them and deplete their funds.

- Work to end poverty, which is an enemy of environmentalism.

14

POLITICS AND MEDIA

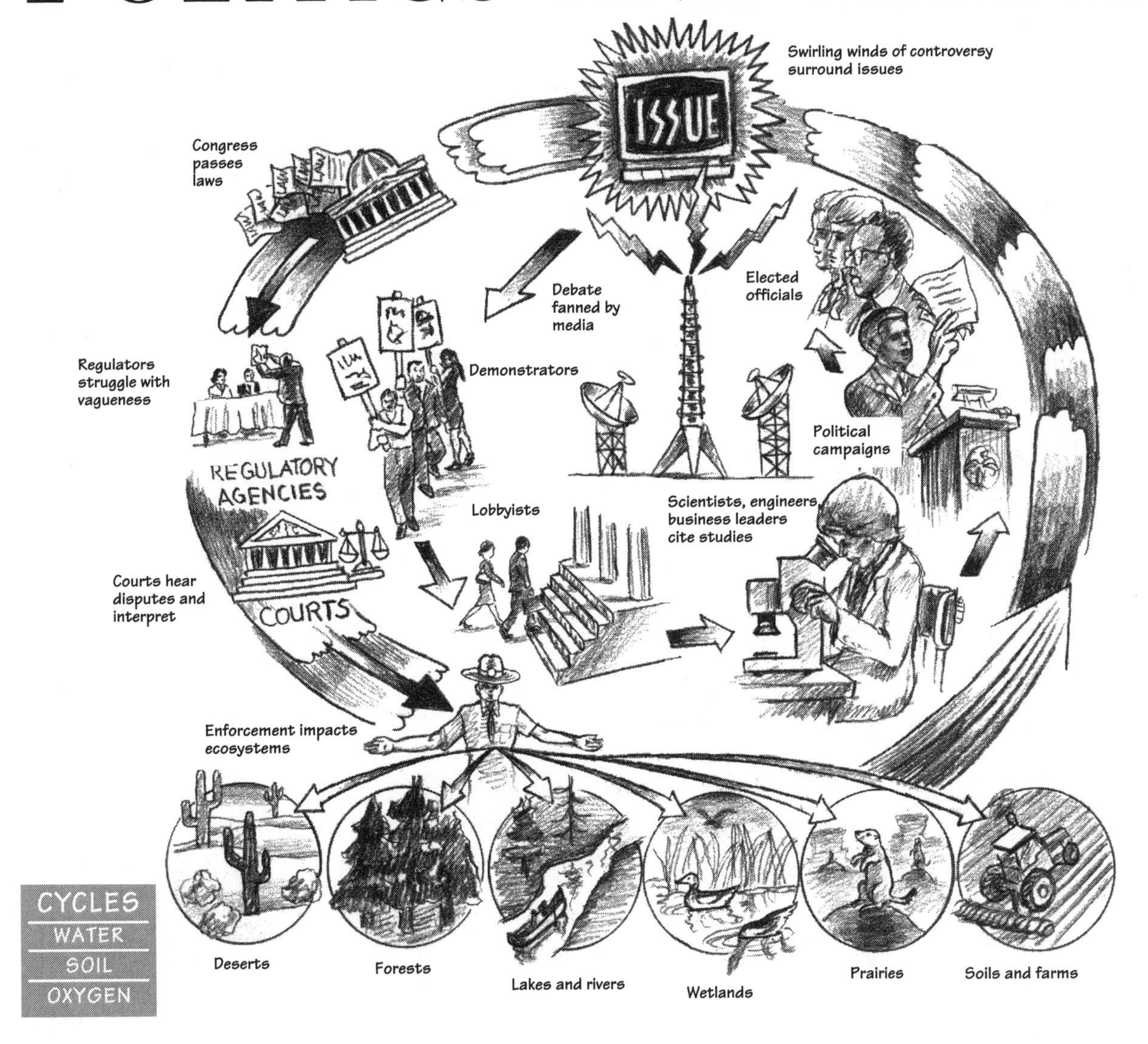

CYCLES
WATER
SOIL
OXYGEN

FROM EARTH DAY TO EARTH SUMMIT

Our country did not get a political conservation movement until the beginning of this century, when the sense of unlimited beauty and abundance of nature began to be replaced by worry over the quality of air, water, and food. As the decades passed, conservationists formed groups and increasingly battled in court over legislation concerning matters such as clean air and water, grazing rights on public lands, toxic waste cleanup, and endangered species. April 22, 1970, the first Earth Day, is regarded as the unofficial birthday of the modern environmental movement in the United States. Twenty-two years later the nations of the world would come together in the first global environmental conference, the Earth Summit in Rio de Janeiro.

Fact: An estimated 10 million people per year are driven from their homes worldwide by environmental degradation.

We are at a turning point. We can reshape our economic and political systems based on how nature works, or nature will do the job for us with an unprecedented and unnecessary increase in human misery and loss of life.

—G. Tyler Miller,
Living in the Environment

The first Earth Day, in 1970, turned out to be the largest mass demonstration around the country since the ending of World War II, and has since been an annual phenomenon. After the first Earth Day, even those U.S. politicians who traditionally courted corporate polluters began to change their stance in order to capture the "green vote." Movie and rock stars gave benefit performances. The League of Conservation Voters scored candidates on their environmental record, and by 1990 a *New York Times*/CBS poll found that:

- 84% of Americans said pollution was a serious problem that was getting worse.
- 74% said that protecting the environment was so important that requirements and standards could not be too high and that improvements had to be made regardless of cost.
- the traditional environmental constituency of conservationists and idealistic young people had grown to include working class and older Americans.

Environmentalism had entered the mainstream, swept along by issues appearing on television, in advertising, and in magazines and newspapers. The state of the Earth improved in some ways but worsened in others. In Europe, especially Eastern Europe, the winds of change blew mightily. The Green Party mobilized to turn environmental values into a political agenda. One of Europe's most respected leaders, Vaclav Havel, then President of Czechoslovakia, declared:

> I live in a country that is one of the greatest polluters in Europe. . . . The past era has taught us, survivors of the totalitarian regime, one very good lesson—man cannot command wind and rain. . . . Man is not an omnipotent master of the universe, allowed to do with impunity whatever he thinks, or whatever suits him at the moment. The world we live in is made of an immensely complex and mysterious tissue about which we know very little and which we must treat with utmost humility.

The year 1992 saw the publication of a book by a major political figure: Al Gore's *Earth in the Balance: Ecology and the Human Spirit.* Gore stated it was necessary to make "the rescue of the environment the central organizing principle for civilization."

In June of 1992 the Earth Summit, officially called the United Nations Conference on Environment and Development, took place. It was attended by over one hundred heads of state, 9000 worldwide organizations, and tens of thousands of citizens. Its most tangible outcome was an 800-page blueprint for the twenty-first century, known as "Agenda 21." Terms such as "biodiversity" and "sustainable development" entered common parlance. People collectively realized that environmental issues crosscut boundaries, that the destruction of nature in one area of the Earth affects other portions, and that a global partnership is needed to protect our ecosphere.

"Agenda 21" covers in detail these major topics: the need for sustainable development to combat poverty, better health standards, improved communities for work and shelter, the need to combat deforestation and the spread of deserts, solutions to solid waste problems, and the protection of mountains, oceans, and biodiversity. The role of women is especially acknowledged. Economic, technological, scientific, and other institutional strategies are summarized. Highly useful as an action guide for the future, a (shortened) Press Summary of this document may be ordered from the United Nations (see Notes). The U.N. has articulated the goals; it is now up to us to bring them to fruition.

Plant and Animal Politics

We must not forget that our compatriots, the animals, are engaged in political activity all the time. Perhaps by looking closely at their behavior, we will discover the roots of our own. If we look even more closely, we may see ingenious ways they solve problems that we as yet bungle.

For example, in a "recession," or scarcity of food opportunities, birds will abandon their independent ways and join together, just as they do to protect themselves from attackers.

Flocking: To Eat or Be Eaten

Frank B. Gill

Whereas stable food resources and defensible spaces promote territoriality, unstable food resources and indefensible areas promote flocking. In certain ways, though, flocks resemble colonies, and flocking behavior has features in common with territoriality. For example, even though a flock member benefits from the group effort, it is subject to a dominance system. . . .

Casual aggregations of individuals at rich feeding grounds are obviously fortuitous, but why do unrelated individuals form stable foraging partnerships? Social tensions and the frequency of fights increase with group size. Subordinate individuals could avoid dominant "bullies" by feeding alone. Competition for conspicuous or rich food items also increases in groups. What then are the advantages of feeding together in organized flocks? The answers lie in the inescapable daily concerns of foraging efficiency and predation risk. Some of the advantages are straightforward, practical ones. Flocks of pelicans encircle and trap schools of fish in shallow water; groups of cormorants and mergansers drive fish toward the shore, where they are more vulnerable. Common Ravens steal Black-legged Kittiwake eggs more easily when hunting in groups than when alone, and subordinate birds profit by moving together onto defended food sources where they can overwhelm the territorial individual. Autumn migrants crossing the Mediterranean may escape the attack of one Eleonora's Falcon, but they have less chance of evading several falcons hunting cooperatively near their colonial breeding areas.

Flock members also benefit from the "beater effect": Prey that is flushed (and missed) by one bird can be grabbed by another. Ground Hornbills in Africa, for example, walk in a line across fields to catch insects flushed by each other. Drongos and flycatchers participate in mixed foraging flocks and specialize in prey flushed by other birds. Flock membership also improves foraging in more subtle ways. Group foraging by pigeons and titmice helps them locate food because members can join successful individuals at rich clumps or concentrate their search efforts nearby. Groups of four titmice in captivity, for example, found more hidden food together than alone. They watched each other's successes and modified the intensity and direction of their searches accordingly. Dominant individuals tend to benefit most because they can usurp the sites discovered by subordinate members of the flock. . . .

Predator detection also improves in flocks; greater individual security is the result. . . .

Flock members warn each other of danger and communicate so that they can take off at the same time. Ducks flush together at the approach of a predator because the individuals synchronize their takeoffs with a series of flight-intention movements that prime every duck's readiness for flight. Flight calls enable longspurs to flush as a group rather than singly. Contact calls enable birds to associate and to maintain a cohesive flock structure even in dense vegetation. Alarm calls serve to alert other members of the social group to possible danger. When one member of a flock spots a predator, it gives an alarm call, and the rest of the flock either freezes or dives for cover. Giving an alarm call would seem advantageous to all but the one that thus revealed its position. Warning calls may seem to be heroic or altruistic acts, but they carry benefits for the caller as well if others in the flock are genetic relatives, such as siblings, parents, or offspring. Each flock member also can count on a certain degree of reciprocity. Most important, by calling loudly the potential victim robs a predator of the element of surprise and thereby reduces the likelihood of attack. The intended victim reduces its own danger as it alerts kin and neighbors.

Animals also have manners that are as disciplined as our diplomacy. Having to behave without the benefit of props other than their own bodies, their protocol centers around gestures. Ear, feather, or tail action expresses dominance, attack, retreat, and submission. Sometimes authority is signaled by badges, such as horns in sheep. Among birds, a darker head and "bib" (throat and upper breast) is a sign of higher prominence.

Even plants have political methods. For instance, if a leaf is wounded by being fed upon by a caterpillar, through subtle chemical changes it sends signals from cell to cell until the message reaches other leaves. Upon receiving the signal, another leaf readies its defenses by producing chemicals that retard an insect's digestion. If the insect stays on the leaf, slowed in its ability to eat and digest, it is more likely to be snapped up by a predator bird.

The plant kingdom is not free from war either. Ponderosa and lodgepole pine trees are subject to attack by mountain pine beetles, each of which is no larger than a grain of rice. The beetle will land on a pine and bore in, hoping to nest and reproduce. The tree will respond by bleeding sticky resin in order to stop the beetle; the resin is also laden with chemicals that poison the air and the beetles' brooding chambers. Beetles call in for support troops. They also, while eating the resin, convert it into an exotic perfume that excites and attracts other beetles by the thousands. While the beetles drill, the tree sends out its poisons. If the tree wins, the beetles move onward or die; if the beetles win, the tree dies, leaving the beetles free to raise their young.

In many instances wildlife must strategize in a defense against the threats imposed by humans.

Endangered Community

Although our lives depend on the survival of wildlife and unstressed nature cycles, we allow many wildlife foraging and breeding grounds to be sacrificed to the sprawl of our activities. Federal regulators will be asked to "compromise" and permit half of an area to be destroyed, although previously the habitat may have already been considerably reduced. Yet in the long run our needs are the same as those of wildlife: we are mutually harmed by global warming, ozone depletion, wetland destruction, and desertification.

The need to protect a regional network of forest preserves and wetlands is readily apparent when we understand how ecology works, how interdependent plants, animals, insects, birds, and trees are on each other for seed dispersal and food.

Rene Dubos' "Think globally, act locally" has become a motto, suggesting that we start with our own neighborhoods, getting to know the flora and fauna of our region as well as the condition of the water, soil, and threatened areas. Working with local conservationists and expanding the network of sanctuaries, privately as well as publicly owned, is an important long-term strategy. Regulations alone do not assure preservation. Ultimately, responsibility falls upon individuals, acting in their own locale.

Paul Ehrlich, Professor of Population Studies at Stanford University, and his wife Anne, a senior biology researcher at Stanford, have put forth their ideas for "an endangered community act":

> The first step in strengthening the act would be to broaden the focus in two directions away from "species." In one direction, more attention must be called to the problem of the extirpation of populations. Saving the last few mem-

bers of an endangered species may be important aesthetically and politically—as it is in the case of the California condor, an "umbrella species," whose presence in the wild helps to protect a very large area of habitat containing many other important organisms. But from the standpoints of the practical value of other organisms and the long-term survival of species, it is numerous genetically differentiated populations within species that must be protected. Without genetic diversity within species, not only is the continued existence of the species jeopardized but so is the species' potential for being developed into a commercially valuable domesticated one.

Since distinct populations of most organisms, especially invertebrates and plants, have not even been delimited, they will be virtually impossible to protect one by one. This leads to the other direction that modification of the Endangered Species Act should take—toward an "Endangered Community" or "Endangered Habitat" Act. That is, it should be deemed as law to maintain the maximum number of relatively undisturbed natural ecosystems of the United States in as pristine a condition as possible. That would automatically promote protection of both populations and ecosystems. . . .

Development of relatively undisturbed land should be forbidden unless the developer can demonstrate a critical, long-term, public benefit. . . .

A cactus wren or gila woodpecker in one's backyard is a wonderful bonus for not putting in that expensive and thirsty plantation of exotic grasses known as lawn. Indeed, many ecologists dream of the day when lawns are again a rarity in North America and homes are surrounded by native vegetation. That step alone would be a major stride towards preserving organic diversity in the United States.

An Endangered Habitat Act could also provide for restoration of habitat, even when that habitat is not occupied by a specific threatened organism. When slums are redeveloped to provide decent low-cost housing, for example, areas should set aside as open space to be planted in native vegetation. Programs should be established to restore stream courses both through cities and farmland to as natural a state as possible; water quality would be enhanced, floods would be less frequent, and important corridors for larger wildlife (and habitat for smaller organisms) would be provided. Obviously, such habitat restoration activities would cost money. But incentives such as tax credits could be applied or, in some situations, penalties for failure to make habitat enhancement part of redevelopment schemes that involve federal funding.

Certainly one of the largest political battles involving all life is war. Often the ecological consequences of war are ignored. Sometimes, it should be noted, wildlife can flourish because humans are more focused on eliminating each other than other creatures. Most agree, however, that the losses outweigh the gains. Below are two commentaries on the toll of war.

Long after the shooting has stopped, haunting images of the environmental devastation of the Persian Gulf war linger—funnels of black smoke billowing from oil-field infernos; oil slicks spreading for miles, fouling the gulf's once-teeming waters; the corpses of birds so drenched with oil they look like lumps of blackened sand on tar-ball beaches. Indeed, with the barrage of media coverage and speculation on the long-term effects of Saddam Hussein's tactics, the American public has been forced to confront—really for the first time—war's ecological toll.

The best lobsters in Asia used to be caught in the shallow water off the South Vietnamese town of Nha Trang, but along with so much else, they were done in by the war. The Nha Trang lobsters fell victim to the enduring principle that war is an activity in which normally rational people behave as if there were no tomorrow.

Before 1965 fishermen who worked the South China Sea set just enough traps to catch a few lobsters for local

(Continued on page 204)

(Continued from page 203)

restaurants catering to the carriage trade—provincial officials, French planters, foreign diplomats, a few tourists. Ordinarily the fishermen were too busy with their major catches to consider lobsters more than a luxury sideline.

But when legions of Americans began arriving in 1965, they delivered an economic jolt from which Vietnam never really recovered. In Nha Trang, appreciative GIs quickly discovered the renowned lobsters, and in a matter of weeks, the price of these delectable crustaceans escalated several thousand percent. Suddenly, lobsters meant big money, and to cash in, the fishermen abandoned their nets and traps, harvesting the lobster beds by exploding grenades and scooping up the casualties.

For six months the GIs gorged themselves. But then the lobsters were irretrievably gone, killed or driven from their breeding areas. Even fish avoided the Nha Trang area, and catches fell drastically. A decade later, when I boarded one of the last helicopters to leave Saigon, the Nha Trang lobsters were no more.

The loss of a lobster population, trivial in comparison with the grander tragedies of war, typifies a category of military insult to habitat I have seen in a half-dozen conflicts—an insult even more devastating than shells or bombs. It has nothing directly to do with combat. It is, rather, a result of the collective action of great masses of men and women for whom war is a dispensation to ignore normal restraints.

Eric Fischer:

While peacetime devastation is usually committed in the name of progress, war damage often results from direct attempts to thwart the enemy. The ten million gallons of the Agent Orange defoliant sprayed on Vietnam in the 1960s was justified as a way to decrease the ability of the enemy to hide and live off the land. The Iraqis purportedly intended the great oil spill during the Gulf War to foul Saudi desalinization plants and foil beachhead landings by U.S. Marines.

The inferno of oilfield fires in Kuwait, the petroleum-steeped coastal marshes and mangroves in the Persian Gulf, and the crushing of the desert's protective crust by armored vehicles show that the environmental effects of military confrontation often last well beyond the end of the conflict. And major unanticipated effects may occur in the aftermath of war, long after the last shots are fired.

The "Just Cause" invasion of Panama by U.S. troops in 1989 led to increased deforestation. The conflict occurred at the beginning of the burning season, the rainless time of year during which farmers clear land and most logging occurs. The invasion and its aftermath further weakened an economy severely damaged by U.S. sanctions. For many citizens low employment prospects in the cities made clearing land for agriculture a compelling alternative to city life. And the new Panamanian government could not protect the forests: The fighting destroyed vehicles and other equipment previously used by wardens to patrol protected areas, and the government had no funds to replace them. The result was devastating. . . .

To prevent environmental war damage in the future we cannot count solely on efforts to avoid armed conflicts. The causes of war run deep. It is a routine activity for both human and animal societies. Some coral colonies attack neighbors that impinge too closely. Social primates engage in raids and border disputes with neighboring groups. And ants and termites attack other colonies with the help of soldiers whose bodies are specially adapted for fighting and defense.

Anthropologists who wish to study peaceful human societies have a hard time finding them. Fewer than one society out of ten can be classified as peaceful, and the modern state is particularly bellicose. In this century there has been only one year of global peace, and on average, three wars are occurring at any given time. No matter how much we may want peace, we can expect many more wars before we find other ways to solve conflicts.

What can we do to guard against environmental damage from the wars that will inevitably occur? For nuclear war the probable outcomes are so horrific that prevention is the only feasible approach. Since the bombing of Hiroshima and Nagasaki, the behavior of states with nuclear weapons has exhibited a restraint that would be admirable if it were applied to conventional conflicts.

Dealing with the environmental impact must become an integral part of any war planning. Failure to do so may lead to additional loss of life from such problems as lack of food and potable water. Environmental organizations and government agencies should bring their expertise to bear.

We also must enhance environmental protection during the conflict itself. Like other human social activities, war has accepted rules and conventions. We need to find ways to make environmental protection a stronger part of the laws of war. The United States could take a significant first step by declaring support for provisions in the 1977 Protocols to the Geneva Conventions that outlaw wartime activities having severe environmental effects. However, these protocols are vague and purely admonitory, and stronger measures are needed.

Nations must hold each other accountable for the environmental devastation they inflict in war. Environmental damage should no more be accepted as a cost of waging war than pollution should be accepted as a cost of doing business.

Indeed, only when we begin to give these ideas their rightful place in the international ethos will we finally be moving toward making peace with nature.

How do ecosystems and their inhabitants respond to devastations, such as war and storms that cut a path of destruction?

THINKING LIKE AN

ECOSYSTEM

Violent Disturbances: Volcanoes, Earthquakes, Hurricanes, Floods, Fire

Ecosystems are vulnerable to destruction by violent acts of nature. Biologic patterns are set back. Some wildlife never recover, especially in estuaries. Some do adapt under certain conditions. We must remember that the most stable environments are generally the most diverse.

Earthquakes and volcanoes are the result of the Earth's internal processes. Stress can cause solid rock to fracture, producing a fault line. Abrupt movement along this line results in an earthquake. Humans have increased the likelihood of earthquakes by building dams which create heavy backloads of water, by underground nuclear testing, and by deep-well disposal of liquid wastes. Volcanic activity occurs where magma reaches Earth's surface through a central vent or a long crack. It can release debris ranging from large chunks of lava rock to ash, which may be glowing hot, liquid lava, and gases. It tends to happen in the same areas that have seismic activity and where there are plate boundaries. They may be scary to humans but produce beneficial effects: majestic mountains, lakes, geysers and hot springs, geothermal energy, and the rich fertilization of soil by the weathering of lava.

Near coastlines hurricanes, caused by circular air patterns, draw up large amounts of water and flood surrounding areas. Flooding has been the most common type of natural disaster because so many areas are susceptible to it. The main way humans have increased the probability of flooding is by removing vegetation—through logging, overgrazing, construction, and mining.

Fires and tornadoes that uproot trees clear patches of vegetation that flora and fauna can adapt to. When the disturbance happens frequently, organisms can develop resistance to it. It becomes part of the rhythm of a place, fixed in the genetic memory of its inhabitants. Sometimes when ecosystems are subject to spontaneous fires, they can recover if they are fundamentally sound and strong. Periodic cutbacks can be helpful because they clear and free land for vigorous new growth.

If the violent incidents are infrequent, adaptation is more difficult. There is always a critical point at which the damage becomes catastrophic, beyond what the organisms can bear, and the land becomes degraded, with only lower-grade species. For example, destruction of buffalo, sodbusting, and then excessive, chronic cattle grazing obliterated prairie culture. When the biodiversity of a region is destroyed, the gene pool is curtailed. Nature needs a large pool of genes to reweave genetic combinations. The more variety, the better nature can adapt to stresses and strains.

When ecosystems are devastated, the length of time needed for them to recover is quite uncertain. The slowness of genetic change and adaptation is a factor. So is the principle of succession, as species from an earlier stage of succession reenter an area first and dominate for a period. The rate at which succession occurs is unpredictable because of the variables involved as to when new members of species arrive and are able to establish populations.

THE EMPOWERED ACTIVIST

Blame and Guilt

Reflecting human nature, environmentalists range from aggressive to conservative in their views and actions. Greenpeace activists have risked their lives by placing themselves and their small boats between whales and the harpoon guns of Icelandic and Russian whaling ships. Earth Firsters practice civil disobedience—nonviolence toward living things—and strategic violence against inanimate objects, such as bulldozers and power lines. To divide groups into wrong or right is to perpetuate the habit of projecting onto others what we don't agree with. This blindness leads to a loss of "caritas," the caring for each other so sorely needed in our world.

An unfortunate result of pointing the finger of blame at corporations, homemakers, government—at everyone but ourselves—is that legislative and public meetings often end up as confrontations. In commenting on the use of blame in order to make others feel guilty, historian Ted Roszak said, "Shame has always been one of the worst and most unpredictable motivators in politics; it too easily laps over into resentment. Call people's entire way of life into question and what you are apt to get is defensive rigidity."

Although we all admire the uncompromising idealist who forces society to face the truth, outraged idealists can end up doing more harm than good. Locked into a refusal to consider the needs of the adversary, they sometimes prevent a dialogue from unfolding that could lead them to salvaging more of what they want. It is difficult to know where to draw the line. Brock Evans, distinguished lobbyist for the Sierra Club and later the Audubon Society, cites an example of how he faced this dilemma. Instinctively an uncompromiser, he said that his work lobbying for a Mount Saint Helens National Volcanic Monument of 216,000 acres forced him to learn differently.

> It became very, very apparent for a host of factors that we could never get anything close to that at all because the other side was talking about 45,000, or 80,000 at most. And so I took it on myself, after consulting with our local people, to say, "What must we have? What was going to be lost if we didn't get it? What must we do?" And we drew a boundary of 115,000 acres, which eliminated most of the problems of the other 100,000 and saved most of what we had to save, that we could feasibly get at this time. That's what I had to do, and it was painful to do it, because good places were left out. (The bill ended up with 110,000.) . . . We could have said, "No, we won't take it; we're going to fight for our 200,000 acres," and let it go four or five years, but we would have lost every place in between times. So that's the nature of the bitter choices you have to make.

Dialogue and negotiation are being tested by activists as supplementary tools to legislative combat. Industry leaders become more at ease when environmentalists are interested in market-based incentive solutions as opposed to reliance on regulatory enforcement. More frequently than in the past politicians, government representatives, environmentalists, and business leaders gather together to spell out their positions and see where flexibility can be found before bills are introduced.

As an example, General Motors and the Environmental Defense Fund formally agreed to hold discussions on their conflicts. One lawyer said, "Over the last twenty years, it's been each side preaching to the heathen on the other side. We've decided to skip the great conversion speeches and talk about what we can talk about." The agreement specifies that the EDF will not accept funds from GM and that GM cannot publicize the dialogue in advertising without permission. One result has been a program for scrapping old cars to help cut pollution.

Scientist Jan Beyea speaks about replacing the politics of blame with the politics of vision, which to him means giving people a practical vision that they can accomplish. He says:

> I look at the environmental movement as an ecosystem in itself, with lots of different organizations with different talents, different abilities, competing for resources and evolving. I believe there are certain stages to the process of change. The fact-finding stage. . . . You have to find out what the problems are and what their causes are the best you can. Then you have an educational stage, a consensus-building stage involving your constituency, and then the public education campaign to try to change laws or to change people's actions voluntarily. . . .
>
> We are all ignorant of the natural world. A little humility would go a long way. Occasionally there are glimpses of clarity, insights, that we are meant to hold on to in order not to do so much damage. . . .
>
> [In finding solutions to the environment, cities, income inequities, jobs] people are locked into yelling and screaming, saying "You do this, you suffer, you pay the price, you're to blame." We don't hear about trying to find ways out of the box rather than fighting in the box. . . . When you start conflict resolution, you find out a lot about your adversary and that there might be a door you hadn't realized was there. The goal of negotiated conflict resolution is to find new solutions whereby each side gets eighty to ninety per cent of what they want.

THE POLITICAL ECOSYSTEM

Environmental law has become so complex that it has become a profession in itself. It is an extremely important area because laws have tremendous impact on nature cycles. Here is an explanation of how our laws are made.

When we think of how federal laws are made, most of us picture the drama of a roll-call vote on the floor of Congress. But the real work of shaping legislation occurs in the congressional committees. Before a bill introduced by a member of Congress becomes law, it must survive the scrutiny of the appropriate committee (and sometimes several), where details are hammered out and compromises forged. The process is conducted in the formal committee rooms of the House and Senate office buildings, and is open to the public.

Because of the importance of the committees in determining the outcome of a bill, lobbyists tend to focus their efforts on committee members.

Activists whose representatives or senators sit on committees that deal with conservation issues are very important in helping pass good environmental legislation. Many members of Congress carefully consider what way the wind is blowing before casting a vote. By letting your representatives and senators know your opinions about a bill being considered in committee, you can influence the process.

The Key House Committees

HOUSE ENERGY AND COMMERCE

This committee has jurisdiction over a number of important environmental issues, including air pollution, energy production, and energy conservation.

HOUSE NATURAL RESOURCES

A broad spectrum of public lands issues, including those relating to wildlife and the use of natural resources, falls into the jurisdiction of House Natural Resources.

HOUSE PUBLIC WORKS AND TRANSPORTATION

This committee authorizes public works projects like dams, highways, and flood control projects. It also deals with pollution issues as they relate to navigable waters, and has jurisdiction over the Clean Water Act, including Section 404, which regulates wetlands alteration.

HOUSE MERCHANT MARINE AND FISHERIES

Key environmental jurisdiction of this committee includes marine wildlife, fisheries, coastal management, and wetlands.

HOUSE AGRICULTURE

House Agriculture has jurisdiction over a variety of agricultural issues, as well as pesticides, conservation, and forestry.

HOUSE FOREIGN AFFAIRS

The bulk of the responsibility for direct authorization of conservation projects in developing countries, such as in the Foreign Assistance Act, is under the jurisdiction of this committee.

The Key Senate Committees

SENATE ENVIRONMENT AND PUBLIC WORKS

This committee has jurisdiction over a wide range of environmental concerns: toxic and solid waste, air and water pollution, fish and wildlife, endangered species, wetlands, and ocean dumping.

SENATE ENERGY AND NATURAL RESOURCES

Senate Energy has jurisdiction over such areas as energy development and conservation; the mining of coal, oil, and gas; irrigation and water power; and national parks, forests, wilderness areas, and wild and scenic rivers.

SENATE AGRICULTURE, NUTRITION, AND FORESTRY

Senate Agriculture has jurisdiction over many issues relating to crop production and forestry.

SENATE FOREIGN RELATIONS

This committee oversees the Foreign Assistance Act and its various conservation-related issues.

A Committee Primer

There are sixteen Senate *standing* (permanently authorized) *committees* and twenty-two House standing committees. Most Senate committees have 14 to 20 members; most House committees have 34 to 45. The ratio of majority to minority party members is roughly the same as the ratio in each chamber.

Committees with environmental jurisdiction, such as House Interior and Insular Affairs, House Energy and Commerce, Senate Environment and Public Works, and Senate Energy and Natural Resources, are *authorizing committees*; that is, they are empowered to initiate a government program or activity. When such a committee approves legislation that authorizes $50 million for a new program in a specific year, it is setting a ceiling for spending for that program, but the legislation carries no guarantee that the program will get any money.

Actually committing money to a program is the province of the *appropriations committees,* which annually approve appropriations bills for all discretionary government programs (as opposed to trust funds such as Social Security). In theory, all programs that receive appropriations should have been previously authorized; in fact, many may continue to get money even though their authorizations have expired.

Authorizing committees receive bills that are within their jurisdiction; for example, House Interior and Senate Energy could get a bill to create a new national park. Once a bill is introduced, it can be acted on at any point during the two-year Congress. In most cases, a bill is first considered by the appropriate *subcommittee.*

If the subcommittee chairperson (or possibly other members of the subcommittee or full committee) is favorably disposed toward the bill, a *hearing* may be held. At a hearing on a bill, experts on the issue, representatives of groups supporting or opposing the legislation, and other interested parties present testimony before the committee or subcommittee members.

Thereafter, the subcommittee may *mark up* the bill (make changes, which can range from a few technical amendments to an entire new substitute bill) and approve it. If the bill is approved, a similar markup is then held in the full committee. The bill may then go to the floor for a vote by the full House or Senate.

Committee procedures vary considerably. Some committees, such as Senate Energy and Natural Resources, mark up all bills at the full committee level. This reduces the power of subcommittee chairs. Other committees, such as House Appropriations, tend to cede virtually complete control over bills to the subcommittees.

Committees also vary in style. House Interior, for example, tends to be combative, often approving legislation along a party-line vote. Other committees try to develop legislation with bipartisan support.

Bipartisan legislating is probably stronger in the Senate because of the different procedures followed in the two chambers in bringing bills to the floor for a vote. When minority House members are violently displeased with a bill, there is not much they can do to stop it from coming to the floor or to defeat it on the floor. But a single senator or a determined coalition of senators can sometimes prevent a bill from coming to the floor (by placing a *hold* on the bill, which can be an implied threat that a member may filibuster the bill if it comes up), or can filibuster a bill on the floor.

HOW A BILL BECOMES A LAW

(if introduced in the House)

House of Representatives

Introduction of Bill by Member
We will assume this is an appropriations bill so the Constitution specifies that it be introduced in the House

Referral to Standing Committee
by Leadership and Parliamentarian

Committee Action
Possible referral to subcommittee
Hearings on major bills common
Committee Decisions:
–Table –Defeat –Accept and report
–Amend and report –Rewrite

Calendar placement

Rules Committee (Major Bills)
Hearings to decide if bill will go to the floor earlier than the calendar date

Floor Action
- Reading, general debate
- Second reading
  Amendment(s)
  Report to the House
- Third reading
- Passage or defeat

Senate

Referral to Standing Committee
by Leadership and Parliamentarian

Committee Action
Possible referral to Subcommittee
Alternatives similar to those of the House

Calendar

Senate Floor Action
Alternatives similar to those of the House including rejection, acceptance or additional amendments

Conference Committee
If the Senate approves a bill that is not identical to the one passed in the House, a conference committee is requested. This committee consists of appointed members from both houses who compromise on a final version of the bill. This compromise version is then sent to each house for final approval.

Back to the Senate Floor
Bill signed by Speaker and Vice–President

President
- Approve
- Veto
- "Pocket Veto"
- Permit bill to become law without his or her signature

LAW

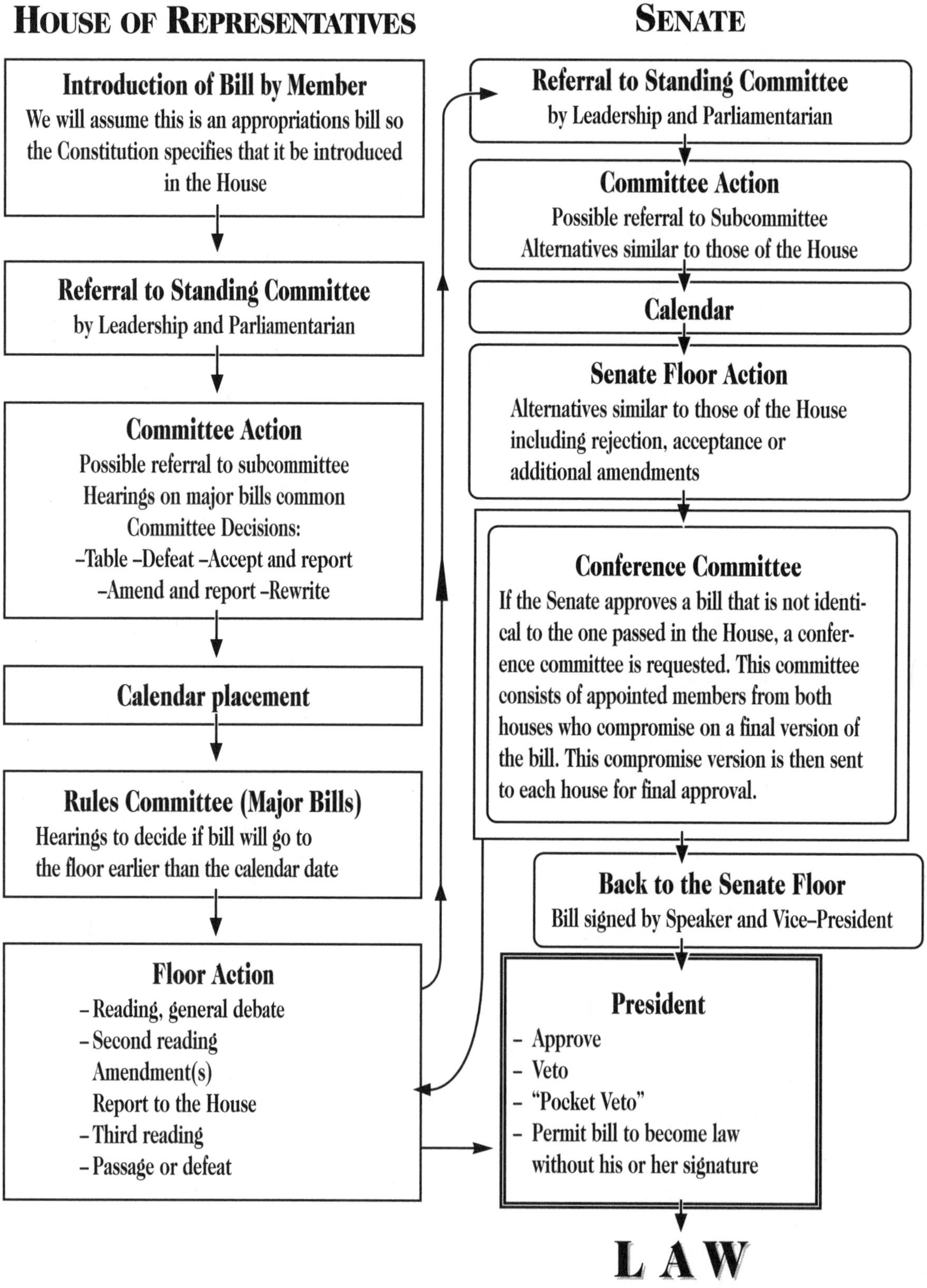

"Endless Pressure Endlessly Applied"

This phrase, voiced by Brock Evans years ago, has become the motto of many activists. Having lobbied for wilderness for decades, Evans described the special feelings that many activists share after long labor on behalf of a cause:

> One reward that came early in my career was in the North Cascades in Washington State, . . . mountains often called the Wilderness Alps, and very beautiful country indeed. Back in the 1950's and 60's the Forest Service . . . was embarked on a program of full-scale logging in all the valleys that we didn't think should be logged, opening up the area to mining. . . . A small band of people got the idea of creating a large national park and stopping that forever. As usual in the case of these issues, almost all the press and politicians were hostile to us and the public was basically apathetic.
>
> We just fought on. I have lots of memories of giving speeches in logging towns on rainy nights and getting hooted down, and lots of memories of licking, stuffing, stamping, and folding parties, and lots of memories of aches and tears when it all seemed lost and hopeless and that we never really could win. But, to make a long and very beautiful story short, finally after years of this kind of effort and struggle, we won. In 1968, Congress created a 700,000-acre North Cascades National Park; safe forever.
>
> It wasn't until about six months after the park was established that I had my first occasion to call up the superintendent of the new park. . . . [When] the receptionist picked up the phone and said, "North Cascades National Park," I held that phone and wept.

Media and Doublespeak

We have to constantly examine how well our media serve us. Most of us get our information from newspapers, radio, television, films, and computer networks. We also use media to broadcast our messages, and we are dependent on journalists reporting events at which we cannot be present. Journalists are supposed to stick to the facts, but often because of limited time and space, reports are reduced to soundbites. Insubstantial visual and aural moments may be misleading. So may the slant of the words.

Problems of adequate coverage are endemic to the profession. Lack of funding, terrorism, and inadequate scientific training may result in limited coverage. Media people tend to pick up a trend or hot political issue, only to drop it when they think their audience won't be interested. Since many ecological disasters come about very slowly, they may very well go unrecorded. TV commercials show environmentally damaging products, such as cars, surrounded by pristine mountains or lakes. Saving the rain forests or planting trees is sometimes marketed like a product, perpetuating the consumerist habits that are part of the problem. Feature films project images of Western landscapes that look magnificent but are trashed in the process of making the film.

To many of us the ecological habitats of rain forests, elephants, tigers, alligators, and rare birds are experienced on television, not in the wild. How can we possibly comprehend the textures, scents, sounds, and complexities of ecosystems on celluloid, while sitting on our couches? Bill McKibben, in writing about a year spent studying television programs, observed that we believe we live in an age of exploding information, but in many cases "we also live at a moment of deep ignorance, when vital knowledge that human beings have always possessed about who we are and where we live seems beyond our reach: we live in an age of missing information."

Nevertheless, stellar media activities abound. Ted Turner's development of Turner Broadcasting System and CNN to bring about the "global village" to help save the planet is certainly impressive. CNN, which enables eighty-nine countries to get the same news at the same time, is regularly watched by intelligence agencies and heads of state. Programs from the National Audubon Society, National Geographic, and the Cousteau Society have long been staples. Turner has also formed (along with Russ Peterson, Jean-Michel Cousteau, and Lester Brown) the Better World Society, dedicated to fostering public awareness and education on global issues that affect the sustainability of life on Earth.

Another innovative organization is the Earth Communications Office, a grass-roots, nonpartisan group, whose board of influential media leaders is led by Dr. Thomas Lovejoy, of the Smithsonian Institution. It focuses on inspiring Americans to attain measurable results in recycling, reducing energy consumption, and buying environmentally sound products.

The Environmental Media Association is the brainchild of Norman Lear and his wife. It brings environmental experts to meet with the creative staffs of motion picture and television studios to come up with ideas for characters, dialogue, and plots that dramatize the environmental crisis. When Roseanne or Murphy Brown discuss issues, they have tremendous influence on viewers.

Well and good, but what are we to do when the message communicated is confusing or even deceptive? For instance, in a political battle, what if one group says the environmental cause will cost people jobs and the other says it will create jobs, or one group says their way represents "wise use" and the other says "abuse"? It's important to have meaningful debate on the consequences of people's positions and to be cautious in deciphering the true intent of organizations, perhaps by looking at a group's board of directors and a breakdown of its spending.

About some of the attacks on environmentalists, Jay D. Hair, president of the National Wildlife Federation, said, "Look behind the fancy rhetoric and you realize this is a movement to bash environmental groups that is funded by the oil and gas industry, the mining industry and the development community." In order to resist unwanted change, industries can instill fear in people by threatening the loss of jobs. New jobs, however, may appear around the corner.

To nature lovers a disturbing trend is the so-called wise-use movement, which seeks to overturn and weaken environmental protection and open public lands and natural resources to economic development. The movement has gained significant numbers of supporters with its stress on development for human consumption. It appeals to individuals who believe environmental protection measures damage their short-term economic interests and therefore such measures should be weakened or eliminated.

Just as the "wise-use" name is misleading, so too are the names some groups choose to call themselves, such as People for the West! and the Blue Ribbon Coalition. What is termed the wise-use "movement" is actually an amorphous collection of independent organizations that share one or more common goals and may at times cooperate and support each financially or strategically. Some deny being part of the wise-use movement, although their policies and objectives are consistent with the thrust of the "wise-use agenda."

While the movement's goals are often couched in environmentally sensitive rhetoric and make strong appeals to traditional American values, in Audubon's opinion they would seriously threaten this country's ecological resources. The movement's platform includes the following goals as quoted from *The Wise-Use Agenda:*

- "Immediate wise development of the petroleum resources of the Arctic National Wildlife Refuge (ANWR) in Alaska as a model project showing careful development with full protection of environmental values."
- "Passage of the Global Warming Prevention Act to convert in a systematic manner all decaying and oxygen-using forest growth on the National Forests into young stands of oxygen-producing, carbon-dioxide-absorbing trees to help ameliorate the rate of global warming and prevent the greenhouse effect." [This suggestion is ecologically disastrous once you understand the necessary function of all irregular stages of a forest. —*author*]
- ". . . that all public lands including wilderness and national parks shall be open to mineral and energy production under wise-use technologies in the interest of domestic economies and in the interest of national security."

Other wise-use goals would:

- Severely weaken the Endangered Species Act by placing legal emphasis on program cost rather than species protection and by reclassifying some species as "relicts" [sic] not worthy of protection;
- Undertake a massive program of building concessions in the national park system;
- Reverse protection offered by the Wilderness Act by allowing "all commodity industry uses on an as-needed basis in times of high demand" on 10 million acres of the present wilderness system, and allow "motorized trail travel and limited commercial development" on another 30 million acres of current wilderness lands;
- Propose laws that would put the courts out of reach for the vast majority of citizen suits aimed at enforcing environmental statutes. Under what it terms "obstructionism liability," anyone who litigated "an economic action or development on federal lands" and lost in court would be required to pay the winner "the increase in costs for completing the project plus money damages for loss of economic opportunity."

The net result of the movement's agenda, however, would not benefit the average citizen, but would help industry increase short-term profits. Rather than encourage multiple use of public land, as its public relations campaigns claim, many "wise-use" proposals would allow ranchers, logging interests, and the mining industry to take precedence over recreational uses, such as hiking and fishing, on public land.

Policy Guide for the Future

In the final analysis we could use a general framework for handling political and ecological problems as they arise. Here are some pointers:

1. In general, support public officials who favor global environmental platforms. Join organizations that act as partners with nature. Try to make our politicians and business leaders understand the importance of biodiversity to the maintenence of all life on Earth. They should be aware that biodiversity is essential for perpetuating a large gene pool that will serve as the source for future adaptation, and that it is necessary for pharmaceuticals, fibers, sources of food and oil, and soil regeneration. They should realize that to let one supposedly insignificant species go under in order to carve out another mile of road is not acceptable.
2. In deciding on which side you stand in public debates, look for evidence of ecological studies. Watch out for arguments that have as their by-product the removal of regulatory protections and open up America's resources to industrial exploitation.
3. Remember, the people who suffer the most from pollution and degradation are poor ethnic groups. Incinerators, landfills, lead paint, and hazardous waste facilities are usually found in or near their neighborhoods because corporations figure the residents are too poor to legally oppose them. Tons of radioactive uranium have been dumped on Native American lands, and numerous Hispanic farm laborers suffer from illnesses related to pesticides. Municipalities often overlook how much setting up recycling programs could contribute to revitalizing a community by providing jobs and cleaner air and water.
4. Habitats need to be protected from fragmentation. We need to be aware of the importance of keeping networks open and accessible, just as our highways enable us to move from place to place. We must make room for migratory species to rest, feed, and breed in order to maintain their populations. Degraded areas must be restored.
5. We need to implement practices that enable us to live lightly on the land, that reduce our stress on supportive ecosystems. We must look for better options in energy, waste disposal, agriculture, population growth, and construction—subjects that are discussed in this book. Political change rests more on personal commitment than on money.
6. We can eliminate many problems by being more responsible about pollution at its source.

   I believe in the two-percent solution to things. It can be just too overwhelming to deal, for instance, with the whole pollution problem. On the other hand, if you think about doing it at two percent a year, that's doable. Although I believe in the incremental approach of the two-percent-a-year solution, you have to calculate and make sure you really are getting two percent and you really will reach your goal at the end. —Jan Beyea, Audubon's Chief Scientist

7. Adequate funding of scientific research is vital in planning ahead to determine how ecosystems will respond to global warming: for example, how we will deal with competition over water supplies, and how populations of species can be preserved in the face of this threat.

8. Scientists should be more helpful in predicting the consequences of governmental policies. Their reluctance to speak out can be used politically to permit ongoing destruction and ignorance, as in the U.S. government's foot-dragging on global warming and its policy of leaving wetlands open to attack. Environmentally oriented policy must be made even when total certainty of the results is not available.
9. Citizens can play the role of environmental watchdogs in their communities.

Local Hero Keeps Golf Course off Wetlands

In November 1988 the Environmental Protection Agency drew a line through a tract of coastal mangroves fifteen miles south of Miami. The owner of the land wanted to build a 253-home resort community, the centerpiece of which was to be a "championship" golf course designed by the Golden Bear himself—Jack Nicklaus. The developer was told not to cross east of the line into fragile wetlands.

Miami's Tropical Audubon Society had spent two years fighting the golf course. Inspired by the tireless efforts of Karsten Rist, conservation chair at the time, a legion of Auduboners and other activists had succeeded in halving the acreage of wetlands to be filled for development. Rist and the others saw the line as a reasonable compromise.

But by the end of 1989, the developer had declared he needed "only" eight acres east of the line to make a profit—the additional space was indispensable for the Nicklaus-designed course, he said. The Army Corps of Engineers relented and prepared to issue a Section 404 permit for the project. The state and county had caved in earlier, and the EPA stood on the sidelines. Nobody wanted to squabble over a few acres—except Rist.

Sitting behind his desk in the office of his modest plastics manufacturing business, surrounded by photographs of wildlife, Rist picks up the story in his gentle German accent. "I had gotten a copy of the developer's maps and they just did not seem to match the EPA's description of the development line," says the fifty-nine-year-old former mining engineer. "Kevin Sarsfield, president of Tropical, and I went back to the site with a 100-foot tape measure, located the EPA's red flags, and marked them off. Lo and behold, the line on the developer's map was drawn in error."

Rist notified the EPA of his discovery in February 1990. The agency asked the developer to draw a new map. It showed almost sixteen acres of golf course east of the line—enough to renew the EPA's interest. The project was temporarily halted.

The developer was forced to recognize the EPA's line. He bought more land west of the line and redesigned the golf course. Although about fifty acres of wetlands will be lost, an equivalent amount will be restored as compensation. The wetlands east of the line will eventually be turned over to the National Park Service as part of Biscayne National Park, which is adjacent to the area.

South Florida owes a great debt to Rist for his vigilance. The mangrove wetlands he saved are part of the vanishing Florida swampland vital to the state's wildlife. Herons, egrets, and ibis nest in mangrove swamps, and the Interior Department has identified some sixteen species of sport fish that depend on the swamps, as well as eighteen species of birds, reptiles, and mammals that are threatened or endangered. Not one to rest on his accomplishments, Rist has continued to monitor Florida wetland issues.

Quick Camera Work Halts Destruction of Michigan Marsh

Jack Smiley wasn't smiling when he heard that a local developer wanted to build a car wash on a cattail marsh in a Detroit suburb. As executive director of the Detroit Audubon Society, he had already seen too many of Michigan's wetlands erased from the landscape.

He registered his concern with the city council in Westland, Michigan, site of the wetland. He advised them that the marsh was regulated by the Michigan Department of Natural Resources under the state Wetlands Protection Act and that the developer needed a permit to fill it. But the city ignored Smiley's counsel and approved the developer's plans for a car wash.

Fearing that the marsh, covered in cattails and red osier dogwood and threaded by a small stream, would soon be gone, he borrowed his parents' video camera and headed for the site. He wanted to record the vegetation and wildlife. By the time he arrived, the bulldozer was already at work. "It was a shock and a surprise," he says. "They'd dredged most of it. They'd bulldozed a lot of the cattails and pushed a lot of water around."

Undaunted, he began to document on tape the filling of the wetland. He filmed the bulldozer scraping at the soggy earth and then interviewed the construction supervisor, who admitted the site was a wetland. "It won't be a wetlands for very long," he told Smiley on camera. "We're going to fill it in."

Smiley shot ten minutes of footage. He made a copy of the tape and took it immediately to the Department of Natural Resources, which has authority over wetlands in the state. After viewing the tape, agency officials ordered the developer to stop work at the site.

Eco Quiz

Q: Why should citizens monitor government and industry officials?

A: Citizens who oppose big polluting projects exert significant leverage in forcing officials to pursue better environmental policies. Strong opposition costs polluting industries money, time, and damage to their image and thus can ultimately make pollution prevention and waste reduction more economically attractive to government and industry.

Q: Why should some lands that are seldom wet be protected as wetlands?

A: Even wetlands that are wet only a few weeks of the year perform important biological and hydrological functions. The prairie potholes of North Dakota, the vernal wetlands of San Francisco Bay, and the bottomland hardwoods of the Gulf Coast region recharge groundwater supplies, filter pollutants from runoff, and provide essential habitat for migrating waterfowl and shorebird populations.

Q: What is a swampbuster?

A: After 1985, the term applied to a person who converts a wetland in order to produce an agricultural commodity.

Q: What is a conservation easement?

A: It allows a landowner to donate all or some development rights to a nonprofit charitable organization. The land can still be sold yet its important ecological features are protected.

Q: Will you take the Green Pledge for Earth Day 1990, as written by Denis Hayes?

Because our planet today faces severe environmental crises such as global warming, rain forest devastation, growing world population, and water and air pollution; because the planet's future depends on the commitment of every nation as well as every individual, I pledge to do my share in saving the planet by letting my concern for the environment shape HOW I...

ACT. I pledge to do my utmost to recycle, conserve energy, save water, use efficient transportation, and try to adopt a lifestyle as if every day were Earth Day.

PURCHASE. I pledge to buy and use those products least harmful to the environment. Moreover, I will do business with corporations that promote global environmental responsibility.

VOTE. I pledge to vote and support those candidates who demonstrate an abiding concern for the environment.

SUPPORT. I pledge to support the passage of local, state, and federal laws and international treaties that protect the environment.

E C O S Y S T E M   P A T H W A Y S

Have you ever envied nature for existing without money? Are birds, trees, and animals luckier or smarter to thrive solely on the cycles of the ecosphere than we who have organized systems of money? Because the ecosphere is sending numerous signs of endangerment, we must change the way we manage money and do business. Here is why:

- Our economic system has emphasized producing and consuming, while neglecting the decomposing and recycling phases that we see inherent in the way nature functions. In nature every waste discharge is a feedstock for another part of the ecosystem. Given our rapidly multiplying population, we are in danger of depleting our resources unless we plan to replenish what we take out.

- We think of "capital" and "wealth" in terms of dollars, but ecology teaches us that the most important wealth is the biodiversity of life. When a great number of species are destroyed, as is happening now, ecosystems decline visibly. The power of species to restore an area depends on a large enough gene pool to draw upon. As E. O. Wilson writes, "The richest ecosystems build slowly, over millions of years. . . . A panda or a sequoia represents a magnitude of evolution that comes along only rarely. . . . Such a creation is part of deep history, and the planet does not have the means nor we the time to see it repeated."

- Energy drives industry. The extraction, transportation, processing, and utilization of energy accounts for the largest disturbances to nature cycles.

- Commerce can be thought of as the extraction from, processing, packaging, and selling of the environment. Without environmental security, we cannot have economic security.

- Our Gross National Product is misleading in terms of economic well-being. While the GNP in the United States has grown, the slice per person has shrunk. The GNP index does not account for the depletion or degradation of natural resources. It can hide harmful effects. For instance, cigarettes and terribly wasteful and polluting accidents, such as *Exxon Valdez* and Three Mile Island, increase the GNP because of the flow of cash generated in cleanup and legalities.

- Many, many people are unhappy in their lifestyles. Psychologists assert that happiness is related not to consuming more and more products, but in having meaningful jobs, love and friendship, and a sense of belonging to a community or place.

- The implications of poverty are ignored. Poor people lack access to the land or jobs they need to pay their way. Many of them, it should be noted, have learned how to recycle well, since they have been forced to reuse materials and not waste a thing. Being poor leads to poorer health, though, which decreases the ability to work and results in more poverty. Persons trapped in degrading conditions feel powerless and are the first to turn to crime and addiction. All over the world, when humans are faced with starvation, they are forced to overexploit their vital resource base. Thus, poverty becomes an environmental concern for us all. Such poor people should help construct the programs needed in their own countries to end the destruction of resources.

To achieve a sustainable world, we must make a population transition to low birthrates and death rates; emphasize high-efficiency and renewable energy sources; recycle and restore degraded areas; require environmental audits on products used from cradle to grave; protect forests, wetlands, grasslands, soil, wildlife, and aquatic ecosystems; and preserve biological diversity at local, national, and global levels.

15

BUSINESS AND MONEY

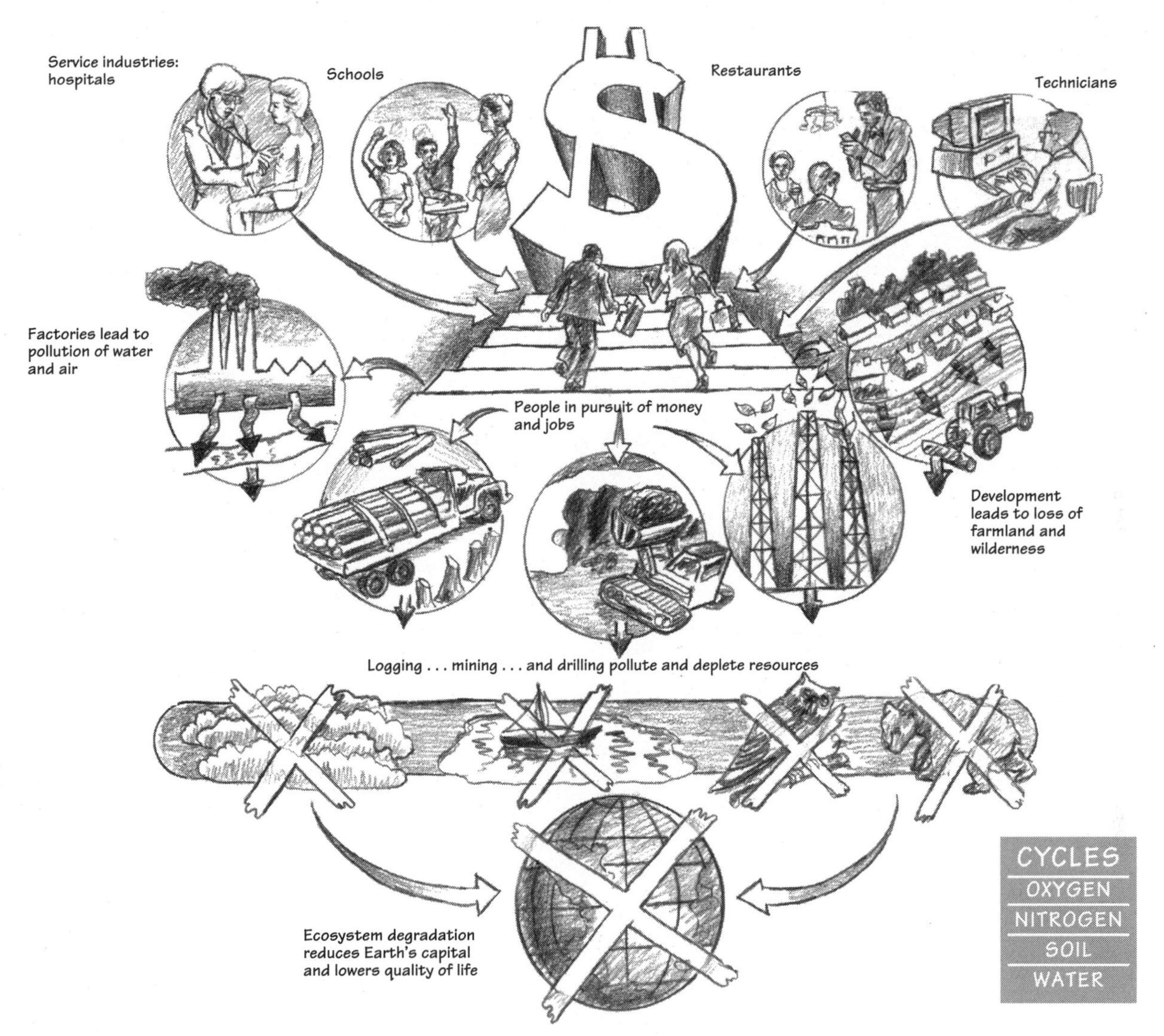

A NEW ECONOMIC ETHIC

A new ethic that weds business and conservation is on the horizon. It involves you and the way you spend money, as well as companies (which are big groups of people) and their practices. We voice our good intentions, but all of us are busy and want to pay our bills. But keeping in mind that producers depend on consumers, and vice versa, perhaps we can make headway together.

First, some basics. The economy is basically a system of production, distribution, and consumption of products and resources, designed to meet our needs for food, clothing, water, shelter, health care, recreation, and education. The difference between our "needs" and "wants" is largely a function of habit and culture. The goal of advertising is to increase our "wants," not to inform.

The economy depends on three basic factors: (1) raw materials, for example, natural nutrients, energy sources, minerals, plants, and animals, (2) capital equipment, for example, manufacturing machinery, equipment, buildings, and distribution facilities, and (3) labor. In a purely capitalistic economy, the market determines how

THINKING LIKE AN

ECOSYSTEM

Producers, Consumers, Decomposers

Organisms that make up an ecosystem are usually classified as producers, consumers, or decomposers, depending on how they get the food they need to survive. Producers can manufacture the organic compounds they need. In most terrestrial ecosystems, green plants are the producers. In the water, most producers are phytoplankton. Only producers make their own food. All other organisms are consumers, and live directly or indirectly on the food provided by producers.

Decomposers, which feed on the dead, are paradoxically the life-givers of the planet. Microorganisms sheltered in roots not only break down and make ammonia, nitrogen, and phosphorus, but also can break down some forms of industrial chemicals and pesticides into simple compounds that plants can absorb. Without decomposers, litter, wastes, and garbage would not be made available for new uses. Ecosystems are dependent on this cycling between death and life. Barren lands can be restored at nature's own pace of production, consumption, and decomposition.

An amazing example of regeneration is the story of an island in the East Indies, which in 1883 was sterilized and reduced to ash by a volcano equal to a 10,000-megaton H-bomb. Nine months after the eruption the only sign of life was a single spider spinning a web. The island was twenty-five miles from the nearest source of life, but after three years eleven species of ferns and fifteen flowering plants had arrived. After ten years a carpet of green covered the scarred land. Twenty-five years later 263 species of animals—mostly insects, birds, and reptiles—had made their way to the island. Fifty years later a dense forest covered the island. Yet even then, the species' populations were fluctuating wildly. In some years rats overran the vegetation, then disappeared.

As with ecosystems, our economic growth experiences wide fluctuations, is a mixture of competition and symbiosis, and depends on the trinity of consumers, producers, and decomposers working in concert.

these factors interact and who survives (if you can't produce cheaply enough, you perish). In a purely command economy, such as communism, the government allocates resources. If the economy is based solely on free enterprise, there is no obligation to society: the aim is to make the most profit in the shortest time; profit alone is the goal unless stockholders demand that the company establish other priorities. A more realistic approach, which has become the norm, is a free-enterprise economy regulated to some extent for the public good.

But how much regulation? Viewpoints on how to proceed vary, of course. In general, one approach is that there is an infinite supply of materials on the Earth and an infinite sink for wastes; this viewpoint places no value on cycles of nature or ecosystems: technology will find alternatives when a resource is exhausted. Among those who want to redirect economic growth so that it supports the ecosphere, a common view is that technology and people are clever enough to solve any problem involving the Earth without waiting for the exhaustion of resources. The purer ecological view is to see the economy as totally dependent on resources provided by the sun and the planet's natural processes and to make sure that business is conducted in such a way that resources are replenished as fast as they are subtracted.

Q: What do ecologists want?
A: Room for all species to do their dance.

If the GNP continues to grow at 2% per year, assuming growth as it has been, in fifty years we will have three times today's amount of pollution. Individuals and corporations would have to reduce their environmental impact by a factor of 10, just to stay even with the level of disturbance to ecosystems that we now have and accounting for other countries catching up with our lifestyle. Since corporations are larger entities than households, they should be required to carry more of the responsibility. So how can we do better than staying even or slipping behind?

New laws will be made. It is hard to foresee the effect they will have, but it is important to assess the potential consequences in advance as best we can.

A crucial issue is how society spends its money. If we place all our bets on technological solutions, what happens if they fail? What will the costs be, and how will we pay them? Don't we want insurance against risking our lives, or that of the planet, in case technologies cause unexpected damage, as they have in the past? Can we integrate the business of technology with the goals of conservation so that the world is kept in balance?

The nuclear industry is a good example of unexpected consequences. It was rushed onto the scene without concern for its side effects, and received a great deal of investment and development prior to consideration of its safety or waste disposal problem. Only a few decades after its introduction it suffered disfavor and decline. Technologies can be made to be part of the solution and not the pollution if we are cautious enough to see that they are not unleashed upon the world without environmental accountability.

In making the transition to a new kind of economy, businesses usually go through several stages. First, pioneers, who want to change the way things are, are foolish or daring enough to try out new procedures. Then, more established companies see that things can in fact change and follow the lead. Finally, regulators step in and make the remaining businesses come along. The food industry provides numerous examples. It was not so long ago that demand for organic foods led to "health food" stores. Now most supermarkets carry leading organic brands, while traditional brands scramble to catch up with the latest trends.

John and Nancy Todd were two pioneers. In 1969 they founded the New Alchemy Institute, a research and education center focusing on organic agriculture and sustainable living, and more recently they began Ocean Arks International, which focuses on using the principles of ecology to purify water. They have promoted the use of living organisms assembled within structures to produce fuels and food and to treat wastes, purify air, and regulate climates. When asked what has sustained them through the frustrations of the decades, Nancy says, "When you work with the *transforming* energy, it's as though life isn't worth living unless you're on its side. When you have work and a cause you believe in, it is such a profound and satisfying way to live."

Another pioneer is Anita Roddick, who started the Body Shop. Roddick figured out a network in which people who franchised her stores could be environmental activists as well as earn a living. The products could help indigenous people live sustainably in the rain forests. She blazed a trail in "doing well by doing good." Here, from her book, are her standards:

> I am still looking for the modern-day equivalent of those Quakers who ran successful businesses, made money because they offered honest products

and treated their people decently, worked hard, spent honestly, saved honestly, gave honest value for money, put back more than they took out and told no lies. This business creed, sadly, seems long forgotten.

I have no intention or desire to stack up a pile of accumulated wealth, which goes on and on, ad infinitum, for generation after generation. We believe it would be obscene to die rich and we intend to die poor by giving away all our personal wealth, through a foundation of some kind.

Jobs to Save the Earth

Polls show that many people would feel much better if they had jobs that did not degrade the Earth or deplete its resources. Although workers suffer when an industry in transition cuts back on jobs, it is not true that in the long run environmental regulations lead to more jobs lost than gained. In an evolving economy, new employment opportunities emerge. Let us remember that many of today's jobs did not exist 100 years ago and, presumably, will be different 100 years from now.

Some careers that conservation opens up are:

sustainable forestry
range management
parks and recreation
air and water quality control
solid waste management
hazardous-waste management
urban and rural landscape planning
soil and water conservation
fishery and wildlife conservation
environmental education
environmental health, geology, ecology, biology
environmental chemistry
climatology
population dynamics
environmental law, journalism, engineering
design and architecture
marketing of sustainable products
business related to cleanup and reuse of materials
politics with environmental platforms

In addition to creating jobs, protecting our natural resources preserves the foundation for economic progress that will enable the United States to remain a leader in the international marketplace. Policies, technologies, and processes that prevent pollution at the source offer numerous business opportunities.

An example is the coaltion of labor, industry, and environmentalists that formed to promote a Clean Water/Jobs platform. The program grew out of efforts to clean up Long Island Sound, and calls for a sewage treatment infrastructure in national estuaries and the Great Lakes. Billions of dollars invested in public works projects will provide jobs to the unemployed.

Green Design and Pricing

An essential component of a conservation economy would be "green design," or the designing of products made with recycled materials and manufactured without harmful chemicals. Engineers could be educated ecologically so that they would design safer products from the beginning. Pollution would be seen as an unfinished and imperfect aspect of design.

> If individuals were personally liable for the half billion tons of toxic chemicals and the 11,000 different organochlorines (for example, dioxins and PCBs) produced by chemical companies every year, would they still sell them? Would they ship them without foreign-language labels to Third World countries, where they end up poisoning children, workers, and watersheds?
>
> I would suggest that no company be allowed to make a manufactured product unless it is willing to take it back—all of it. . . . We don't consume what we produce. That is why our landfills are full, our water is polluted, our skies are acidified.
>
> A principle of good industrial and ecological design should be that every appliance, car, and file cabinet be assembled in such a way as to be completely reusable. When you buy a new refrigerator, the manufacturer would be responsible for reclaiming it when you are done with it, and for remanufacturing and reusing it so that there is no waste. If you have twenty televisions in a truck, you are technically a toxic-waste hauler by Environmental Protection Agency standards and require a license. The one and a half ounces of mercury in every TV picture tube should belong to Sony, not you or me or our landfill.
>
> —Paul Hawken, the founder of Erewhon, one of the first natural foods companies, and Smith & Hawken, the horticultural catalog company

Another component of a conservation economy would be "green pricing." This would enable consumers to have a direct role in influencing decisions. For instance, they would be offered an optional electricity product that uses a renewable rather than a fossil-based mix. The rate is likely to be higher at first, but public opinion polls show that many consumers are willing to pay higher electricity bills in return for greater reliance on cleaner and safer generating technologies.

Bringing economics and ecology together means figuring all environmental-damage and health costs for the whole of its life cycle into the price of a product. As mentioned earlier, the GNP rate does not reflect depletion or degradation of resources. Government action is required because businesses find it unprofitable to volunteer to include "externalities" on their own, when the guy next door doesn't.

The inventor of the socially responsible credit card idea, Peter Barnes, suggests, "By act of Congress, a new line should be inserted in business and financial statements just below the tax line and above the bottom line. This line would be called the E-line, for 'externalities' or for 'earth.' It would incorporate social and environmental costs. It would also reward businesses for performing social and environmental tasks."

Take the price of a car, for example. Traditionally, it includes factory costs, raw materials, labor, marketing, shipping, and company and dealer profits. Also, consumers pay for gas and maintenance once they own the car. The pollution caused by driving the car causes health costs to rise, as well as increasing town and city costs for cleanup and protection. These costs are not paid for. If they were included in the cost of making a car, its price would rise, but not necessarily if efficient techniques were incorporated upfront. Manufacturers would end up competing to reduce pollution damage. If environmental damage costs were included in the price, consumers would be able to make more informed decisions about the impact of their lifestyle on the planet's life-support systems.

Market forces have been engines of economic progress, but sustainable growth requires that we try to employ them to create incentives for environmental improvement. Controlling pollution via tradable emissions permits is one way. This approach has been used to reduce both CFCs and sulfur dioxide, and could be used to cut carbon dioxide emissions. The way it works is that the government specifies an overall level of pollution that will be tolerated. This total quantity is allotted in the form of permits among polluting companies. Companies that can control pollution most cheaply will be able to sell permits to those with higher costs and will therefore do most of the control. Everyone has an incentive to invent low-cost control technologies in order to sell permits. So far businesses have not made extensive use of the Emissions Trading Program, partly because states are not requiring them, and there is uncertainty about their future.

Another way to get the market to work in favor of the environment is to replace subsidies with taxes on polluters and depleters, then reward businesses that practice conservation, waste reduction, recycling, and pollution prevention.

It takes leadership to make environmental considerations and concerns part of design and pricing right from the start. But, if consumers and investors exercise their considerable powers, visionaries tell us that a sustainable economy could be established in twenty years. Moreover, it could be done by using the profit motive and self-interest to work for the Earth, not against it.

Business Accountability

In 1969 the passage of the National Environmental Policy Act (NEPA) declared that the federal government has a responsibility to restore and maintain environmental quality, and that all federal agencies are required to file an Environmental Impact Statement (EIS) or assessment for any proposed legislation or project that has a significant effect on environmental quality. The EIS must include short- and long-term effects, backup references, and objections raised by reviewers. It is also possible for private citizens to request to see the files on any such project.

Problems exist in carrying out the intention of the Act. An EIS can be ineffective if it is issued after a project has already been started instead of in the planning phase. It can be weakened by the use of obscure, meaningless generalities; inadequate preparatory studies; and by deliberately using the process to delay and raise the costs of legitimate projects.

In 1988 the Coalition for Environmentally Responsible Economies (CERES) put together a guide (originally known as the Valdez—now CERES—Principles) for individuals, industry, and the government. Included in sponsorship were the National Audubon Society, the Sierra Club, public pension trustees, the U.S. Public Interest Research Group, and the AFL-CIO Industrial Union Department. Aimed at corporations that once were blamed for most environmental insults but that were now seen as a critical part of the solution, the Coalition's objectives were to encourage industries to work toward improving their environmental performance in the areas defined by the Principles. They have been asked to make the following pledge:

THE CERES PRINCIPLES

By adopting these Principles, we publicly affirm our belief that corporations and their shareholders have a direct responsibility for the environment. We believe that corporations must conduct their business as responsible stewards of the environment and seek profits only in a manner that leaves the Earth healthy and safe. We believe that corporations must not compromise the ability of future generations to sustain their needs.

We recognize this to be a long-term commitment to update our practices continually in light of advances in technology and new understandings in health and environmental science. We intend to make consistent, measurable progress in implementing these Principles and to apply them wherever we operate throughout the world.

1. Protection of the Biosphere

WE WILL minimize and strive to eliminate the release of any pollutant that may cause environmental damage to the air, water, or Earth or its inhabitants. We will safeguard habitats in rivers, lakes, wetlands, coastal zones and oceans and will minimize contributing to the greenhouse effect, depletion of the ozone layer, acid rain or smog.

2. Sustainable Use of Natural Resources

WE WILL make sustainable use of renewable natural resources, such as water, soils, and forests. We will conserve nonrenewable natural resources through efficient use and careful planning. We will protect wildlife habitat, open spaces and wilderness, while preserving biodiversity.

3. Reduction and Disposal of Wastes

WE WILL minimize the creation of waste, especially hazardous waste, and wherever possible recycle materials. We will dispose of all wastes through safe and responsible methods.

4. Wise Use of Energy

WE WILL make every effort to use environmentally safe and sustainable energy sources to meet our needs. We will invest in improved energy efficiency and conservation in our operations. We will maximize the energy efficiency of the products we produce and sell.

5. Risk Reduction

WE WILL minimize the environmental, health and safety risks to our employees and the communities in which we operate by employing safe technologies and operating procedures and by being constantly prepared for emergencies.

6. Marketing of Safe Products and Services

WE WILL sell products or services that minimize adverse environmental impacts and that are safe as consumers commonly use them. We will inform consumers of the environmental impacts of our products or services.

7. Damage Compensation

WE WILL take responsibility for any harm we cause to the environment by making every effort to fully restore the environment and to compensate those persons who are adversely affected.

8. Disclosure

WE WILL disclose to our employees and to the public incidents relating to our operations that cause environmental harm or pose health or safety hazards. We will disclose potential environmental, health or safety hazards posed by our operations, and we will not take any action against employees who report any condition that creates a danger to the environment or poses health and safety hazards.

9. Environmental Directors and Managers

WE WILL commit management resources to implement the CERES Principles, to monitor and report upon our implementation efforts, and to sustain a process to ensure that the Board of Directors and Chief Executive Officer are kept informed of and are fully responsible for all environmental matters. We will establish a Committee of the Board of Directors with responsibility for environmental affairs. At least one member of the Board of Directors will be a person qualified to represent environmental interests to come before the company.

10. Assessment and Annual Audit

WE WILL conduct and make public an annual self-evaluation of our progress in implementing these Principles and in complying with applicable laws and regulations throughout our worldwide operations. We will work toward the timely creation of independent environmental audit procedures which we will complete annually and make available to the public.

For the most part privately owned companies have signed on or conducted environmental audits. Pressure to get the big corporations to cooperate needs to come from those who sit on their boards, own their stocks, and buy their products.

While the creators of the CERES Principles did not want to resort to cost-benefit analysis to argue for the adoption of the principles, there are savings to be made. Reducing the use of toxic resources, recycling, and reusing raw materials lessens the creation and disposal of wastes and improves overall efficiency. Anticipating governmental regulation is proving cost-effective in avoiding legal battles. Since it is estimated that we spend over $100 billion annually to comply with regulations, many corporations are seeing such investments to be profitable over the long term.

Who Owns Biodiversity (i.e., Life)?

> Present day economics does not and cannot take into account the value and diversity of biological wealth. . . . Economists can assist in the bioeconomic analysis of ecosystems, but they should avoid fruitless and dangerous exercises in cost-benefit analysis, which is far beyond their reach. They should instead join in the ethic of saving every scrap of biological diversity possible, the approach of the so-called safe minimum standard, until humanity figures out how to understand and evaluate this most valuable part of its natural heritage. Anything else would be tragic and, in the deepest and purest sense of the word, unethical.
>
> —Pulitzer prize–winning biologist E. O. Wilson

Biological diversity—complex beyond understanding and valuable beyond measure—is of fundamental importance to all ecosystems and economies. At least half the planet's species reside in tropical forests. Brazil, for instance, has more tropical forest—and probably more species—than any other nation. The range of products—for food, medicine, and cosmetics—hidden in forests, reefs, and other ecosystems is a powerful argument for their conservation. Surely, however, the removal of commercially driven quantities of fruits, nuts, and oil seeds will impact animal populations; continual harvesting of the best leaves and fruits will leave only the inferior species to survive over time.

We don't know enough about wise management to ensure against such losses to biodiversity. Commercial conservation has been known to be extremely hazardous to ecosystems. Attempts to farm sea turtles or other wildlife or to exploit whole ecosystems invariably introduces unforeseen problems. Ecologists must be brought in to determine what is a sustainable commercial pursuit. It is too risky to let business alone decide what is to be saved and lost.

The tensions between the highly industrialized continent of North America and the developing countries of South America are based on issues of what belongs to regional economies and what belongs to the good of the world. The United States, which has a higher rate of consumption per capita than people in the South, is perceived as dictating to the South about reducing its consumption of forests. South Americans think that the North wants to turn developing nations into parks where forests and biological diversity are preserved, keeping the local citizenry in poverty. They believe that international treaties infringe on their rights to use the natural resources within their borders.

Population control is another bone of contention. Northerners believe the population explosion in the Southern Hemisphere will keep living standards there low and cause environmental degradation. Southerners say that the North's high rate of consumption stresses global resources more. But, while some passengers on the planetary boat have been traveling first class, and others in steerage, if the boat sinks, we will all drown together.

Species and habitat preservation could be enhanced by an international agreement to create property rights in wild genetic resources for countries in which they reside. This would give rain forest countries more incentive to sustain rather than mine their forests. Developing fair (not free) trade practices is a goal of conservation economists.

Here is an example of the commercial benefits derived from the protection of biodiversity:

> An example of agricultural crop products and the importance of biodiversity is the U.S. grape industry. Wild species of domesticated crops are very vital for the genetic involvement of domesticated species. In 1860 plant lice were accidently introduced into Europe, and the entire grape-growing regions throughout Europe were in danger of ruin. The French grape industry was saved by native American graperoot stock, and eventually the old French varieties were crossed with the disease-resistant American grapes to develop nearly immune hybrids. 95% of European wine grapes are grown on root stocks of the wild N. American grape species, and the annual value of the U.S. root stock to the European wine industry is $6.2 billion. Money talks and it is important to maintain those wild species. In the U.S. in 1990 the grape wine industry added $1.66 billion to our economy.
>
> —Claudine Schneider, former Congresswoman from Rhode Island, director of the Artemis Project, which is documenting the financial value of biodiversity, and Energia Global, a value-driven company marketing appropriate energy technologies in lesser-developed countries

However, one of the more alarming complexities is the world market in the gene pool. Corporations scramble to grab patents in a Gold Rush fever, and they will soon be in control of the biological wealth. Although making patents available to free enterprise provides an incentive for research, do we want the future directions for life to be in the hands of corporate directors? We must decide to what extent we will permit gene-patenting, and, if permitted, how we will regulate it.

Another problem devastating to biodiversity is that in traditional economics future resources are held less valuable than current ones. On spreadsheets the value of resources is discounted in the future. Oil, natural gas, clean water, and healthy forests can be considered without significant value if conserved for more than a few decades. This practice obviously encourages perilous consumption today rather than conserving resources for tomorrow.

YOUR LIFESTYLE

We have discussed what companies can do to curtail heedless destruction of the Earth. Now to your part. Although people have been admonished for decades to lead a simpler lifestyle, the problem is that the message doesn't take hold. Our government is now faced with reducing the national debt,

The use of money is all the advantage there is in having money.

—Benjamin Franklin

which has kept growing, along with Earth-plundering, and is forcing us to make reductions. Everyone wants to lower the debt, but when it comes to concrete measures, resistance arises.

Will your personal debt induce you to adopt a lifestyle that is lighter on the planet? Money is the crux for us as well as for business and government.

The consumptive, throwaway lifestyle is a trap that leaves us, as well as nature, the loser. Poorly served by inefficient and costly appliances, cars, and homes, we know the anguish of getting little for our dollar. Most of us resort to credit cards to pay for "things," and most of us cannot keep up with the monthly balance. Many of us are severely in debt. Others cannot afford the big expenses of health, college for our children, income in old age. Some are trapped in jobs that pay below or near the poverty line. Families today often require two earners just to get by, and single parents slide onto welfare that they hate.

Because people like us make this economic system happen, we can turn it around, for the system depends on the symbiotic connection between producer and consumer. We householders buy products that companies make, and companies buy natural resources, equipment, and labor to supply these products. Currently, when we go into a store and buy food, the product we get is likely to contribute to the loss of topsoil, to groundwater depletion or contamination, to stream destruction, and to higher health costs for farmers. If manufacturers and sellers are not made responsible for these effects, the cost is handed to us one way or another. In order to make the environmental improvements we all seek, we are going to have to eliminate the wasteful aspects of our own lifestyles.

We can change the system—with our voices and money. Instead of being the victims of endless consumption and debt, let us use our money for real value.

When you go into a store, demand healthful foods and lasting hardware. Read labels carefully. Use your power as a consumer to demand better products and less packaging. (Some people send wrappings back to the manufacturer.)

Our personal budgets—as well as state and national—are signaling us to cut back and simplify our appetites. We can do it by making sure our money goes for what we value. As the chapters in this book have shown, when we spend our money, every item has pathways to ecosystems somewhere. Think of them wherever you go. Think of how you can cause less stress on nature and less financial stress for yourself, your community, country, and world. You have the opportunity to break out of the present system, get out of debt, and live a more environmentally benign lifestyle.

Let us make saving the Earth our personal mission. Too many of us have felt intimidated by bulky multinational corporations. We can use the tools of rejection, boycott, and organizing. We can be conscientious and demand better performance and service. Remember, money casts a vote every time we open our wallets to buy a product, make an investment, or contribute to a cause. Every financial transaction we make has a direct connection to values promoted throughout the world.

Using credit cards that donate a percentage to conservation is one way. You can make a gift of cash or securities that will pay you a lifetime income and provide generous tax savings.

One way you can cast votes for a sustainable world is by investing in companies that are planning to make the world a better place for generations to come. There are now more than a dozen "social conscience" mutual funds that pick securities according to a company's record in protecting the environment, or based on other ethical or social criteria. Some even donate their profits to environmental groups. Here is a list of well-known funds and their returns for a period of three years ending March 31, 1992.

Calvert-Ariel Growth, 9.5%
Calvert-Managed Growth, 10.5%
Dreyfus Third Century, 15%
New Alternatives, 10.7%
Parnassus, 9.4%
Pax World, 16.6%

You better believe that money talks. Any-size investment makes a positive difference. According to *Fortune* magazine and other monitors, business executives and investment bankers are especially sensitive to trends in the flow of cash that funds new ventures, products, and plant expansion. When industry leaders observe investors taking a serious view of the environmental records of companies, they respond rapidly.

A number of financial books offer advice on investments and companies. One of the best known is *The Better World Investment Guide*, put out by the Council on Economic Priorities, which also provides detailed reports on the environmental practices and policies of companies. Among other things, these fascinating reports list summonses for violations a company has received and instances of health risks to workers. They include appendixes that discuss issues pertinent to the industry. Here is a sample graph that compares the toxic releases in the forest/paper industry.

FOREST PRODUCTS INDUSTRY, TOXIC RELEASES—RAW INDUSTRY DATA
(millions of pounds of releases)
The Toxic Release Inventory is compiled annually by EPA. Data is self-reported by companies. "0" represents no pollution.

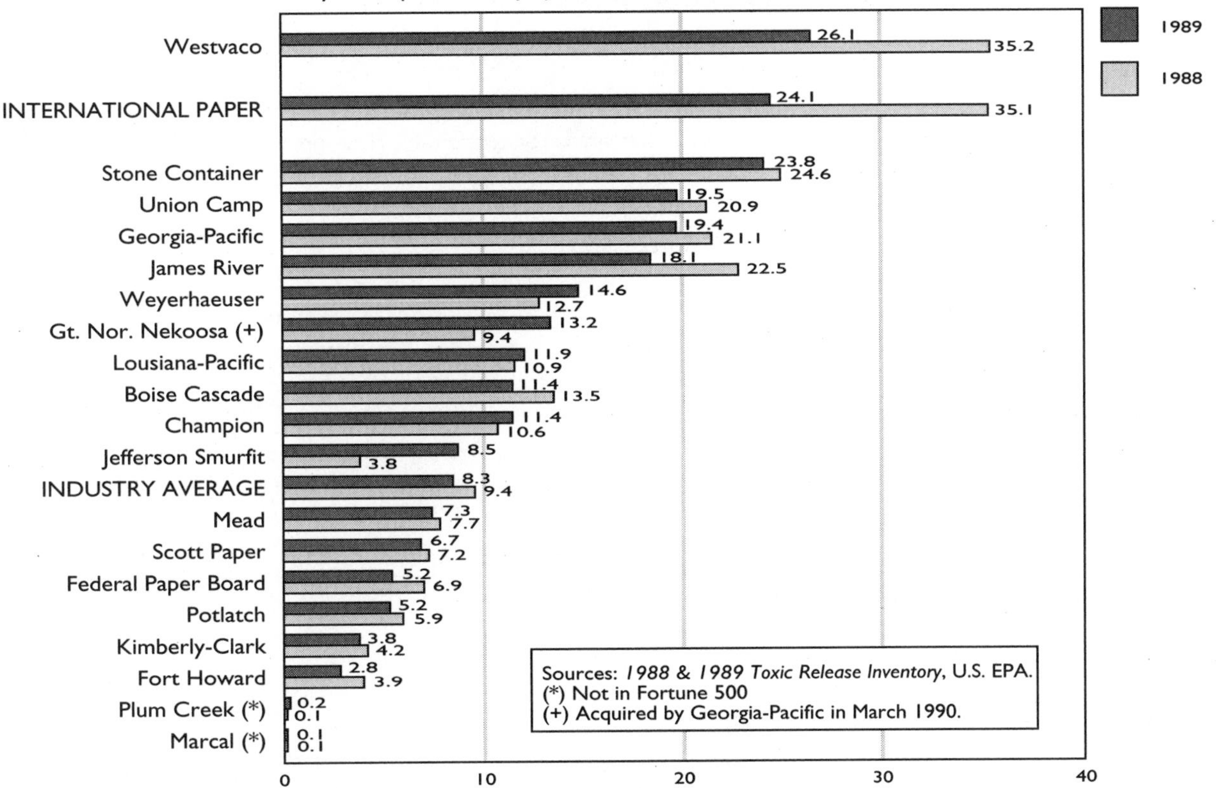

Besides how you spend your money, another factor to think about is where you live. Many of us are no longer rooted in a place, and as a result fail to care for the areas we inhabit. Every place on Earth is made up of ecosystems to cherish, every region has flora and fauna native to it. Some conservationists believe that if we had an economy based more on regional food, building materials, and clothing, we would stress the environment less by eliminating transportation costs. Recently some people have created business opportunities in food co-ops and farms that grow specialty foods for the region.

Some architects are building communities that conserve resources and promote human interaction. Cohousing projects are popping up all over the United States. The basic concept is the building of homes around a central dining/recreation area. Everyone owns their own home but shares the common facilities. Some designers have planned eco-villages, so that people can work near where they live, thus minimizing transportation. Seven such villages are under construction, with a number of others in the works.

In the following excerpt, Wendell Berry, author, farmer, and well-known advocate for ecological living, lists some precepts for living more centered on our communities.

- If we could think locally, we would do far better than we are doing now. The right local questions and answers will be the right global ones. The Amish question "What will this do to our community?" tends toward the right answer for the world.
- If we want to keep our thoughts and acts from destroying the globe, then we must see to it that we do not ask too much of the globe or of any part of it. To make sure that we do not ask too much, we must learn to live at home, as independently and self-sufficiently as we can. That is the only way we can keep the land we are using, and its ecological limits, always in sight.
- The only sustainable city—and this, to me, is the indispensable ideal and goal—is a city in balance with its countryside: a city, that is, that would live off the net ecological income of its supporting region, paying as it goes all its ecological and human debts.
- The cities we now have are living off ecological principal, by economic assumptions that seem certain to destroy them. They do not live at home. They do not have their own supporting regions. They are out of balance with their supports, wherever on the globe their supports are.
- Industrial procedures have been imposed on the countryside pretty much to the extent that country people have been seduced or forced into dependence on the money economy. By encouraging this dependence, corporations have increased their ability to rob the people of their property and their labor. The result is that a very small number of people now own all the usable property in the country, and workers are increasingly the hostages of their employers.
- Ecological good sense will be opposed by all the most powerful economic entities of our time because ecological good sense requires the reduction or replacement of those entities. If ecological good sense is to prevail, it can do so only through the work and the will of the people and of the local communities.
- The right scale in work gives power to affection. When one works beyond the reach of one's love for the place one is working in, and for the things and creatures one is working with and among, then destruction inevitably results. . . .
- The real work of planet-saving will be small, humble, and humbling, and (insofar as it involves love) pleasing and rewarding. Its jobs will be too many to count, too many to report, too many to be publicly noticed or rewarded, too small to make anyone rich or famous.

- To make a sustainable city, one must begin somehow, and I think the beginning must be small and economic. A beginning could be made, for example, by increasing the amount of food bought from farmers in the local countryside by consumers in the city. As the food economy became more local, local farming would become more diverse; the farms would become smaller, more complex in structure, more productive; and some city people would be needed to work on the farms. Sooner or later, as a means of reducing expenses both ways, organic wastes from the city would go out to fertilize the farms of the supporting region. . . . The increase of economic intimacy between a city and its sources would change minds (assuming, of course, that the minds in question would stay put long enough to be changed). It would improve minds. The locality, by becoming partly sustainable, would produce the thought it would need to become more sustainable.

EcoQuiz

Q: What can you do to help be a green consumer?

A: You can buy products that are durable and reusable, that are made from recycled materials and will be recycled after use and that have minimal packaging. Avoid harmful products.

Q: According to John Thompson, the stimulating author of *The Environmental Entrepreneur*, all you have to do is find the solution to any environmental problem, and the world will beat a path to your door. What are opportunities for doing well and doing good that others have found?

A: *Business* magazine recently listed the top ten businesses that it evaluated for vitality, continuity, originality of technology, profitability, and impact on industry. Maybe these can give you ideas:

—Calvert Group of Funds, for connecting social responsibilities with environmental venturing;
—Clivus Multrum, for its compost toilets;
—Croxton Collaborative, for its design of buildings such as for the National Resources Defense Council and Audubon's headquarters;
—Deja, Inc., for its making shoes out of mostly recycled materials;
—The Hummers, for its woodcraft products;
—NaturaLawn, Inc., for its safer ways to care for lawns and control pests;
—Natural Cotton Colours, Inc., for its "green cotton" line;
—Resource Conservation Services, Inc. for its recycling of municipal and industrial wastes;
—Stonyfield Farm, for its yogurt products;
—Terra Verde, for its "green" retail stores.

Q: What are your basic material needs, and to what extent do you exceed them? Do you do enough to tap into the abundance of nonmaterial needs, such as being close to people and nature?

E C O S Y S T E M   P A T H W A Y S

One World, One Goal

"I consider nothing human alien to me," the Roman playwright Terence wrote more than 2,000 years ago. Yet today's headlines prove that much of the world still divides itself into humans ("us") and aliens ("those people"). What, then, are the chances that the planet's 5.5 billion residents will come to understand, before it is too late, that *no* living thing is alien to them?

National boundaries fade into insignificance in the face of planet Earth's environmental disintegration. The easy passage of wild plants and animals, as well as humans and their pollutants, from one country, even one continent, to another now demands planetary solutions. Some basics:

- Ecological thinking rules out isolationism, both in politics and in wildlife conservation.
- The loss of biodiversity threatens the underpinnings of life itself. All plants and animals exchange gases with their environments, thus maintaining the atmosphere that sustains all living things, ourselves included, because as Earth loses more species, climates will change and agriculture be put at risk. The destruction of forests creates floods and leads to the loss of freshwater sources. The destruction of insects is by no means a positive event: a very large number of insect species are beneficial to mankind, pollinating crops or keeping noxious pests under control.
- Every key environmental initiative, including the protection of marine mammals and migratory birds, rain forest preservation, restrictions on the trade in wild animals either as pets or for their skins, and halting the growth of human populations, will require an enduring network of international partnerships.
- Only mutually binding, multinational treaties can adequately deal with global warming, ozone depletion, and the allocation of scarce water, food, mineral, and fiber resources.
- In sum, let the word *alien* be confined to science fiction.

16

OUR GLOBAL FAMILY

Guest author: Frank Graham, Jr.

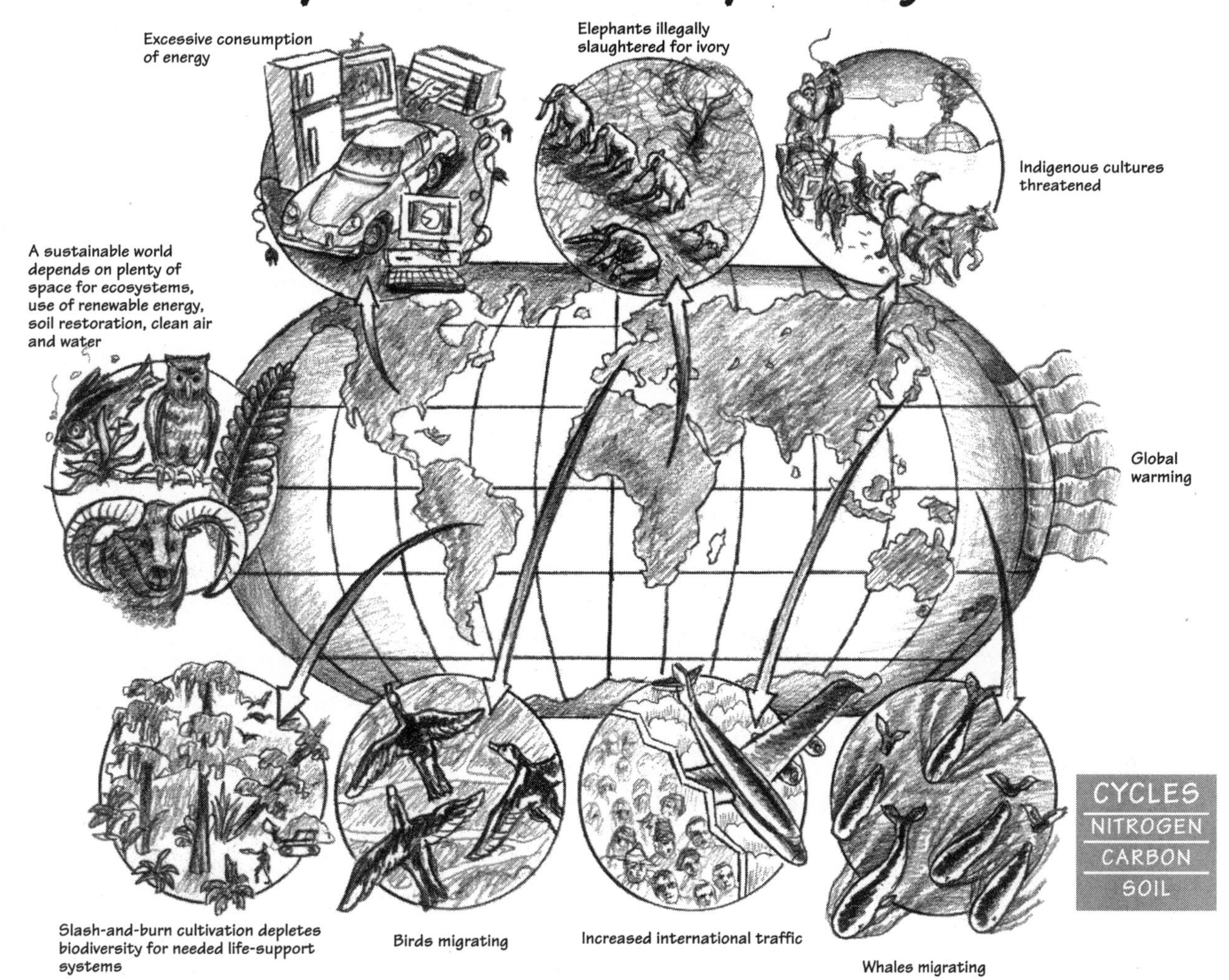

CYCLES
NITROGEN
CARBON
SOIL

EVOLUTION AND EXTINCTION

During the nineteenth century, geologists and paleontologists accumulated convincing evidence of unending and even cataclysmic change on Earth over long periods of time. For instance, they learned from the fossil record—from marks and petrified bones left in age-old rocks—of the past existence of hundreds of kinds of large mammals. Dinosaurs and hairy mammoths once grazed and fought and reproduced here. But they have all vanished. They have become extinct.

These discoveries altered the way that men and women think about life. Educated people came to know that Earth and its living things are not static, as earlier generations had believed. Volcanoes erupt, glaciers expand and retreat, species are constantly coming into being and passing away.

Extinctions and the evolution of new species have been taking place since life first appeared on our planet. But until very recently the mix of species was endlessly replenished out of the ferment of evolution. Today Earth holds a teeming variety of living forms that caring and observant humans have come to cherish. Orchids and roses, apes and butterflies, brighten our world and, many scientists insist, make up an integral part of that vast complement of living things that we may obliterate only at our own peril.

Now plants and animals are disappearing at a rate rivaling that of the era of great extinctions, when the fate of the dinosaurs was sealed. One of the crucial questions of our time is: Why is this happening? There are no simple answers, but scientists agree that in essence humans are introducing profound change so rapidly that most other forms of life can no longer deal with the new and artificial conditions.

Life thrives on change, but Earth and its living things (perhaps including humans themselves) have reached the outer limits of their capacity to adapt. The word *extinction* now looms larger than ever before in our vocabulary.

> Hidden among the western Andean foothills of Ecuador, a few kilometers from Rio Palenque, there is a small ridge called Centinela. Its name deserves to be synonymous with the silent hemorrhaging of biological diversity. When the forest on the ridge was cut a decade ago, a large number of rare species were extinguished. They went just like that, from full healthy populations to nothing, in a few months. . . . [Isolated mountain places like that] tend to evolve their own species of plants and animals, found nowhere else or at most in a few nearby localities. . . . Around the world such anonymous extinctions—call them "centinelan extinctions"— are occurring, not open wounds for all to see and rush to stanch but unfelt internal events, leakages from vital tissue out of sight. —E. O. Wilson, in *The Diversity of Life*

Species—No Photocopies!

What is a species? The answer to this question keeps changing with the advance and refinement of scientific knowledge, but a definition widely accepted among today's biologists is the one put forward by the most eminent evolutionary biologist of our time, Ernst Mayr, of the Museum of Comparative Biology at Harvard:

"A species is a reproductive community of populations (reproductively isolated from others) that occupies a specific niche in nature."

Let's take, for example, a kind of beetle that we shall call Species A. Members of this species may be scattered in different places all over a large continent such as Africa or South America, and wherever it lives it may feed and reproduce alongside other kinds of beetles that look very much like it.

But each individual of Species A maintains the capacity to mate and produce offspring with other individuals of its own species. At the same time, it maintains individuality in its genetic makeup that leads to differences in its form and behavior, however small, from other, somewhat similar, kinds of beetles.

Under normal conditions in nature, such differences will prevent members of Species A from mating with members of any other distinct species. This individual and all other members of its species are "reproductively isolated." However scattered, they go on forming a community of populations that is *unique*.

But the planet, and consequently life itself, is continually changing. A population of Species A may undergo significant changes. If a mountain chain rises, or a new body of water forms, cutting off one population of Species A from the rest, this newly isolated population may acquire in the

THINKING LIKE AN

ECOSYSTEM

Biomes

Most people tend to think of the Earth and its plant and animal populations in patterns that conform to the distribution of the continental land masses: elephants in Africa, bison in North America, sloths in South America, kangaroos and platypuses in Australia, and so on. But ecologists also see the patterns of life on our planet in terms of biomes.

A biome is a major ecological unit, classed according to its key plant and animal life. A savanna grassland is a biome, as is a tropical rain forest and a desert. This type of floral and faunal community may appear on more than one continent because it is created by climate: average annual temperatures and rainfall, as well as latitude and altitude. Biomes exist and change over time within the complex worldwide pattern of air currents.

Climate then determines the kind of plants that grow in any region. A plant fitted for life in the harsh, stony soil of a blazing hot desert will not be found growing in the cold, wet lowlands along a northern coast. The plant community, in turn, determines the makeup of resident animal populations. A bird fitted for opening seed cones in a northern spruce forest will not be at home among the large, fleshy fruits produced by trees in a tropical rain forest.

Although the species inhabiting distinct biomes may differ from one part of the world to another, they share many characteristics. The cacti of the North American desert, for instance, are unrelated to the euphorbias of dry African regions. But their reduced leaves, swollen stems, and spherical structure often make them lookalikes and help both groups conserve precious water. Again, the large, hoofed mammals of temperate grasslands on the steppes of Asia, the prairies of North America, and the pampas of South America share many physical characteristics, and all tap the same kinds of resources.

The marvelous diversity of life, everywhere evolving and adapting through time, nevertheless displays unexpected kinships in the pockets of space represented by the Earth's biomes and their inhabitants.

course of time a series of wholly new genetic characteristics.

These isolated beetles may become so different from members of other individuals of Species A that they are no longer recognizable by them as "suitable" mates. They are reproductively as well as physically isolated from all other beetles. They have evolved into a *unique* population—Species B.

Yet, triggered by changes in Earth's topography, this evolutionary event only added to the diversity of life. A new life-form, Species B, has evolved, keeping much of the genetic advantages of Species A but adding variations of its own. Nothing has been lost.

What concerns modern environmentalists is another scenario that brings about change. Human interference, in the form of clear-cutting forests, filling and paving over wetlands for business and housing developments, or submerging vast areas of formerly habitable land to build large dams and reservoirs, now threatens the habitat of many kinds of plants and animals. Species that cannot move or adapt to changing conditions are doomed.

Biologists estimate that as many as 50,000 species are now lost annually because of human interference in natural environments. Fully one fourth of all species on Earth may disappear within the next fifty years. An unknown number are reservoirs of irreplaceable genetic material, with a potential for making priceless contributions to other forms of life, now and in the future. In rain forest plants alone, more than 200 species have proved to be important as food crops for humans. Another 2,000, says the National Cancer Institute, have anti-cancer properties.

These doomed species will never be reconstituted, even if planet Earth were to continue in existence for a hundred billion years. They are extinct—and no photocopies are left behind.

VALUES AND DEFINITIONS

What is the value of the biological diversity of the planet? . . . It is one thing to treat the valuation of biodiversity as a guessing game or as a set of very interesting theoretical problems in welfare economics. It is quite another thing to suggest that the guesses we make are to be the basis of decision making that will affect the functioning of the ecosystems on which we and our children will depend for life.

If we are not taken seriously unless we quantify our answer, I would like to suggest some new units of measurement. An *oops* is the smallest unit of chagrin that we would feel if we willfully extinguish a species we need later on. A *boggle* is the amount of ignorance encountered when an economist asks a biologist a question about species and ecosystems, and the biologist answers: "I don't know, and I'm so far from knowing, it boggles the mind." If I understand what the economists are saying, irreversible oopses and boggles of uncertainty are the main factors in decisions affecting biodiversity. In the passion to express the values of a species in dollar figures, it will be unfortunate if we forget to count oopses and boggles as well.

—Bryan Norton, "Commodity, Amenity, and Morality," in *Biodiversity*

FROGS IN HAZARD

The extinction of a species can be viewed as an isolated tragedy. The disruption of an entire class of animals, some scientists fear, may be the harbinger of planetary catastrophe.

In recent years many scientists who study amphibians—the class of animals that includes frogs, toads, and salamanders—began to realize that these animals are suffering a mysterious decline in many parts of the world. A conference held on the subject in New Orleans in 1990 concluded that there was not yet any "hard data" to back up their suspicions. Yet one scientist after another spoke of the disappearance of certain frogs in areas where they had been studied for years.

The pattern of decline is confusing. Many once-familiar species are gone from large areas of our Pacific states. Researchers report major losses in Latin America, Europe, and Australia. Yet there are few similar reports from our Southeastern states, except locally where developers have destroyed good amphibian habitat. All sorts of reasons, including pesticides, heavy metal contamination of waterways, acid rain, predation, and ultraviolet radiation, have been advanced as reasons for the losses.

"In Oregon we looked at 30 populations of the Cascades frog and found an overall 80 percent drop in numbers," Andrew Balustein, a herpetologist at Oregon State University, said. "But one population near several decimated ones were doing very well. In one study area 600,000 eggs laid by western toads failed to develop. The only exceptions were eggs I took into my lab. They did fine, and the tadpoles that hatched from them also did fine when I put them back into their habitat."

David Wake, director of the Museum of Vertebrate Zoology at the University of California, Berkeley, also emphasized the fact that amphibians are not disappearing en masse. He pointed out that frogs are known to be killed off in areas of heavy pesticide spraying, but he also said there is evidence that a thinning of the ozone layer may be responsible in part for the decline. Studies have shown that frog eggs and tadpoles are extremely vulnerable to increased levels of UV-B radiation, the kind of rays that reach Earth when the ozone layer is pierced.

"Birds are good indicators of environmental problems, but amphibians are even better at the local level," Wake said. "Birds can fly away, but frogs must face the problem. They are extremely susceptible, being herbivores when they are tadpoles and predators as adults. Many of them readily absorb contaminants through the skin. Whatever happens to them happens right here—not in Belize or Brazil, as may be the case with birds."

Scientists have speculated that the decline may have been masked for some years because frogs are often a "top carnivore" in their ecosystems. ("A lot of frogs die of old age," one herpetologist says.) Whatever is striking at their populations may be affecting only the early life stages, so the problem did not appear until the adults began dying of old age and there were no new generations to replace them.

"Frogs are telling us the ecosystem is sick," David Wake said. "If something bad is going on out there it may be hard to turn around at this stage, and all forms of life may be affected."

FROGSONG

I have always liked frogs. I liked them before I ever took up zoology as a profession; and nothing I have ever had to learn about them since has marred the attachment. I like the looks of frogs, and their outlook, and especially the way they get together in wet places on warm nights and sing about sex. The music frogs make at night is a pleasant thing, full of optimism and inner meaning. . . . The male sits in the pond edge and croaks or whistles or buzzes or roars, according to his kind, when something mystic outside and inside tells him it is time for new frogs, and his calling draws the female there, and this way frogs are provided for the new year. The singing of the hermit thrush is a sweet blustering. The song of the frog is a shout for the flowing on of the life stream.

—Archie Carr, in *The Windward Road*

BIRDS KNOW NO BORDERS

There is much about the phenomenon of bird migration that is mysterious. But the reason for migration is comparatively simple: birds fly off to other climates to find enough food to raise a family.

We are apt to speak of the warblers, vireos, flycatchers, tanagers, and other small species we see in the woods and fields every summer as "*our* birds." But they are with us for a very short time each year. They may spend only three or four months in North America, and the rest of the year on their wintering grounds in the tropics and traveling back and forth on migration.

The wintering grounds for most of them are the forests of Central America or island outposts in the Caribbean. Several thousand species of birds live in those areas, and when the breeding season begins and there are many chicks to feed, food is in short supply. So these birds take off in early spring each year on the long and hazardous flight to North America, with its sprawling land mass and (as every summer vacationer knows!) an incredibly rich hatch of mosquitoes, caterpillars, spiders, and other small creatures that make perfect food for growing chicks.

This strategy, despite heavy losses to weather, predators, and obstacles such as tall, man-made structures, has served migrant birds well for millions of years. There are about 250 species of birds that breed in North America but spend their winters to the south, some flying as far as Argentina. The blackpoll warbler, a species that doesn't weigh any more than a dime, migrates from Alaska to the Amazon Basin!

But the warblers and other migrants have fallen on hard times. The Breeding Bird Survey conducted annually by the Fish and Wildlife Service shows that migrants from tropical forests declined in eastern North America during the 1978–1987 decade at a rate of 1% to 3% a year. It is now clear that millions of birds are vanishing.

The problem boils down mainly to loss of habitat. Migrant birds take a terrible beating in all phases of their life cycle:

- Unrestricted logging for fuel and farm clearings has swept away huge areas of tropical forests, especially in Central America, where many North American breeders spend the winter.
- Once-hospitable feeding and resting places along migratory routes, such as the shorelines of Delaware Bay, are now beset by domestic and industrial sewage, oil spills, and development. The birds have a hard time "fueling up" for the final stages of their long journeys.
- Many of the northern woodlands where migrants formerly bred have been fragmented by roads, malls, and housing developments. A large percentage of the breeders there, nesting on or close to the ground, are more vulnerable to predators such as raccoons, house cats, jays, and crows, than they were in extensive forests of the past.

When bird-watchers used to get together, they would regale each other with stories of the big "waves" of migrating warblers they had found themselves in the midst of, or the numbers of rarities spotted in the underbrush. Too often the talk today is about the song and color fast vanishing from America's woodlands.

> Whatever most delights us and makes us glad to live on so fair a planet; whatever enhances our existence by its grace and beauty; whatever intensifies our appreciation of nature's vastness and bounty; whatever stirs our imagination and stimulates us to exercise all our powers, mental and physical, in the effort to learn and understand; whatever stiffens our determination to protect the natural world from devastating exploitation—whatever does this for us has a special claim to our love and gratitude. . . . For a large and growing number of people, birds are the strongest bond with the living world of nature. They charm us with lovely plumage and melodious songs; our quest of them takes us to the fairest places; to find them and uncover some of their well-guarded secrets we exert ourselves greatly and live intensely. In the measure that we appreciate and understand them and are grateful for our coexistence with them,

we help to bring to fruition the agelong travail that made them and us. This, I am convinced, is the highest significance of our relationship with birds.

—Alexander Skutch, *A Bird Watcher's Adventures in Tropical America*

Beringia: A Bridge Linking Two Worlds

At their summit meeting in June 1990, presidents George Bush of the United States and Mikhail Gorbachev of what was then the Soviet Union called for the creation of a Beringian Heritage National Park. The agreement was greeted as a hopeful break in East-West hostility, a forerunner of the contacts to come.

On hearing the news, many citizens of both countries must have asked, "Where in the world is Beringia?"—a response that emphasized only the region's forbidding remoteness from the centers of civilization. It was hardly a measure of its enormous impact on the history of the New World and the vital movements of plant and animal populations.

The word Beringia describes an ecological entity, not a political one, for this vast area of the Far Northern Pacific links extensive Arctic and subarctic lands of the Old World with those of the Americas. It includes the Bering and Chukchi seas, with adjacent areas of Alaska and far Russia, and a sliver of Arctic Canada. Asia and North America were once joined there, but the sea rushed in when the last of the giant glaciers retreated and the two continents now confront each other at the Arctic Circle across the Bering Strait.

The idea of an international park for this region is belated recognition that it remains united by common human cultures and a ceaseless interflow of wild plants and animals. Environmentalists and Native people on both sides of the strait are working to make the park concept a first step in the preservation of Beringia's diversity after decades of plunder by oil and mining companies, huge fishing fleets, and those who would exploit its wildlife.

Thousands of years ago, the tremendous amounts of water frozen into glaciers to the north and east of Beringia caused ocean levels to drop and expose an expanse of sea bottom as a dry "land bridge" that connected Asia and North America. This broad plain served as a highway for ancient people and many Old World plants and animals as they invaded the Americas and spread out south of the glaciers. The discovery of the fossilized remains of a mammoth on the Pribilof Islands, now isolated in the Bering Sea, was an early clue to those invasions.

As the glaciers melted and water covered the old land bridge, the continents were once again divided. But contact between them never ceased. Only fifty-five miles separate them at the narrowest point, a short haul for flying and swimming creatures and no deterrent for polar bears, foxes, and others that travel on pack ice.

Humans also found the strait no obstacle. They crossed back and forth in skin-covered boats in search of trade or plunder. Though many of the people we now call Amerindians continued inland to inhabit the plains or forests, others, such as the Eskimos and Aleuts, remained and became superbly adapted to making a living on the rim of the icy seas. That they continued going back and forth is evident from the similarity of goods and traditions found on both sides of the Bering Strait. Anthropologists trace common themes in the clothing and houses, as well as in their religious icons and rituals.

Alaskan Eskimos regularly visited trade fairs in Russia until the sale of Alaska to the United States in 1867. But even today, after years of Cold War isolation, Alaskan Eskimos readily converse in the Yupik and Inupiaq languages with relatives across the strait.

Europeans eventually discovered Beringia and plundered both its people and wildlife. Yet after two centuries of depletion, Beringia's natural resources can only be described as opulent. The Bering Sea's bottom fishery may be the richest on the planet. In an era of declining wildlife, considerable populations of whales, Pacific walruses, seals, sea lions, polar bears, and other marine mammals still come to prey on the smaller life in the nutrient-rich waters.

Perhaps most impressive of all is the birdlife. Nearly three dozen species of seabirds feed and nest there during Beringia's short summer. More than 200 kinds of birds in all stream in and out of Beringia every year on a variety of age-old flyways. For instance, about 80,000 lesser snow geese that winter in California become an international resource as they fly across the Bering Strait to nest on Wrangel Island on the Russian side of Beringia. Songbirds that normally halt at ocean barriers also fly across the strait as if the old land bridge still existed.

Development is a fact of life in modern Beringia. But thousands of men and women in two nations are trying to bring to reality that vision of a great international park where the diversity of human cultures and wildlife communities may flourish beside the compatible use of its natural resources. Only then will the park carry out the promise of its name, preserving the *heritage* of a world that came into being around an ephemeral bridge of dry land and for a time achieved a rare kind of harmony.

Biology in an Icy Land

Several years before the old Soviet Union collapsed, American and Russian biologists were working together on Wrangel Island, a bleak outpost and Russian nature reserve in the Chukchi Sea, north of Siberia.

"I went there in 1989 to work with the Russians on a study of black brant," recalls Dirk Derksen, a specialist on those small geese for the Fish and Wildlife Service. "Brant had been declining for some years in the Pacific, and Wrangel Island is a key area for them. They fly there from nesting grounds on both the Alaska and Russian mainlands to molt and rest before heading south on migration."

The biologists were trucked every day from their village quarters to Jack London Lake, a shallow freshwater basin where brant gather to molt. The Russians set up their nets and the biologists began herding the flightless birds into them for capture and banding.

"But when we got all the geese inside, the nets collapsed!" Derksen said. "The nets and poles were rotten. We had to have nets shipped over from Alaska so we could keep working."

Other American biologists come to the island in early spring to study polar bears. At summer's end, 150 to 200 females excavate dens in huge snowdrifts, often left from the previous year. They spend the winter giving birth and nursing their cubs. The dens are so closely clustered, often only a few yards apart, that when the bears break out of them in spring there are few better places to observe their behavior.

How St. Lawrence Island Appeared in the Bering Sea

Ages and ages ago, the Eskimo people say, a great giant lived in the far north. One day he happened to be standing with one foot on the Siberian coast and the other on the shores of Alaska. The two continents are not very far apart there, so the giant stood comfortably, looking out over the world. As it happened, he chanced to look down at the narrow strip of water between his feet, and nonchalantly reached down and took up a handful of the sand and stone of the ocean's bed. He stood looking at it and idly squeezing the water from it. Then he raised his great arm and threw the handful of dirt and rock out before him into the water. It stayed there—an island between the two continents. —Margaret E. Murie, *Island Between*

Parrots For Sale

The future survival of at least forty-one of the world's 330 parrot species is threatened by the trade in wild pets, according to the International Council for Bird Preservation. Coupled with the loss of habitat that persists in most of their native regions, the pet trade hits parrots especially hard.

The United States is the largest importer of wild birds, contributing to a booming but deadly business that brought 6 to 7 million birds to this country as pets in the last decade.

Parrots have always been prized for their colorful plumage, but the rate of capture has been going up as roads and development make it easy for poachers to reach for them deep into tropical forests. Many Third World countries where the birds are captured have laws to limit bird exports but no money to enforce the regulations. Nor has the United States been very successful at enforcing its own laws. The Endangered Species Act outlaws trading in endangered and threatened species, while the Lacey Act prohibits imports of birds protected in their native countries. But only 20% of bird shipments into this country are inspected.

Statistics show that the United States imports 250,000 parrots legally each year, but diligent investigation by conservationists indicates that at least 150,000 more are smuggled in from Mexico. Stuffed in boxes or the trunks of cars, many die of rough handling, disease, starvation, or dehydration. More than a million birds are known to have died during the past decade even before reaching pet stores in the United States, but the number that succumbed in the smuggling trade is unknown. Environmentalists are pushing for legislation that will help to keep rare parrots where they belong—in the wild. Among the key provisions are ones that

- phase out the import of wild birds except for zoos, captive breeding, and scientific and conservation purposes,
- promote captive-breeding programs to supply pet stores,
- set up licensing and record-keeping requirements for breeders and importers of wild birds,
- set standards for the handling, transport, and identification of birds, and
- establish stiff penalties to deter smuggling.

Outpacing Extinction

In June 1973 two Peruvian hillmen made their way up a trail through an "elfin forest" of stunted trees, 12,000 feet high in the eastern Andes. They were returning from a nearby village with supplies for a party of North American ornithologists they worked for seasonally as field assistants. A group of small birds fed low in the trailside trees.

The men stopped and conferred briefly. Then one of them lifted his gun, fired into the flock, and a little brown bird with a tawny-olive throat fluttered to the ground. When the men reached camp, they handed the dead bird to the ornithologists, calling it by its local name, "pardusco."

"The bird was totally new to us," recalled the party's leader, John O'Neill of Louisiana State University. "It was a species that had never been described before. But on subsequent trips to the region we discovered that the pardusco is fairly common in several isolated valleys."

The pardusco was a revelation almost in the religious sense—a living creature suddenly shown to mortals, its existence made known at a time when biologists tend to talk gloomily of extinctions. Since O'Neill began visiting Peru in 1961, he and his colleagues have discovered more than two dozen new species of birds, as well as several kinds of mammals and reptiles.

According to the International Council for Bird Preservation, the planet has been losing one species or subspecies of bird about every four years since the dodo's extinction some three centuries ago. No wonder then that prior to the findings of the Louisiana State team in the Andes, it was generally believed that the age of ornithological discovery was a thing of the past. In a sense, these scientists "produced" bird species faster than the rest of humanity was obliterating them!

"The amazing thing is that these birds have been living all along in a region that until twenty years ago was thought to have been pretty well explored," O'Neill said of this latter-day form of creation. "In fact, most of them are relatively common in their restricted areas."

Why did this treasure trove of life suddenly come to light in the Peruvian Andes?

"First of all, the region is still intact ecologically," O'Neill replied. "The country is high, cold, and wet. Few people live there. In the past, even dedicated naturalists couldn't stay in the area long enough to get to know all the species that lived in the dense, trackless, forest scrub. But we have the advantage of modern equipment—nylon tents, good tarps, medicines, mist nets in which to trap night-flying birds like owls, and so on. We can stay longer in one area, and for a time we returned year after year so that we knew the common birds and were alert to any strange calls or plumages."

O'Neill pointed out that the role of fluctuating climate in creating species is nicely demonstrated in the diverse land of Peru. The Andes have been subjected to drastic alterations of climate and upheavals of rock, isolating all kinds of animals in remote valleys and bringing on the subtle changes as they adapted that finally resulted in the evolution of new species. Each valley has animals found nowhere else in the world, not even in neighboring valleys.

"A key to new species is high mountains, with their remoteness and diversity," John O'Neill explained. "Africa and most of Asia are quite well known, though such places as the mountains of New Guinea and the Philippines, perhaps even China, must hold undiscovered species. We may have sent men to the moon, but there are still all kinds of things that we don't know about right here on Earth."

Niches and Diversity

For years the accepted estimated figure for the number of living species around the globe has ranged from 1 to 3 million kinds of plants and animals. But in recent years a respected entomologist has come up with a new figure: based on his field studies in the Amazon Basin, he says there may be 30 million species of insects alone!

Terry L. Erwin, of the Smithsonian Institution, sampled insect life in three plots of treetops, each twelve square meters, in five different types of Amazon forest. Some findings:

- He collected 3000 species of beetles.
- The beetles' average length was 2 to 3 millimeters.
- Only 8.7% of the beetles found in one of the plots could be found in another plot of the same forest type only fifty meters away.
- Only 2.6% of the species were shared between the plots surveyed in that study and another one Erwin studied in treetops 1500 kilometers away.

Why such diversity? The treetops of the Amazon forests are an incredibly tangled web of branches, twigs, leaves, vines, aerial plants, seeds, and fruits. Each provides a niche for some kind of creature, and these niches may be occupied by an entirely different set of creatures at varying times of the year. Extrapolating his figures through cumulation curves based on increases of samples, Erwin arrived at a potential figure of 30 to 50 million insect species. His estimate would have come as no surprise to the great evolutionary biologist J. B. S. Haldane. Some years ago he was asked by an acquaintance if his scientific studies had given him any insight to the Creator.

"Yes," Haldane is said to have replied. "I find that He is inordinately fond of beetles."

Rapunzel

Rapunzel lurched to her feet when Jim Doherty walked into the cavernous, richly odorifous building. She stood only four feet tall and had shaggy tufts of reddish-brown hair on her ears, which she twitched in the man's direction when he whistled.

"She's a Sumatran rhino," Doherty, a large, soft-spoken man who is general curator of mammalogy at the Bronx Zoo, remarked to a visitor in the rhino's quarters. "This is the smallest of the five rhino species in Africa and Asia, but it has now nearly disappeared from the wild."

The story Doherty told that day is already too familiar to conservationists. Like populations of many of the world's large animal species in the tropics, that of the Sumatran rhino has been decimated by loss of its habitat and indiscriminate hunting. That they have any prospects for the future has been mainly the result of captive propagation programs carried out by zoos in Europe and North America.

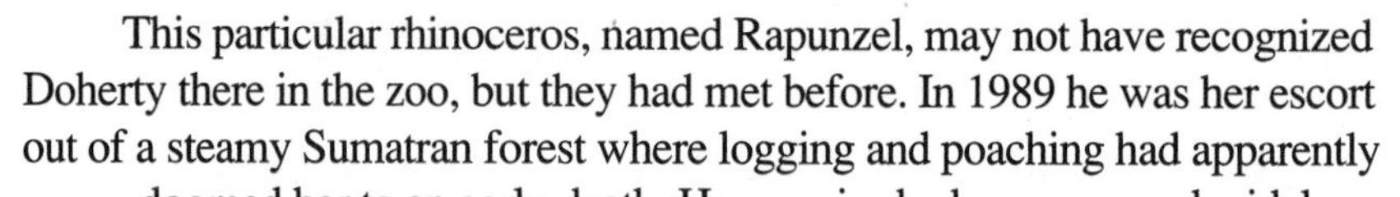

This particular rhinoceros, named Rapunzel, may not have recognized Doherty there in the zoo, but they had met before. In 1989 he was her escort out of a steamy Sumatran forest where logging and poaching had apparently doomed her to an early death. Her species had once roamed widely across northern Burma, the Malay Peninsula, and the Indonesian islands of Borneo and Sumatra. Doherty guessed there may be only 500 to 700 left. Their wet forest habitat is increasingly fragmented by loggers, putting stress on these shy animals and concentrating the survivors for the convenience of poachers.

Rhinos of all species are hunted for their horn because some Asian men are persuaded by folklore that this mass of tightly packed, hairlike material, when ground up and brewed into a dark tea, passes the male rhino's celebrated sexual prowess on to them. As the Sumatran rhino's numbers plummeted, zoos in San Diego, Los Angeles, Cincinnati, and New York worked out a plan. Hoping to establish a breeding program, and with the encouragement of Indonesian conservationists, they hired a professional trapper and obtained government permission to capture rhinos in areas where they weren't likely to survive.

The capture team prepared pit traps in a wet area favored by the rhinos. When Rapunzel wandered by and tumbled into a pit, word reached Doherty in New York at the Bronx Zoo and he left to claim his prize. Arriving in Sumatra, he found Rapunzel in a corral built especially for her, carefully watched by the trapper who was trying to shift her diet from leaves and twigs to one of hay, grain, and alfalfa brought from Australia.

"After a while she adjusted and was ready to travel," Doherty said. "We moved her out of the forest to a paved road and then by truck to the coast, where we put her aboard a small freighter to Singapore. Then we trucked her to the airport and got her on a 747. One third of the plane was a freight compartment, and the passengers didn't even know a rhino was aboard."

When the little rhino arrived at the Bronx Zoo, director William Conway was much taken with her "long hair" and her propensity for singing while she wallows—in reality a cacophony of grunts, snorts, and whistles. Conway came up with the name Rapunzel. He and Doherty arranged for her to be visited by a male rhino which had also been captured in Sumatra.

"We hope to establish a captive population of rhinos here," Doherty said as he watched Rapunzel nose about in some hay. "After the land is cut and abandoned in Sumatra, it will begin to recover. People may decide to treat their forests differently in the future and rhinos can be returned to the wild."

The tragedy of the commons develops in this way. Picture a pasture open to all. It is to be expected that each herdsman will try to keep as many cattle as possible on the commons. . . . The rational herdsman concludes that the only sensible course for him to pursue is to add another animal to his herd. And another; and another. But this is the conclusion reached by each and every rational herdsman sharing a commons. Therein is the tragedy. Each man is locked into a system that compels him to increase his herd without limit—in a world that is limited. Ruin is the destination toward which all men rush, each pursuing his own best interest in a society that believes in the freedom of the commons. Freedom in a commons brings ruin to all. —Garrett Hardin, "The Tragedy of the Commons"

Population: The All-Inclusive Threat

The human population has spun out of control since World War II. New medical technologies and modern agricultural techniques set in motion a spiral that has seen human numbers leap from 2.5 billion in 1950 to 5.3 billion in 1990 and perhaps 15 billion by the year 2100. The ever-increasing numbers of people have had a predictably deleterious effect on the natural world and the quality of human life.

In the absence of effective measures against the population explosion by the world's governments, many private organizations have tried to grapple with the problem. One of the most imaginative and revealing international initiatives in recent years is the Sharing the Earth Project, a joint effort of the National Audubon Society's Population Program and its Sanctuary Department.

In this project, eight Audubon wildlife sanctuaries that were obviously affected by human population pressures were matched with eight sites in developing countries where wildlife and habitat were threatened by rapid population growth. The managers of each of the Audubon sanctuaries visited his or her counterpart and in turn played host to the foreign wildlife manager. These men and women worked together across continents and oceans to understand one another's problems and arrive at possible solutions, and found some real surprises.

"When I went to Pakistan I knew I'd learn a lot, but I have to admit that at one point I was really astounded," said Kenneth Strom, manager of Audubon's Rowe Wildlife Sanctuary on the Platte River in Nebraska. "Along the Indus River, which has run through populated areas for hundreds of years, I saw what the Platte *used* to look like. In some ways they've done less damage to their river than we've done to ours."

Strom expanded on the differences and similarities along the two rivers. For instance, 80% of the world's sandhill cranes gather on migration along the Platte each spring, while Asia's migrant demoiselle and common cranes likewise depend largely for survival on a great river, the Indus. Strom saw that the Pakistanis have, after centuries of abuse, nearly destroyed their soil. Yet the Indus survives because the technology of "development" was not available to them until recent decades.

"The help that developed countries is giving to Pakistan is mostly in engineering," Strom said. "But many Pakistanis now believe that building so many irrigation projects is a problem. The land is becoming quite saturated as irrigation diverts much of the natural flow and shrinks the great river the people have depended on since the Mogul emperors ruled. What we are doing is taking away their Indus River and giving them our Platte."

Another Audubon wildlife manager, Norman Brunswig, of the Francis Beidler Forest Sanctuary in South Carolina, became convinced that unless the rate of global population growth declines, the future is bleak for wildlife habitats of any kind. Finding himself transported from a low-country Carolina swamp to Zimbabwe's spectacular Eastern Highlands, he saw that the area had been preserved much better than his own because of the local people's spiritual values. They respected the forest and indeed were reluctant to venture among its strange mists and tremors.

But this is changing as other people, lacking reverence for the forest, have moved in. The area is now scarred, and its wildlife is depleted. Invited into local homes, he heard disturbing comments.

"In Africa, children are our wealth," a woman told him. "If you do not have lots of children, there may be no one who will care for you when you are old."

Norm Brunswig found a reluctance everywhere to discuss the population problem.

"The cultural momentum related to large families is strong," he said. "Only very, very strong, fully committed government and popular programs are going to stem this surging human tide."

Two worlds are becoming aware that their deficiencies, like their strengths, are very much akin.

A Warning— Spoken Almost Half a Century Ago

The history of our future is already written, at least for some decades. As we are crowded together, two and a quarter billion of us, on the shrinking face of the globe, we have set in motion historical forces that are directed by our total environment.

We might symbolize those forces by graphs. One of them is the curve of human populations that, after centuries of relative equilibrium, suddenly began to mount, and in the past fifty years has been climbing at a vertiginous rate.

The other graph is that of our resources. It represents the area and thickness of our topsoil, the abundance of our forests, available waters, life-giving grasslands, and the biophysical web that holds them together. This curve, except for local depressions, also maintained a high degree of regularity through the centuries. But it, too, has had its direction sharply diverted, especially during the past hundred and fifty years, and it is plunging downward like a rapid.

These two curves—of population and the means of survival—have long since crossed. Ever more rapidly they are drawing apart. The farther they are separated the more difficult it will be to draw them together again.

—William Vogt, *Road to Survival*

EcoQuiz

Q: What is the biggest global threat to wildlife today?

A: Loss of habitat. There are many local forces that take away living space for wild plants and animals: destruction of forests and wetlands, building huge dams and reservoirs, overgrazing grasslands. But ultimately all threats can be traced to a single cause: there are too many of *us*. People need space and natural resources to live, and more and more people leave less and less space and resources for other living things. As conservationist David Brower has said, "The wild places are where we began. When they end so do we."

Q: But until we can get the human population under control, what course can we take to hold our own?

A: The key phrase here is *global sustainability*. That means living within our means, both today and tomorrow. Deserts, eroded landscapes, chronic water shortages and dwindling wildlife testify to the flagrant plunder of natural resources. The whole world, rich and poor alike, has been using slash-and-burn tactics.

Q: But don't we have to keep using up resources to feed and clothe the billions of people already here?

A: The use of sustainable resources at a rate that cannot be sustained leads to ultimate disaster. This is "the tragedy of the commons" that Garrett Hardin talks about. We have to overcome our addiction to growth at any cost and adopt guidelines for an economic policy emphasizing global sustainability.

Q: Is that course *practical* on a global basis?

A: Absolutely. We've got to shift from competition between regions and nations to a cooperative system where we work together to abolish war and help each region sustain itself. Some key points are building the organic content of soil through sound agricultural practices, conserving water by putting a price on it that reflects its true value, and shaping each region's forestry and agriculture to reflect local conditions.

Q: Can you give some suggestions?

A: Don't grow "thirsty" crops or animals in arid areas, don't cut forests on steep slopes where erosion is sure to follow, replant with native trees that don't need pampering, and treat farms with organic fertilizers from local plant and animal wastes. Tap renewable sources of energy, such as wind, sun, and running water wherever possible—and recycle, recycle, recycle!

Q: But is there anything I can do about global population?

A: Yes, urge Congress and the Administration to increase assistance and research for family planning and international population programs. And remember: overpopulation is at the root of environmental problems, but it cannot be separated from other social issues. Population programs have little chance in areas blighted by war, poverty, and illiteracy.

ECOSYSTEM PATHWAYS

The picture asks how our aesthetic and spiritual traditions relate to nature. How have they drawn from the natural world and how do (or don't) they illuminate our relationship to it? Do our arts and spiritual life support, partner, or control nature? Moreover, what are the values behind our decisions and acts regarding education, politics, and money?

The conservation ethic of preserving a balance among living beings and land presupposes a partnership between humans and nature. Values can range from "let it be" to "let's get the most out of it." A burning moral debate surrounds the extent of nature's rights and how much value to give to nature.

The following developments set the stage for the issue of nature's rights:

- The abolition of American slavery in 1863 reflected the understanding that humans could not simply be regarded as property but had inalienable rights.
- Reverence for animals came next, characterized by the values of such people as Albert Schweitzer and John Muir.
- The Endangered Species Act in 1973 gave legal protection to plants and animals.

At the heart of the matter is the place of humans within nature. As numbers of people have increasingly spread over the land, the question of human beings versus nature has gained in urgency. The intensity of the debate has been carried on in a number of periodicals, such as *Environmental Ethics*, *Ecophilosophy*, *The Deep Ecologist*, *Between the Species*, and *Ethics and Animals*.

17
ART AND SPIRIT

RELIGIOUS TRADITIONS

It is a general assumption in Western culture, as shaped by Christianity and Judaism, that man is the crowning glory of creation and has dominion over all of nature. This belief arises from Genesis, wherein it says, "God blessed them and said to them: 'Be fruitful and multiply, and replenish the Earth, and subdue it; and have dominion over the fish of the sea, and over the fowl of the air, and over every living thing that moves on the Earth.'" This human-centered attitude ordains the use of resources solely for the benefit of humans and has governed our politics and policies toward land and wildlife for a long time.

But the quote from Genesis does not reveal the whole picture. Jews believe that although God apportioned the world to humankind, God always remained its Master. In pure practice God owned the land and spewed people out if they didn't "behave." According to the Bible, every seventh year land was supposed to rest and lie fallow. Trees are considered sacred. Jews plant them whereas Catholics light candles.

Jews and Christians both have a strong sense of caring for Earth as the "garden of Eden," and when the environmental crisis mounted, they formed numerous organizations to protect God's sacred creation. Such organizations were among the most numerous at the 1992 Earth Summit. They have pledged to work with science and government toward sustainability. They publicly witness, initiate curriculums, and disseminate materials throughout their networks.

Eastern religions have always held a vision that stressed the oneness of all beings. Much of the focus of contemplation is becoming united with the beings of creation and getting rid of the ego's tendency to separate.

Among many indigenous cultures, shamans are the wise, holy "doctors" who have the ability to enter states of consciousness in which they envision animals, plants, and other beings. They do so in order to access the power for healing that resides in nature. Meditating upon and invoking animal, bird, insect, fire, water, air, and fertility spirits through dance, song, and ritual are common practices.

The myths and ceremonies of American Indians originally centered around the sun and the moon, and around animals such as the bear, eagle, deer, buffalo, snake, and coyote. The Great Spirit dwells in the wind and rain and flowers. In a creation story of the Iroquois, the Earth was created on the back of a giant turtle, and is a living being that must be nurtured. Life came from the sky world, as a woman who fell to the Great Turtle Island. Indians in the past demonstrated an integration with the Earth that the "white man" lacked.

It is ironic to note that the widely quoted speech of Chief Seattle was written for Earth Day 1970 by a white man named Ted Perry in a filmscript for a Southern Baptist Convention. Here is part of it:

> How can you buy or sell the sky, the warmth of the land? The idea is strange to us. If we do not own the freshness of the air and the sparkle of the water, how can you buy them? Every part of this Earth is sacred to my people. . . .
>
> We know that the white man does not understand our ways. One portion of land is the same to him as the next, for he is a stranger who comes in the night and takes from the land whatever he needs. The Earth is not his brother, but his enemy, and when he has conquered it, he moves on. He leaves his father's grave behind, and he does not care. He kidnaps the Earth from his children, and he does not care. His father's grave and his children's birthright are forgotten. He treats his mother, the Earth, and his brother, the sky, as things to be bought, plundered, sold like sheep or bright beads. His appetite will devour the Earth and leave behind a desert. . . . This we know: the Earth does not belong to man; man belongs to the Earth. All things are connected. We may be brothers after all. We shall see. One thing we know which the white man may one day discover: our God is the same God.

Father Thomas Berry, a Passionist priest and author of *Befriending the Earth: A Theology of Reconciliation Between Humans and the Earth* and *The Dream of the Earth*, is an exponent of a theology based on ecology. His view emphasizes the oneness that emerges out of new physics concepts, such as the view of the world as an indivisible dance of atoms.

Trained as a historian of cultures and specialist in Asian religions, he argues, "From its beginning in the galactic system to its earthly expression in human consciousness, the universe carries within itself a psychic as well as a physical dimension." He credits modern science with the discovery that the cosmos is not fixed but self-transcending and expanding in an irreversible process. But he blames the materialist bias of science for focusing only on the physical side of this process and ignoring its numinous dimension. Humans have a primordial genetic attachment to the natural world. He says, "The same atoms that formed the galaxies are in me." The evolving cosmos is our teacher.

The recent attention to goddess myths and ecofeminism is a manifestation of the appeal of the spiritual aspect of ecology. The patriarchal gods, which have been in ascendance for over two centuries, are blamed for the splitting off of humans from nature consciousness. Goddesses such as Artemis-Diana or Persephone-Demeter are associated with healing the wounds of the Earth in the psyches of both men and women. Earth and our physical bodies are one, requiring the same care. Spiritual feminists resonate with the richness of goddess traditions and find inspiration therein for living in harmony with nature. Rituals honoring nature take place in many parts of the world at solstices, equinoxes, and phases of the moon.

Starhawk (author of *Dreaming the Dark, The Spiral Dance, Truth or Dare*) describes this view:

> Earth-based spirituality is rooted in three basic concepts that I call immanence, interconnection, and community. The first—immanence—names our primary understanding that the Earth is alive, part of a living cosmos. What that means is that spirit, sacred, Goddess, God—whatever you want to call it—is not found outside the world somewhere—it's in the world: it *is* the world, and it is us. . . . Our deepest experiences are experiences of connection with the Earth and with the world. When you understand the universe as a living being, then the split between religion and science disappears. . . . Earth-based spirituality makes certain demands. That is, when we start to understand that the Earth is alive, she calls us to act to preserve her life. When we understand that everything is interconnected, we are called to a politics and set of actions that come from compassion, from the ability to literally feel *with* all living beings on the Earth. That feeling is the ground upon which we can build community and come together and take action and find direction.

Communities have formed dedicated to spreading such a message. One example is Genesis Farm, a program of the Dominican Sisters of Caldwell, New Jersey, that practices its principles and offers learning programs. "Rooted in a spirituality that reverences the Earth as a primary revelation of the Divine," it welcomes people who search "for more authentic ways of achieving human and earth security."

The sacred is the emotional force which connects the part to the whole.

—William Irwin Thompson

"Deep" Ecology

The problem of the centrality of humans' place in nature has engendered intense philosophical debate. To some even the word "stewardship" or the idea of "spaceship Earth" assumes dominance and control of nature by humans. The phrase "deep ecology" was taken from Arne Naess, a Norwegian philosopher, who put forth the idea of ecological equality.

The debate emerged out of conflict over the treatment of wilderness. One side allowed wilderness areas to be managed for recreation and tree farms. Some argued that humans have always influenced the direction of nature and that therefore there can be no such thing as pure wilderness. Others claimed that wilderness should be left entirely alone and nature allowed to work in its own time, however impatient people get. As wilderness increasingly

Eco Joke

Student: I am a deep ecologist.

Eco Guru: You mean in contrast to a "shallow" ecologist?

Student: Yes.

Eco Guru: Whatever you do, be sure that you don't become an arrogant ecologist.

A thing is right when it tends to preserve the integrity, stability and beauty of the biotic community.

—Aldo Leopold

disappeared, more people sided with nature against man. Writer Edward Abbey declared that he could no more slice through the tissues of a tree than those of a person. In 1979 David F. Favre, a lawyer, made the radical proposal that an amendment to the Constitution be made to the effect that all wildlife should have the right to a natural life and that humans could not interfere with that right without due process of law.

Gary Snyder, American poet and Buddhist, saw his role as a spokesman for wilderness, by which he meant all nonhuman creatures and things "that are not usually represented in intellectual chambers or in the chambers of government." He said, "Non-harming must be understood as an approach to all of living and being, not just a one-dimensional moral injunction. Eating is truly a sacrament. . . . Looking closer at this world of oneness, we see all these beings as of our own flesh, as our children, our lovers. We see ourselves too as an offering to the continuation of life."

In 1985 Bill Devall and George Sessions published a book called *Deep Ecology*, which furthered the argument against the anthropocentric, human-centered aspects of the conservation movement. They said that the usual conservation approach has as its objective the health and well-being of people, whereas wilderness should have a value independent of human access to it. In the usual approach, concern for natural resources is utilitarian, implying that if some life forms or ecosystems have no value to us, they should be allowed to disappear. Just the act of attributing value to life forms is itself a human activity. They argue that other species have the same rights as humans because they are part of the cosmic web that has evolved together over the millennia.

Shallow ecology was seen as the ineffectual outcome of outmoded conservation efforts. According to Devall and Sessions, people try to defend ecosystems and wildlife in human courts, invariably losing much of it over the years through compromise, abuse, and inefficiencies. Whether "deep ecology" can save more or fewer species than traditional approaches remains to be seen. In any case, "deep ecologists" believe that a more long-term vision has to emerge, one that questions the human right to dominate nature, and includes our organic connections to all of nature. Humanity's highest possibility is participating in the evolutionary destiny of a community in which all components are equal.

Deep ecologists Joanna Macy and John Seed hold workshops called the "Council of All Beings." Through guided visualizations they take members of their workshop through the details of evolution to help them feel more empathetic with nature's processes. In this way people learn to identify the needs of all, from rain forests to reptiles, and to cope with their despair over threats of extinction.

Eco Koan

A student came to the guru and asked, "Shouldn't all species have equal rights to life and habitat?"

The guru responded, "That's a difficult question. I wonder: how much is a cell or a virus to be valued compared to a hawk? What legal rights should each of them have and what will that mean? Should a corn plant have rights? We could duck the difficult issues in your question by giving legal protection to ecosystems, which provide opportunity for all species to survive. On the other hand, what kind of protection should be given to the ecosystem and how much? These are the vital questions we must wrestle with in the years ahead."

Artistry of Nature

We love nature for its beauty, aside from its functions. Every season has its colors, the birds their plumage, the flowers their shapes and scents, the meadows, mountains, moonlight, sun glistening off water, its infinite variety.

Many of nature's designs inspire awe, for instance, the geometrical precision of each snowflake, but most of these designs go unnoticed by any

AURORA BOREALIS

Walter Sullivan

Few natural phenomena are as awesome in scale and grandly beautiful as the ever-changing curtains of light that hang in the sky during an auroral display. For those living in the auroral zone at high latitude, as in Fairbanks, Alaska, Fort Churchill on Hudson Bay, and northern Scandinavia, visible auroras are common, particularly during the long winter nights and when a major eruption on the sun has disrupted the magnetic field of the Earth. It is easy to understand why many northern peoples believed the multicolored displays were sunlight reflected off the polar ice. The Norse regarded the dancing lights as Valkyrie riding across the sky. According to Eskimo legends they represent wild ball games being played in the heavens by departed ancestors. . . .

Today, thanks to recent discoveries from space and on the Earth, we realize that auroras are not supernatural but are to a large extent generated by the sun, 93 million miles away. It has been known for some time that the aurora's rays, glows, and rippling curtains of light are produced by extremely high-energy particles (electrons and protons) striking atoms and molecules of the high atmosphere as they plunge Earthward along field lines of the Earth's magnetism. . . .

At times the cascading of high-energy particles into the auroral zone may carry energy that is equivalent to a billion kilowatts. Not all of this is manifested in visual displays. An intense electric current is generated in the auroral zone—the polar electrojet—that may induce voltages in long electric conductors on the Earth, such as powerlines and pipelines. . . .

The displays occur between sixty and two hundred miles above the Earth, their colors being determined by the nature of the atoms or molecules hit by incoming electrons and by the energies of those collisions. Molecules of oxygen, for example, glow red or green. Hydrogen molecules may likewise produce a red glow, whereas individual hydrogen atoms will radiate green light. Nitrogen atoms generate purple, and nitrogen molecules glow pink.

human eye. Early on, however, American Indians perceived the pattern of the circle in the cosmos:

> Everything the Power of the World does is done in a circle. The sky is round, and I have heard that the Earth is round like a ball, and so are all the stars. The wind, in its greatest power, whirls. Birds make their nests in circles, for theirs is the same religion as ours. The sun comes forth and goes down again in a circle. The moon does the same, and both are round. Even the seasons form a great circle in their changing, and always come back again to where they were. The life of a man is a circle from childhood to childhood, and so it is in everything where power moves. Our tepees were round like the nests of birds, and these were always set in a circle, the nation's hoop, a nest of many nests, where the Great Spirit meant for us to have our children.
>
> —*Black Elk Speaks*

Likewise, we know that planets circle around the sun, and electrons around the center of atoms. Our fingerprints form unique circular patterns. The symmetry in the cycling of nature's elements—water, nitrogen, carbon—dazzles.

The courtesy of the soul is lived only when every man as an individual person has been recognized as kin—and indeed when every animal and thing too is given the rights of kin in the oneness of creation.

—Helen M. Luke

THINKING LIKE AN

ECOSYSTEM

Color and Beauty

Annie Dillard extrapolates from the goldfish to the larger world:

Chloroplasts bear chlorophyll; they give the green world its color, and they carry out the business of photosynthesis. . . . If you analyze a molecule of chlorophyll itself, what you get is one hundred thirty-six atoms of hydrogen, carbon, oxygen, and nitrogen arranged in an exact and complex relationship around a central ring. At the ring's center is a single atom of magnesium. Now: If you remove the atom of magnesium and in its exact place put an atom of iron, you get a molecule of hemoglobin. The iron atom combines with all the other atoms to make red blood, the streaming red dots in the goldfish's tail.

It is then a small world there in the goldfish bowl, and a very large one. Say the nucleus of any atom in the bowl were the size of a cherry pit: its nearest electron would revolve around it one hundred seventy-five yards away. A whirling air in his swim bladder balances the goldfish's weight in the water; his scales overlap, his feathery gills pump and filter; his eyes work, his heart beats, his liver absorbs, his muscles contract in a wave of extending ripples. The daphnias he eats have eyes and jointed legs. The algae the daphnias eat have green cells stacked like checkers or winding in narrow ribbons like spiral staircases up long columns of emptiness. And so on diminishingly down. We have not yet found the dot so small it is uncreated, as it were, like a metal blank, or merely roughed in—and we never shall. We go down landscape after mobile, sculpture after collage, down to molecular structures like a mob dance in Breughel, down to atoms airy and balanced as a canvas by Klee, down to atomic particles, the heart of the matter, as spirited and wild as any El Greco saints. And it all works. 'Nature,' said Thoreau in his journal, 'is mythical and mystical always, and spends her whole genius on the least work.' The creator, I would add, churns out the intricate texture of least works that is the world with a spendthrift genius and an extravagance of care. This is the point.

—Annie Dillard, *Pilgrim at Tinker Creek*

Recognizing the wisdom in nature's designs, some people have tried to adapt them to our purposes. (See "Drawing on Nature," on facing page.)

Artists and Ecology

Nature has been a sublime inspiration to artists always. Here is an excerpt on the use of birds in art and music. (See "Man and Birds," on facing page.)

As the ecological crisis has heated up, artists have made many kinds of statements about the Earth. *Time* magazine featured Christo's globe wrapped in plastic for an issue on the state of the environment. Cornelia Hesse Honegger did a series called "After Chernobyl," consisting of paintings depicting in microscopic detail the deformity of insects. Many artists document or denounce the effects of pollution and waste, and today museums more frequently offer exhibits on ecological fragility.

Some artists have been inspired by textures and shapes in stones and

Drawing on Nature

Delta Willis

The participants investigate the blueprints of nature. The dome of a sea urchin, for example, is used as a model by no fewer than five speakers. There is a debate about the architecture of sand dollars. The finesse of a butterfly's proboscis is compared to the design of an oil drill. The pleat of a poppy demonstrates how folding patterns strengthen even the most delicate material. A barnacle is seen as the epitome of efficient design. "These are lovely things that need nothing extra," says paleontologist Adolf Seilacher. "They have only the perimeters that need to be there."

Frugality is key, or as Charles Darwin noted, "Natural selection is continually trying to economize every part of the organization." Castles of clay built by African termites are better thermoregulated than any of our skyscrapers; a single chamber is maintained at a constant 86 degrees Fahrenheit, no matter the temperature outside; the hotter the climate, the taller the chimney. The shell of a pearly nautilus encases what amounts to a jet propulsion engine with soft parts, less likely to break than rigid ones. The shell itself is capable of enduring the pressure at depths of 1,800 feet, a model of maximum strength employing minimum material.

Trees invite particular envy. Their leaves are highly efficient solar panels, yet they have many other functions. Leaves fold to avoid drag when the winds blow, and by

transpiration move small rivers up trunks as tall as 360 feet. As the oldest living things on Earth, trees also win the prize for fatigue resistance. The oldest in North America are bristlecone pines in the White Mountains near the California-Nevada border. Now 4,600 years old, they were drawing sap via complex hydrological systems before aqueducts were built in Rome. Trees even recycle their wastes: Forest duff is buried, then converted into fossil fuels like charcoal. . . .

Drawing blueprints from nature is nothing new. During the 15th century Leonardo da Vinci designed sleek ship hulls based on the movements of fish in water. His notebooks are rich in comparative anatomy and corresponding machines, several inspired by birds in flight. The Wright brothers devised stabilizers after the way a turkey vulture employs its primary feathers to reduce turbulence at low speeds. The cockpit of the supersonic Concorde is lowered on approach, like the head of a swan. The interior braces and struts of an eagle's wing have inspired bridge designs. Latticework for London's Crystal Palace was inspired by the veins and ribs of the water lily Victoria regia. . . .

With the advantage of millions of years of evolution, natural forms offer us blueprints in efficiency. It's yet another argument for preserving the planet's biological diversity—within every form is a potential blueprint.

Man and Birds

The Strange, Wonderful, Enduring Relationship of Man and Birds

Frank B. Gill and Joseph Ewing

With no other animal has our relationship been so constant, so varied, so enriched by symbol, myth, art, and science, and so contradictory as has our relationship with birds. . . .

Indeed, in almost every primitive culture birds were divine messengers and agents: To understand their language was to understand the gods. To interpret the meaning of the flight of birds was to be able to foretell the future. Our words *augury* and *auspices* literally mean bird talk and bird view. By the time Greek lyric poetry was flourishing (5th and 4th century BC), the words for bird and omen were almost synonymous, and a person seldom undertook an act of consequence without benefit of augury and auspice. This practice prevails in Southeast Asia and the Western Pacific. . . .

Perhaps the greatest influence of birds on art has been in music. The earliest piece of English secular music of which we know, "Summer is Icumen in," is a canon for four voices and the words are those of the 13th century lyric in which the cuckoo welcomes summer with its song. The cuckoo, nightingale, and quail

(Continued on page 256)

(Continued from page 255)

are heard in Beethoven's Sixth Symphony. The 18th-century composer Boccherini wrote a string quartet called "The Aviary," perhaps the first complex composition in which a number of birds are imitated.

Birds as subject and as metaphor are found frequently in opera. Wagner wrote an aria about owls, ravens, jackdaws, and magpies for *Die Meistersinger*. In Puccini's *Madame Butterfly* a character sings of a robin, in *La Bohème*, another sings of swallows. In what is probably the most popular aria in the most popular opera of all time, the "Habanera" in Bizet's *Carmen*, the opening words are "Love is a rebel bird that no one is able to tame." Janáček's *The Cunning Little Vixen* is a 20th-century "nature opera" with animal characters that include numerous birds, and in Gershwin's *Porgy and Bess* there is a buzzard. These examples are only a few of the many that come to mind. . . .

In rock music, the best treatment of birds may be Jimmie Thomas's 1958 "Rockin' Robin," a solid hit when first sung by Bobby Day and revived to equal acclaim by Michael Jackson and The Jackson Five. Swallows, chickadees, and crows urge the robin to "Go, bird, go," a raven teaches him the Charleston, and he turns out to be a better dancer than buzzards or orioles. . . .

Our enjoyment of bird sound in music extends beyond mere imitation. The phonograph recording of the singing of a real nightingale is played in performances of Respighi's *Pines of Rome*, and James Fassett, an American composer, has written a *Symphony of Birds* that consists entirely of the recorded songs and calls of real birds.

The role of birds in paintings and sculpture is impressively large. Birds appear in paleolithic cave paintings in France and Spain as early as 14,000 BC, and as neolithic cave paintings in Eastern Turkey, 8000 years later. In Egyptian tombs at Thebes, very accurate bird paintings appear before 2000 BC. . . .

A few centuries later birds begin to abound in Greek vase painting, some depicted with a quite modern realism. In Roman frescoes and mosaics, bird are sometimes stylized, but many are not. Among the most vibrant and brilliantly colored are those in mosaics from Pompeii that are now in the National Museum of Naples.

In much of medieval art, birds became so highly stylized that it is often impossible to identify the species. A remarkable exception is an assemblage of species in a 13th-century illuminated manuscript of the Book of Revelations. The composition, in which the birds sit before a man preaching, is strikingly similar to Giotto's famous painting of St. Francis preaching to the birds, with one latecomer arriving after the sermon has begun. Hieronymus Bosch's *Garden of Delights* (about 1500), is filled with birds, some realistic, though sinister, and some monstrous hybrids.

Birds appear frequently in English and Dutch nature paintings and still lifes of the 17th century, but the Romantic painters of the 18th century showed little interest in birds despite great enthusiasm in Europe at the time for Japanese prints in which birds abound. The impressionists and postimpressionists also showed little interest in birds, though van Gogh's last work, painted on the day of his suicide, is of a flight of rooks across a somber sky. . . .

Birds are ubiquitous in literature. In his comedy *The Birds*, Aristophanes mentioned 79 species and was well informed about their habits and appearance, as his Athenian audience must have been; otherwise they would have missed much of the playwright's wit. For its perfect matching of avian and human characteristics, the play has been described as an "ornithomorphic view of man." One of the earliest lyric poems in English describes a quarrel between an owl and a nightingale. The cuckoo, nightingale, and lark probably appear in poetry more often than other species.

wood and land contours, making these the focus of their work. Some "earthwork" artists throw the beauty of nature into high relief and invite us to appreciate it more. Sculptor Harvey Fite turned an abandoned bluestone quarry into six and a half acres of curving stone walls, terraces, and paths culminating in a central monolith. He enclosed springs to make pools and fountains and preserved existing trees. Having worked on restorations of the Mayan cities of Copán and Tikal, he used ancient techniques and let his own intuition dictate how to reveal the stone with its presence of glacial scars and fossils. The aim of such works is not to glorify the artist but to make recompense to corrupted land.

For two decades Helen and Newton Harrison, a painter and sculptor from California, have created projects that reclaim rivers, restore riparian habitat, and replenish groundwater tables. Artists have turned garbage landfills into parks. Such artists, working with engineers and designers, give a new twist to "public art."

Artists have long been of value for enabling us to study wildlife in its particulars. One of the masters is Roger Tory Peterson, whose art and field guides have served us for decades.

A Field Guide to Roger Tory Peterson

William Zinsser

"Very often a single bird will get a person started birdwatching," Peterson told me. "In my case it was a flicker that I saw when I was eleven, which I thought was dead—it was just a bundle of brown feathers. All of a sudden it exploded into life. That was the crucial moment of my life. I was overwhelmed by the contrast between something that was suddenly so vital and something I had taken for dead. Ever since that day I've felt that birds are the most vivid expression of life. Birds symbolize freedom, and I think that's why birdwatching is so important to so many people."

The person who got Peterson out and looking at birds was his seventh-grade teacher, Blanche Hornbeck. "We'll find out what they are," he recalls her saying. She thereby set him on a path so obsessive that, as he once wrote, "birds have been the focus of my existence; they have occupied my daily thoughts, filled my dreams, dominated my reading." Miss Hornbeck had organized a Junior Audubon Club, where children were given a 10-cent leaflet that contained a description of a bird, a color plate of that bird, and an outline drawing to be colored in with crayons.

"Miss Hornbeck soon decided that coloring an outlined bird wasn't the way to learn to draw," Peterson told me. "One day she brought in a portfolio of The Birds of New York State, by the great bird painter Louis Agassiz Fuertes. Each of us was given a small box of watercolors and a color plate from Fuertes' book to copy. I got the blue jay, and that was my first bird painting. People often ask me what my favorite bird is, and I have to say it's the blue jay, though many people dislike it, and of course my other favorite is the flicker. Locally, here in Connecticut, my favorite is the osprey, and among sea birds it's the wandering albatross. My favorite family are the penguins. . . .

"I grouped birds that look alike and therefore might be mistaken for each other, instead of grouping them by species, and I painted them in similar positions so that they could be compared. I made my paintings schematic and two-dimensional, and I drew little arrows to point out the 'field marks' that are the main information you need to identify a bird. Those arrows were my invention."

The result was an innovative book submitted by an unknown artist in a Depression year, and it was rejected by four publishers before Houghton Mifflin took a chance on a printing of 2,000, stipulating, as a financial hedge, that Peterson would receive no royalties on the first 1,000 copies. That was the last time Peterson would be considered a gamble. The first printing sold out in a week, and the 58 years since then have brought no hand-wringing at bookstores that sell the Field Guide or its many companion volumes, which have Peterson as series editor and which operate on the same principle, identifying such natural forms as butterflies, moths, beetles, wildflowers, shells, and rocks and minerals. . . .

I asked Peterson, "What do you know at eighty-four that you didn't know when you were seventy?"

"What I've come to know," he said, "is how much I don't know."

"About painting?" I asked.

"No," he said. "About birds."

I am no longer trying to 'express myself,' to impose my own personality on nature, but without prejudice, without falsification, to become identified with nature, to see or *know* things as they are, their very essence, so that what I record is not an *interpretation—my idea* of what nature *should be*— but a *revelation*, a piercing of the smoke screen artificially cast over life by neurosis, into an absolute, impersonal recognition.

—Edward Weston, *Daybooks*

Photographers such as Ansel Adams and many others who have devoted themselves to the study of nature have played an important role in protecting land and wildlife. Their images have graced the pages of nature magazines, highlighting scenes and creatures that many of us do not have a chance to see.

Visual artists use materials in their work that are often a danger to themselves. Photographers stand for hours over trays of developer, stop bath, and fixing solutions. Color-film processing involves even more chemicals that require careful handling. Dyes and toners may contain selenium, uranium, iron, gold, and platinum. Kodak recommends that workrooms have ten changes of air per hour as well as exhaust ventilation. Acids can burn and blind. Benzene, a petroleum distillate found in varnishes, lacquers, and solvents for waxes and oils, poses a cancer risk.

Painters use cadmium paints. Cadmium is a silver-white metal that can poison and cause cancer. They use solvents with irritants such as acetone and turpentine, a mixture of resin and oil, with vapors that can cause tissue death, kidney damage, depression of the nervous system, or death. Formaldehyde, a colorless liquid or gas with a pungent odor, is found in adhesives, paints, textiles, fiberboard, and foam particles. It is toxic and carcinogenic. The higher the concentration, the higher the risk. The Center for Safety in the Arts offers artists extensive guidance on potential dangers.

The realm of music has seen a surge of environmental works. Paul Winter and others draw attention to the sounds of whales and perform ceremonies to honor the solstices and equinoxes and the majesty of canyons and rain forests. There has also been an increase in the dissemination of the music of indigenous people.

The Personal Is Political

While many of us have not understood ecosystems, we have sought a connection with them in a myriad of ways. Those who cultivate gardens spend hours learning about the propagation of seeds, soil conditioning, the rhythms of flowers and vegetables, disease and pests, harvesting, and the beauty not just of a patch of lilies but of systems. Gardens are miniature ecosystems that we cultivate.

We walk in woods or hike in mountains, resonating to the trees, ferns, stones, and meadows even if we don't know their names. We travel long distances to watch birds or butterflies. We fish, paddle down rivers, go to beaches, meditate, and keep journals. Our spirits are nourished by such contact, perhaps much more deeply than we know or can articulate.

Here are some journal excerpts. The one that follows is part of a contemporary American Indian's attempt to connect with the energy of the Earth.

> *May 3, 1977.* I went to the Vapor Caves and spent hours there, underground. I walked down the old stairs into stone beneath the ground. In the long tunnel I began to feel the steaming heat, to smell the sulphur and minerals of Earth condensing on my body. The slow drips of water wearing holes into the stone. Walls coated in white crystal, salty to taste, and green algae. I lay down on a warm marble slab, feeling a small amount of fear at being alone underground with only the sounds of Earth. Even objects placed inside the cave by people seemed primitive and were becoming overgrown with natural elements. Some bricks were coated with a green deposit. The marble slabs

placed inside as seats were slowly being shaped by the moving water. I liked the feeling of balance and naturalness, as if nothing could withstand the slow passing of time. Also the feeling of being inside the powerful subconscious of the physical world. I kept thinking that certain spirits or spirit thoughts were stored here. Old people, years ago, healed in these caves, left something of themselves. Alone, I felt these energies. There was the heat, the vapor and steam, the odor of being inside ground, buried in the center of a potent vibrance. Then the water steamed up, bubbled from some core inside the stones and made gurgling noises. Occasionally the rock itself would open, groaning. And myself a part of all that, detached in some way, not static, not long-lasting in the physical world but still somehow connected to that energy and it to me, as if we fed each other. —Linda Hogan, Chikasaw

Here is the journal entry of Gretel Ehrlich, author of *The Solace of Open Spaces*, when her beloved horse had to be shot. In it echoes our love for animals.

> Bluebird is dead. My horse who has been with me since I came to Wyoming, who has herded sheep, followed me from town to town, tracked cattle . . . who, clownishly, shared pans of dogfood with Rusty, clambered up stairs to the veranda and looked in the house, who often stuck his head in the kitchen door on summer days to beg for cookies, who would not be caught the first six months I owned him just to show who had the upper hand, who never bucked, who hated most men, who was so ugly, big-headed, hairy-legged that some ranchers refused to haul him in their trailers, who was the butt of many jokes, who laughed silently at all of us, who knew things but wouldn't tell me except when I wasn't asking, who gave me a look one day that told me something about the pain of not having language, of the longing to have a communication that was not muddled (on my part), that said, "Animals know things and want things too," who comforted me because we shared hard winters and harrowing lonely years. —Gretel Ehrlich

Here Thoreau comments on how vital to personal wisdom deceptively mindless hours in nature could be:

> I love a broad margin to my life. Sometimes, in a summer morning, having taken my accustomed bath, I sat in my sunny doorway from sunrise till noon, rapt in a revery, amidst the pines and hickories and sumachs, in undisturbed solitude and stillness, while the birds sang around or flitted noiseless through the house, until by the sun falling in at my west window, or the noise of some traveller's wagon on the distant highway, I was reminded of the lapse of time. I grew in those seasons like corn in the night, and they were far better than any work of the hands would have been. They were not time subtracted from my life, but so much over and above my usual allowance. I realized what the Orientals mean by contemplation and the forsaking of works. For the most part, I minded not how the hours went. The day advanced as if to light some work of mine; it was morning, and lo, now it is evening, and nothing memorable is accomplished. Instead of singing like the birds, I silently smiled at my incessant good fortune. As the sparrow had its trill, sitting on the hickory before my door, so had I my chuckle or suppressed warble which he might hear out of my nest. —Henry D. Thoreau, "Sounds", *Walden*

In these examples, nature evokes strong personal responses of caring and connection. Yet many of our Earth's problems stem from a lack of such affection. The overall poverty of spirit in our culture parallels the poverty in nature to such an extent that a movement to establish an "eco-psychology" is afoot. Conferences focused on "Healing the Earth, Healing the Self" are based on a recognition of how the suffering and victimization of the Earth and our selves is mutual, that the problems facing all of us are not just solved in legal or economic battles but must happen in our psyches, and that we humans need the presence and partnership of wilderness and animals above and beyond survival purposes.

Such a psychology, as put forth by Theodore Roszak in *The Voice of the Earth*, focuses on the issues of greed and anxiety that lead to the abuse of nature.

Many have seen humans' habits of consumption as an addiction that needs to be treated like overdrinking, overeating, overworking, or overspending. The addiction itself cuts us off from experiencing the vivid and vibrant natural world. Denial of addiction leads to more and more consumption, along with a sense of futility at ever being able to change. Psychology has taught us that greed has roots in emotional anxiety and narcissism, and that we won't change our ways unless those tendencies are addressed.

One way is to reflect on your habits as a consumer. When did you start to shop and buy as a child? Does money give you a sense of power? Do you know why you buy the things you do? Many of us don't know what necessities are.

Anxiety appears in two distinct forms. One is fear about possible illness (even fatal) from radiation, toxic chemicals, polluted food, water, and air. The psychological toll of our environmentally related illnesses and deaths has not been adequately assessed. Nor has the sadness and shame we feel at the devastation of forests, rivers, and endangered species. The other form of anxiety is fear of nature itself: earthquakes, hurricanes, unfamiliar insects, the immensity of mountains. When people feel lost among what they do not know, they see nature as the enemy, and may want to subdue it, plunder it, rape it before it hurts them.

As science and philosophy have shown us, mind and matter are not separate. Thus, the trashing of nature originates in our psyches and brings us pain in return. Perhaps our saving grace lies in our soul's yearning to know its larger ecological habitat, to be aware of ourselves as an organism in the matrix of the cosmos.

Can we transform ourselves in time to save our nest?

Eco Quiz

Q: What goes into a nature journal?

A: You can make it whatever you like! You can use it to record dates, such as flower blooming times and bird appearances. When you go back to places year after year, you will be able to document changes to an ecosystem. Or you might use it to sketch a favorite tree. Even if you are no artist, you can learn a lot simply by concentrating on the details of a leaf or trunk. Express your thoughts and feelings, just as they are, and see where they take you.

Q: How do I think like an ecosystem?

A: You can start anywhere. For example, when you sit down to eat, think not only about where the food came from, how it was grown, and what materials went into the packaging, but also where the fork, knife, spoon, table, and napkin originated. Then consider the walls, floor, heat, and water. If you don't know, make a point of finding out.

When you are outside, try putting yourself imaginatively into the place of a tree or a bird or the soil, and sensing how life goes for it and what its needs are. If it had a voice, what would it say? This exercise can extend your empathy for nature's forms.

AFTERWORD

ON WHY IT'S IMPORTANT TO THINK LIKE AN ECOSYSTEM

Dr. Jan Beyea
National Audubon Society's Chief Scientist

Several years ago, at an Audubon sanctuary, I noticed a wasp trying to escape from a building through a windowpane, bumping its head constantly against the glass as it followed its biological instinct to move upwards and towards the light. I thought nothing of this until, passing by an hour later, I found the wasp still trying to force its way through the same impassable barrier. "Stupid," I thought, until I realized that we humans aren't much brighter. When confronted with difficult problems, we too tend to bang our heads on impassable barriers. The wasp was confronted by something evolution had never prepared it for: a transparent solid. We too are confronted with something evolution never prepared us for: reaching the limits of the biosphere to gratify our expansionist nature. We too keep trying to force ourselves through a barrier we just can't believe is there.

While contemplating the similarities between wasps and humans, my eye caught sight of the lower windowpane. It was half open. If only the wasp could have gone in a direction opposite to that dictated by its instinct, it could easily have flown to freedom. The utter simplicity of its escape path made me sad as I pursued the analogy with humankind.

Is there not a corresponding way for us to effortlessly reverse our degradation of the planet? This Almanac explores that question. Its vision is that of a planet and its ecosystems protected by an informed citizenry that understands ecological and ecosystem concepts. Fortunately for the reader's patience, the path we travel in this book is fun and fascinating, not an ascetic one.

The underlying theme of the Almanac is the inescapable web of connection between ourselves and the Earth's ecosystems. Taking the ecosystem approach is crucial to solving environmental problems without causing new ones to emerge. There are few readers alive today who are not aware that this planet's water, air, and soil are in danger, that open spaces are being gobbled up by development, that animals, plants, birds, and insects are disappearing. Too often these problems are tackled as they arise, without comprehension of the big picture. To practice environmentalism symptomatically is like prescribing medicine for one part of the body without consideration of the whole, or practicing politics without understanding how the branches of government work.

Years ago as a college professor, I used to teach Earth Science. Over and over again, in those days, we talked about how the physical world shaped life. We talked less about how life had correspondingly shaped physical processes. The biological and physical were supposed to be separate worlds.

It is wiser not to separate the two worlds. It is wiser to think of the biological and physical components of our world as locked in an intimate dance—an eco-dance, if you will—that defines an ecosphere. It is amazing to think that so many of the rules of the dance are governed by the codes locked up in DNA and RNA found inside biological cells.

I first learned about the importance of merging the biological and physical worlds as part of my participation as a professor in a team-taught interdisciplinary seminar. Virtually nothing could be ruled out as irrelevant, because solutions to environmental problems demanded contributions from all major academic disciplines: economics, psychology, political science, philosophy, physics, chemistry, biology. Undergirding the whole enterprise was the notion of the great cycles of nature.

Matter, in the form of atoms and molecules, cycles from physical reservoirs into biological reservoirs (the bodies of living organisms) and back again. When we know this, when we think like an ecosystem, it comes as no surprise that the large-scale pollution we humans eject into the system must eventually end up in our own bodies and that of other living creatures.

Thinking like an ecosystem also requires knowing something about the rather abstract concept of energy, a term we all use in daily life, but one that has a more precise definition in science. Would that it were easy to explain energy in a sentence or two. I'm still not sure I understand it, and I've been studying it for thirty years.

Fortunately, for our purposes we have to know only a few basic facts about the concept before we turn to its role in ecosystems. Without fresh supplies of energy, nothing would move or grow. Economists would probably relate energy to money. Like energy, money can be converted into many forms: land, investments, bonds, all of which can be exchanged among each other according to certain rules. Under the "rules" of energy, however, conversions cannot always go back and forth; sometimes they are allowed only in one direction, which is why energy cannot be constantly recycled and why we speak of its "flowing" through a system, not recycling.

Biological systems need energy to overcome resistance to the friction of motion internally and in the world in which they move about; they need energy to overcome gravity when climbing, or moving parts of their bodies upwards; they need it to assemble molecules into cells, and to make the chemicals they can't find ready-made in their environment. Small creatures need energy to overcome their tendency to stick to surfaces.

From where does this energy come? Ninety-nine percent of it comes from the sun in the form of "photons," bundles of energy that flow across the void. Energy may be stored for a while—most significantly in biological molecules—and then released for later activity, but ultimately almost all of the solar energy falling on the Earth is sent back into space. We use it for a while, transform it, and send it back to the void. The number of photons that make up the radiant solar energy that impinges on us is unbelievably high, so great that it defies comprehension. Our eyes have evolved to detect these photons. Green plants have evolved to capture them, also. As this captured solar energy is used up for activity by biological systems, it ends up as heat, which is radiated away by the surface of the Earth, again as photons, but in a form that we have not evolved to see with our eyes (although we can feel it with our skin in the form of warmth). Looked at from space, if you had magical eyes to see photons of all frequencies, and could see their numbers, you would instantaneously see the pattern that governs the large-scale flow of energy. You would see visible photons striking the Earth, some being reflected from the lit side of the globe, but most disappearing as they are absorbed. At the same time, about twenty times as many "infrared" photons would be leaving the globe in all directions. Although you couldn't see it from this scale, a tiny fraction of the incoming photons would be stolen and stored for use by life's creatures—by orchids to grow, wolves to hunt, and children to play. A leaf, with stored energy, might drift over, and fall into, a shallow sea and be destined to end up as coal one hundred million years from now.

A lot happens that is of crucial interest to us and other creatures in the process of visible photons being used by biological systems. The initial photons interact with green plants, including algae on the ocean waters. The energy content of less than 1% of the photons is stored, using carbon as the storage agent, and is prevented from being immediately turned into heat. Carbon in the form of starch or other biological molecules is analogous to a storage battery in that energy can be withdrawn later, when the right conditions are met. Carbon bonds are the "carrier" of energy throughout the biological system. If we see a squirrel or chipmunk or songbird, we know that these creatures have gotten their energy from converting carbon from plants to carbon dioxide as an outgrowth of the breathing process. If we raise our arms, we are seeing the same energy conversion process taking place. As we exercise and feel our bodies heat up, we are experiencing the final stage in the energy-flow process. Once we contribute to heating the air, infrared photons will soon be leaving the planet for outer space.

Of all the cycles that shape life on the planet, and are in turn shaped by it, the carbon cycle is most important. Except for a small number of life forms found near hot ocean springs that depend on sulfur for energy storage, all life on the planet draws from the nonliving carbon reservoir in the biosphere, and then cycles carbon back to this reservoir after death. Over enormously long time frames, carbon also cycles from the atmosphere through rock back to atmosphere. This longer cycle has many components. Carbon moves first from air into the oceans; from the oceans into carbon-containing rock that is drawn down into the Earth by the motion of tectonic plates; from its temporary resting place under continents upwards to the surface as a component of new mountains; from surface rock to soil as wind and rain wrestle it into small pieces; from soil consumed by plants; and finally back to the atmosphere in the form of carbon dioxide exhaled or converted by life forms—an intricate mix of physical and biological interactions.

This cycle can be made less abstract by imagining that we can track an individual carbon atom as it runs through the cycle. This atom, which we may name "wanderer" if we are of a poetic bent, starts out in the atmosphere bound to oxygen as carbon dioxide. It wanders around, bouncing off oxygen and other molecules in the atmosphere for several hundred years. Sometimes it bounces off the ground; decades later it bounces off the ocean. Finally, it sticks to a part of the ocean on one of its many passes, where it remains in solution for thousands of years until it forms into calcium carbonate and is precipitated out onto the ocean floor. Hundreds of millions of years later it will appear again at the surface as rock, having been forced towards the continent by plate tectonics, driven down under the continental shelf, eventually pushed upwards. As a constituent of rock, it eventually is eroded away by rain, carried into streams, then underground to be absorbed by a plant.

It is this longer cycle that we interfere with when we burn fossil fuels and send extra carbon dioxide into the atmosphere. The oceans are the bottleneck—they can't absorb the excess carbon fast enough to prevent a buildup; hence higher concentrations of carbon dioxide result in the atmosphere, with the potential of climate change taking place so rapidly in terms of geological and biological time that life forms can't adapt. Even human survival may be at stake, should a runaway greenhouse effect materialize. Thinking like an ecosystem in this context means understanding that the carbon cycle is the basis of everything we know as life on the planet.

We can't see the long-term carbon cycle in action—it takes decades, centuries, hundreds of millions of years to reveal its workings. We can, however, see the shorter-term cycles, in which life withdraws carbon from the biosphere's carbon reservoir and gives it back as decomposers go to work on dead organisms. We can pick up a plant and feel the cellulose that was made from carbon being drawn from the soil; we can run our foot over decaying vegetation, knowing that bacteria are changing carbon in the plant to carbon dioxide. If we return, weeks later, to see the vegetation decreased in size, we know that the carbon has risen into the atmosphere.

The carbon cycle is complex, with many more variations than I have discussed. There are many "niches" for life forms to exploit, each of which modifies the cycle somewhat. The carbon cycle is but one of many cycles that we must understand to survive. Another is the water cycle.

As with the carbon cycle, let's follow an individual water molecule on its rounds, letting it start out in the middle of a breath from a child's mouth, whereupon it begins bouncing around in the air as it is hit by a constant stream of air molecules. Because of the barrage and collisions, its net progress off the ground is very slow. We imagine it taken in by an insect for a while, but not forever. Ejected once more, it continues its wanderings. Over a period of days it rises a few thousand feet in the air until it gets so cold that it can condense (bind) with other water molecules into a tiny water droplet. Now the water droplet drifts hundreds of miles with the wind until it passes down below the condensation level and starts to evaporate. Our initial water molecule may evaporate from droplets many times until it finally ends up in a droplet heavy enough to fall to the ground without completely evaporating.

Having been moving around inside this large raindrop, our water molecule falls on top of a mountain, rushes downward, helping in the process to wear down the mountain. Billions and billions of such raindrops over the centuries erode the mountain rock, turning it into sediment. The water from the rain passes from a small stream into a river several hundred miles downstream. A human mother drinks water from the river to sustain her life, and two hours later she urinates our molecule into a septic system, from which it goes underground, is picked up by a plant root, is carried up into the plant, and is evaporated into the air again. The rapid collisions with air molecules begin again, causing another wildly erratic trip that eventually takes the water molecule back to the "water vapor line," where it again helps to form another tiny droplet.

As before, it travels hundreds or thousands of miles inside droplets until it collects enough companions to fall down through the layer and to reach the ground before evaporating. Back we go hitting the ground, contributing ever so slightly to weathering rock, helping to make soil. This time, as the water molecule enters a river, it makes it all the way to the ocean, spending eons there before it again reaches the surface and is evaporated once more into the air. Over and over the process continues: evaporation from the ocean, transport by wind over land, precipitation as rain, runoff to the sea, with some fraction absorbed by biological systems and used as part of the fabric of life.

The water cycle provides an artistic bonus: rivers can be the most beautiful of spectacles if they are not tamed by the engineer. Rivers are centers of biological diversity, with life forms adapted to the seasonal variations that rule the collection of water from watersheds. Think for a moment of a mighty river in Canada, still unspoiled, that begins to roar as the snow melts in the early summer. As it sweeps over its bed, it picks up nutrients for life that are carried down to an inland sea and thrust underneath the sea ice that has yet to be broken. The onslaught of water goes far out under the ice and helps to break it up, making it ready for summer melting. The nutrients have been spread out, and phytoplankton can grab them for growth. As sunlight pierces the ice, the ice clears, and photosynthesis takes place.

Thinking like an ecosystem: the phytoplankton bloom is the base of the regional food chain. Mess with it and the system will crash.

Biological creatures need other chemicals to function. Nitrogen from the air is an example. Having accompanied a carbon atom and a water molecule in their wanderings, it is time to learn the life cycle of a nitrogen atom as it cycles between the biological and physical reservoirs. We start our journey on a particularly lucky atmospheric nitrogen molecule that is grabbed by nitrogen-fixing bacteria living in a nodule on a clover plant somewhere in England. In its new chemical form it can be taken up by the clover and used in protein synthesis. Not too many years later, the clover is eaten by an unseen animal and our nitrogen atom is utilized by yet another biological system. Ultimately, decomposers will have their way and specialized bacteria will return the nitrogen to gaseous form where it can return to the atmosphere. There are many other pathways by which nitrogen can cycle, but we turn instead to a limiting role nitrogen plays in biological growth. Growth of an organism ultimately pushes up against limiting factors. For some species, nitrogen is such a limit. As a result, we humans intercede when extra growth is perceived to help us, as is the case with agriculture. Perhaps this is only a form of symbiosis between one form of mammal and certain plant species. Hopefully, we will be wise enough to utilize our agriculture sustainably, so our soils can last even with the five billion, and ever increasing, numbers of us.

Some of the major cycles, like the phosphorus cycle, involve the ocean and land as major physical reservoirs, not the atmosphere. Phosphorus, for instance, is lifted up from the sea bottoms and exposed as part of geologic processes. Thereupon, it is weathered away by the rain cycle and brought to soils, where plants and the organisms that depend on plants can use it. Ultimately it passes to the oceans and is precipitated out on the sea bottom as rock, starting the cycle all over again. Phosphorus is another limiting growth factor for certain organisms, so once again humans are tempted to intervene to improve production of plants valuable to the economy.

In fact, to enhance plant growth, we may bring to our farms as fertilizer phosphorus deposited by birds on islands millions of years ago.

There are many other elements and minerals that life has made use of in its evolution. All of them have a story to tell as they cycle in and out of the biological reservoirs that use them. None of them can be ignored if we are to think like an ecosystem, if we are to save the planet and the natural resources that we treasure.

Like many nature lovers, I first fell in love with nature in New Hampshire, where I spent summers as a child in camp. With a small mountain in the background that could be climbed, fields in which blueberries could be picked, snow caves that could be marveled at, and a sandy river that could be camped near, I felt at peace. But I saw so little of what was really there. I knew nothing of most of the major classification of species. I knew not of single-cell bacteria, many of which play crucial roles as decomposers (nature's garbagemen). I knew nothing of "protists," simple membraned organisms like diatoms and slime molds, which also get their energy from other creatures and can help to recycle dead organisms. I would never have guessed at the variability of such creatures; certainly not that slime molds can have thirteen sexes.

Fungi, another major class of species, I did know about; had knocked off many of them from their visible perches on trees. Following my human instinct, I struck and shaped without understanding, without knowing in this case that fungi play crucial roles as decomposers of dead organisms. There was no way I could think like an ecosystem in those days.

I certainly had no hint of ecological change back then. I had to wait to read Aldo Leopold to learn about concepts of biological succession: wave after wave of ecological revolutions and how humans can inadvertently set off these waves without having the slightest idea of what they are doing, and barely noticing the results. It takes many generations to see the pain. The destruction of 99% of our tallgrass prairies; the reduction of wilderness to 8% of the U.S. land area, only half of which is legally protected; the virtual destruction of virgin forest throughout the land and the wiping out of the large predators that we once feared. I have always remembered a line of Ian McHarg's in *Design with Nature* to the effect that, had we done our planning right, we could have had eagles in New York City. Yes, had we been thinking like an ecosystem all these years.

Even though I saw so little of what was around me in New Hampshire, there was still an emotional bond formed between me and the land that made me want to one day save prairies and ancient forests. Such bonds are the hope, the motivators to understand ecosystems.

First of all, ecosystems have boundaries. The example we are most familiar with is our own body, which is very "open," depending as it does so starkly on inputs of food and water from outside the boundary, and on disposal of its wastes outside as well. In contrast, the goings-on in a pond ecosystem may be fairly self-contained, with the inputs and outputs of the pond having relatively small impacts if removed—at least for some period of time. However, it is a rare ecosystem that is so closed that it can remain isolated for long from the impacts of changes to what is coming in or going out, from what is going on elsewhere.

To many scientists, an ecosystem is a general concept, one that is helpful for understanding and thinking about the interdependency of biological and physical systems like the Everglades or James Bay in Canada. There is no precise definition of the concept, just as there is no single definition of the word "system." Obviously, there is no physical line that precisely separates one ecosystem from another. Yet, to a newcomer to the ecological world, the concept is often revolutionary and magical. To understand how the elements of a country pond interact and depend on each other is as grand an insight as the symmetries of mathematics and physics, or as overwhelming a personal discovery as a great symphony.

The global ecosystem is essentially "closed" to biological and material transfers. Radiant-light energy does cross the boundary and "drive" the whole system. Smaller ecosystems may be "open," with some matter and creatures passing in and out. All involve the interaction of living and nonliving matter in a complex ritual that is both repetitive, cyclical, yet ever-changing (succession). Change one part of an ecosystem and the rest of the system that depends on that part, or is kept in check by it, reacts as well.

On the other hand, change is an essential part of ecology. This is a paradox for those of us who are desperately trying to preserve parts of the natural world that are being lost to development and who are trying to head off global climate disruption. Yet, the truth is that the "poisoning" of the first planetary atmosphere, which was a catastrophe for many organisms, made our own lives possible. The geological record shows a dramatic shift 3 billion years ago from a primarily carbon dioxide atmosphere to the primarily oxygen atmosphere that we have today—attributed to life's emergence and transformation of its environment on a global scale, primarily through the activity of plants.

Since that time the atmospheric composition has stayed relatively uniform, and millions of species have evolved depending upon it. Millions of species are thought to exist on the planet today, somewhere between 10% and 50% of which are projected to go extinct over the next fifty years because of human development. Clearly we are at a critical period in the history of the planet. Humans have become so technologically powerful that we are disrupting both the physical and biological state of the planet. Unrestrained, with our ever-expanding technological arsenal—now even including the ability to manipulate and swap genes with ease—we humans follow the normal biological pattern of species in altering the biosphere on which they depend.

Should we lie back and let our species make the planet safe for some new kind of creature, possibly a genetically engineered cross between a computer chip and our own DNA-based life forms? Each reader will have to decide that question on his or her own.

We should, however, never, never think that we can have our way with the planet without forever changing it. Our only realistic option is to walk as lightly as we can, stabilize our impact, and protect as much of the original physical and biological diversity as we still can. Fortunately, the species that is disrupting the planet this time has the power to prevent irreversible change. All that is needed is the will to save the natural world and a little restraint.

The vision of a protected planet must rely heavily on integrating our activities with the natural cycles. I see future households run by solar and renewable energy, with equipment that is technologically advanced so that just a little sunlight can power it. I see one child, two children, or no children in that household, with adults working in businesses that add to the quality of life without requiring the use of a great deal of matter, and the matter that is used having been recycled in factories also powered by solar energy. I see food waste from meals consumed in the household being turned back into soil components through the process of composting. I see members of the household traveling in vehicles that get thousands of miles to the gallon; using communication equipment to reduce the amount of travel that is needed. I see highways that are biologically integrated into the landscape with under and overpasses for nonhuman life forms, no longer serving as enormously long biological barriers that fragment habitat.

I see protected areas expanded and degraded lands restored. Most of all, I have a dream that all people have learned about ecology.

I have been in the environmental business for over twenty years and have finally learned some lessons that are worth passing on about the human-ecosystem interaction. Some of them can be succinctly told in the argot of rural life:

- *Don't spit into the wind.*
- *Don't spit into the well.*
- *Don't sit on a bird's nest; sit somewhere else.*

Other lessons are more abstract:

- *Changes to a complex system produce unanticipated results.*

Ecosystems are incredibly complex. If we remove one species, we affect what it eats, and what eats it, or what depends on it symbiotically.

Economists talk of the "invisible hand" of the marketplace allocating resources efficiently, and they are skeptical of planners who try to improve matters by using "command and control" techniques. Ecologists have their own "invisible hand"—the feedback of natural selection—and are skeptical when developers and economists try to improve the world by interfering significantly with nature.

That lesson can be taken to the local level. If my town allows a new insecticide to be sprayed over the community to kill mosquitoes, some other creature or plant will be affected in ways nobody knows or will tell me. Knowledge of this unanticipated harvest should make town leaders look for alternative ways to deal with mosquitoes by using known techniques whose consequences have already been learned.

At a minimum, experience should teach us to get an independent assessment of those likely impacts that will be obvious to an ecologist.

- *The cumulative impacts of many small changes is one of the most sinister ways that ecosystems are destroyed, and one of the most difficult threats to guard against.*
- *Careful anticipation of the consequences of past mistakes can reduce the number of unanticipated consequences.*

Ecological research, paying attention to past mistakes, environmental impact studies, knowledge of ecosystem principles—they all are very important. They can tell us where the bird's nest is hidden so we don't sit on it; they can tell us the paths that carry pollution to the well; they can tell us which way the wind will be blowing.

- *The first step in assessing the impact of a proposed change on the ecosystem is to look at which of the basic cycles are being interfered with.*

Is it the water cycle? The carbon cycle? The variability of the solar cycle? Once I determine this, I have an idea as to which creatures and plants will be affected, whether it will affect human health or wildlife, whether the impacts will be long-term or short.

- *Attention to ecological sanity does not mean personal pain and suffering.*

It means learning new habits individually, in industry, and in government. It means self-restraint, which can be annoying at times, but does not have to be painful. It means learning not to sit on the bird's nest.

- *Many of the human actors making changes to the natural ecosystem don't care.*

Humans have a wide range of values and degrees of concern about the future, and those who take a devil-may-care attitude often rise to positions of power as developers and entrepreneurs. On an individual basis, there will always be a large number of our fellow *Homo sapiens* who will refuse to adopt a particular set of values. Thus, regulations are necessary to save the planet; but even so, we may not be able to anticipate all the consequences of our laws. Humility, refinement, and trial and error are the neccesary approaches to environmental problem-solving. Doctrinaire approaches lead to paralysis.

- *"Symbiosis" is a key word to memorize.*

Symbiosis—mutual dependency—includes parasites that destroy their hosts as well as ones that coexist or benefit one another. Two major themes in ecology have been those of predators and symbiosis. The former has given birth to ideologies such as Social Darwinism—"survival of the fittest" in a human context. Symbiosis is the recent lesson that we have learned from nature; a lesson that is feeding our environmental consciousness.

- *Most humans are satisfied with easy answers.*

Using a paper bag instead of a plastic one is supposed to make us feel self-righteous, even when one gallon of gasoline is equivalent to 600 plastic bags and even when the environmental impacts of paper bags are equally as serious in their own way as the impacts of plastic bags.

It is going to take us a long time to understand our impacts on the global ecosystem, for we are engaged in an ecological learning process at the same time we have to halt the damage. Conviction and determination, tempered by humility, are the appropriate attitudes for environmentalists. Some of the solutions we recommend this year we will reject next year—but in time we will get things right.

- *A minority of caring and innovative people (like the readers of this book, hopefully) set the trend that others ultimately follow.*

  Individual action, although at first relatively minor in impact, has the paradoxical ability to cascade into a movement and change attitudes.

  The dark side of moral leadership is that purity can lead to self-righteousness and overconfidence. Leaders do not always go in the right direction; willingness to change and admit mistakes is crucial when trying to modify complex systems.

- *As the magnitude of the perturbation to a complex system increases, the number of unanticipated results and their scope increases rapidly.*

  If you take 20% of the deadwood from a forest for energy purposes, you shift wildlife populations somewhat. If you take 95% out, you devastate communities that depend on the wood-decay cycle.

- *Humans love to smooth out the variability in nature to which biological creatures have adapted.*

  By our smoothing out natural cycles, ecological niches are destroyed.

- *And, finally, the law of large numbers: The changes taking place in my town or locality are probably going on all over the country, and will ultimately likely take place in the developing world as its inhabitants seek to imitate American luxury.*

  When my community considers the ecological damage that may be inflicted by a new type of project or a new ordinance, such as the granting of a variance, I know that the extent of the damage will be global as others make similar changes. Local changes are really global changes, since there are so many of us.

Our ecological problems are not going to get any easier to solve, but the same principles will apply to all the changes that will be proposed in the years ahead. Remember that wasps bang their heads against windowpanes; humans bang their heads against the wall of materialism. Remember to think like an ecosystem.

THE VOCABULARY OF ECOLOGY

Adaptation: a characteristic of an organism that improves its chance of surviving in its environment; for example, the water-storing capacity of cacti.

Agroforestry: a land management system developed to solve the intertwined problems of deforestation and the need for increased production of food and fuelwood. The idea is to grow trees as crops among other agricultural crops or on pasture land. This way not only are more trees produced, but crops are benefited by enrichment of the soil, the prevention of erosion, increased water retention, and valuable shade. About 100 developing nations are practicing agroforestry.

Algae: one-celled or many-celled plants that usually carry out photosynthesis in streams, lakes, oceans, and other waters. They form the base of the food chain. It is the microscopic blue-green algae that often become too abundant in eutrophic waters.

Anaerobic organism: an organism that does not need oxygen to stay alive. Aerobic organisms need oxygen.

Atom: the smallest particle of an element.

Bacteria: one-celled microscopic organisms; those that lack a membrane-bounded nucleus are called prokaryotes; those with a membrane-bounded nucleus are eukaryotes. Some cause diseases; many are decomposers.

Biological diversity: the variety and variability of living organisms—all species of plants, animals, and microorganisms and the ecosystems they comprise. Biodiversity is generally described three ways: (1) diversity of species, the different types of living organisms, (2) diversity of ecosystems, as well as the variety of ecological processes and interrelationships within each type of system, (3) diversity within species as well as the variety of genetic information held in the genes of individuals of a species.

Biomass: any matter from biological systems; more specifically, an agricultural or animal resource utilized to produce solid, liquid, or gas fuels.

Biome: a major land ecosystem that has a distinct kind of plant life, such as grassland or tropical rain forest.

Biosphere: the thin skin of soil, water, and air where life exists on the Earth.

Biotic potential: the ability of a species or a population to increase its numbers. Factors in the environment such as disease and predators usually keep a population from reaching its biotic potential.

Cell: the fundamental structural unit that makes up all tissues of the body and carries on all the body's functions. Each cell is enclosed in a cell membrane and contains a nucleus. Cells reproduce by division.

Chaparral: an ecosystem consisting of thickets of low, thorny shrubs, characteristic of much of the Western United States.

Chlorinated hydrocarbon: an organic compound made up of atoms of carbon, hydrogen, and chlorine. Examples are DDT and PCBs.

Chlorofluorocarbons (CFCs): organic compounds made up of atoms of carbon, chlorine, and fluorine. Gaseous CFCs can deplete the ozone layer when they slowly rise into the stratosphere and their chlorine atoms react with ozone molecules.

Chromosomes: long threads containing DNA, found within the nucleus of a cell. Chromosomes contain the cell's hereditary material.

Climax: the last and most stable stage in succession.

Collapse: an uncontrolled decline in a population or economy induced when that population or economy overshoots the sustainable limits to its environment and in the process reduces or erodes those limits. Collapse is especially likely to occur when there are positive feedback loops of erosion, so that a degradation of the environment sets in motion processes that further degrade it.

Conservation easements: restrictions on a property that legally bind present and future owners. Some, not all, of the owner's rights, such as the right to develop, are surrendered to a conservation organization or governmental organization for the purposes of permanent land protection. Tax benefits are often involved.

Decomposers: plants and animals that feed on once-living material and cause it to mechanically or chemically break down.

Desertification: conversion of rangeland, rain-fed cropland, or irrigated cropland to desertlike land, with a drop in agricultural productivity of 10% or more. It is usually caused by a combination of overgrazing, soil erosion, prolonged drought, and climate change.

DNA: an abbreviation for deoxyribonucleic acid, a substance within the chromosomes of a cell. It carries the genetic information necessary for the replication of the cell and directs the building of proteins.

Ecotone: the place where two ecosystems or biomes meet and blend together; for example, the shore of a pond.

El Niño–Southern oscillation: recurrent fluctuation in the atmospheric pressures and surface water temperature in the tropical Pacific.

Endangered species: wild species with so few individual survivors that the species could soon become extinct in all or most of its natural range. A "threatened" species is more numerous but potentially at risk.

Epidemiology: study of the patterns of disease or other harmful effects from toxic exposure within defined groups of people to find out why some people get sick and some do not.

Estivation: a prolonged sleeplike state that enables an animal to survive during the summer months in a hot climate. As in hibernation, the animal's body processes, such as breathing and heartbeat, slow down drastically.

Estuary: a place where saltwater and freshwater mix, usually where a river enters an ocean.

Eutrophication: physical, chemical, and biological changes that take place after a lake, an estuary, or a slow-flowing stream receives inputs of plant nutrients—mostly nitrates and phosphates—from natural erosion and runoff from the surrounding land basin.

Exponential growth: growth by a constant fraction of the growing quantity during a constant time period. Money in the bank grows exponentially when interest is added at the rate, say, of 4% of whatever is already in the bank every year. Populations grow exponentially when they multiply by a fraction of themselves every year, or every month, or, in the case of microbes, every few minutes. When something grows exponentially, it continuously doubles—2, 4, 8, 16, 32—with a characteristic doubling time.

Fungi: a group of organisms lacking roots, stems, leaves, and the green coloring substance, chlorophyll. Fungi include yeasts, molds, and mushrooms. They aid the decay of dead plants and animals. Fungi and bacteria are primary decomposers in nature.

Gaia hypothesis: the idea put forth by James Lovelock and Lynn Margulis in the 1970s that the biosphere is self-regulating, that matter, air, oceans, and land surface form a complex system that can be seen as a single organism and that has the capacity to keep our planet a fit place for life. According to Margulis, "the concept states that the surface of planet Earth (= planet Water) and the lower part of the atmosphere are actively modulated with respect to temperature, acidity, and chemical composition. The modulation is due to the growth, metabolism, gas emission and other activities of live organisms, especially microbes."

Genes: molecules or groups of molecules that are part of the sex cells (and other cells) of organisms, and that contain chemical "directions" that determine the characteristics of the individual that develops from a fertilized egg. Genomes are the sum of all the genes of a cell or organism. Genetic engineering is techniques used to manipulate the genome of a cell to produce compounds normally not made by that cell.

Greenhouse effect: The way in which the Earth's atmosphere acts as a blanket to keep temperatures at a higher level than would otherwise be the case. The carbon monoxide and water vapor in the atmosphere transmit solar radiation but reflect the longer-wavelength heat radiation from the Earth. The increasing carbon dioxide in the atmosphere, caused by large-scale burning of fossil fuel, will probably produce higher temperatures and have widespread climatic effects unless control measures are instituted.

Hazardous waste: any solid, liquid, or containerized gas that can catch fire easily, is corrosive to skin tissue or metals, is unstable and can explode or release toxic fumes, or has harmful concentrations of one or more toxic materials that can leach out.

Humus: slightly soluble residue of undigested or partially decomposed organic material in topsoil. This material helps retain water and water-soluble nutrients, which can be taken up by plant roots.

Hydropower: electrical energy produced by falling or flowing water.

Immigrant species: Species that migrate into an ecosystem or that are deliberately or accidentally introduced into an ecosystem by humans. Some of these species are beneficial, while others can take over and eliminate many native species.

Indicator species: species that serve as early warnings that a community or an ecosystem is being degraded. A keystone species is one that plays roles affecting many other organisms in an ecosystem.

Integrated pest management (IPM): combined use of biological, chemical, and cultivation methods in proper sequence and timing to keep the size of a pest population below that which causes economically unacceptable loss of a crop or livestock animal.

Limiting factor: a single factor that limits the growth, abundance, or distribution of the population of a species in an ecosystem.

Molecule: the smallest possible amount of a compound that still has the characteristics of larger amounts of that substance. For example, a molecule of water is made up of just one atom of oxygen and two of hydrogen.

Mutation: an abrupt change in the genetic material of a cell that will be transmitted at cell division to descendant cells.

Natural resources: an area of the Earth's solid surface, nutrients and minerals in the soil and deeper layers of crust, water, wild and domesticated plants and animals, air, and other resources produced by the Earth's natural processes.

Niche: the role an organism plays in nature; its "occupation" in the community.

Nonrenewable resource: a resource that exists in a fixed amount (stock) in various places in the Earth's crust and has the potential for renewal only by geological, physical, and chemical processes taking place over hundreds of millions to billions of years. Examples are copper, aluminum, coal, and oil. We classify these resources as exhaustible because we are extracting and using them at a much faster rate than the geological time scale on which they were formed.

Nuclear energy: energy released when atomic nuclei undergo a nuclear reaction such as the spontaneous emission of radioactivity, nuclear fission, or nuclear fusion.

Old-growth ecosystem: natural forest virtually free of all human activity. It has reached a dynamic, steady-state condition characterized by canopy gaps and a wide range of tree species and sizes. Most often used in reference to a stand of trees more than 100 years old.

Old-growth forest: uncut, virgin forest containing trees that are often hundreds, sometimes thousands, of years old. Examples include forests of Douglas fir, Western hemlock, giant sequoia, and coastal redwoods in the Western United States.

Organic compound: a molecule that contains atoms of the element carbon, usually combined with each other and with atoms of one or more other elements, such as hydrogen, oxygen, nitrogen, sulfur, phosphorus, chlorine, and fluorine. Compounds not classified as organic are called inorganic.

Organism: a simple living member of one of the five biological kingdoms: bacteria, protoctists, fungi, plants, and animals.

Passive solar heating system: a system that captures sunlight directly within a structure and without the use of mechanical devices converts it into low-temperature heat for space heating or for heating water for domestic use.

Permafrost: permanently frozen ground occurring in the Arctic.

Petrochemicals: chemicals obtained by refining (distilling) crude oil. They are used as raw materials in the manufacture of most industrial chemicals, fertilizers, pesticides, plastics, synthetic fibers, paints, medicines, and many other products.

Photosynthesis: the process by which green plants convert carbon dioxide and water into simple sugars. The process is powered by radiant energy from the sun and occurs only in the presence of the green-coloring substance, chlorophyll.

Plankton: tiny, drifting plants and animals that live in saltwater and freshwater. The plant, or phytoplankton, includes diatoms and algae. The animal, or zooplankton, includes rotifers and copepods.

Producers: organisms in a community that convert radiant energy from the sun into food energy that can be used by consumers.

Radiant energy: energy given off by the sun. Only part of it is visible, but it is this visible part (sunlight) that provides the energy on which all living things depend.

Radioactivity: a property of some elements by which the center (nucleus) of an atom breaks up by emitting particles. These radioactive particles can't be seen or felt but may harm living things.

Radon: a radioactive inert gas occurring in nature and found everywhere in the environment. It is a pollutant that can contribute significantly to health risks as ventilation rates are reduced in energy-efficient buildings.

Renewable energy resources: Energy sources that can be used without depletion if carefully managed. Most of these sources are derived directly or indirectly from the sun (e.g., direct solar, wind, biomass, and hydroelectric power). Not all renewable energy is environmentally benign, for example, hydropower.

Respiration: the process of "burning" food, in which bound energy is released for use by an organism. In this cellular process, oxygen is consumed and carbon dioxide is given off as a waste. Respiration occurs in both plants and animals.

RNA: ribonucleic acid—nucleic acid composed of phosphate chains, ribose sugar molecules, and nucleotide bases. Involved in protein synthesis.

Silviculture: the science and art of cultivating and managing forests to produce a renewable supply of timber.

Sink: the ultimate destination of material or energy flows used by a system. The atmosphere is the sink for carbon dioxide generated by burning coal. A municipal landfill is often the sink for paper made from wood from a forest.

Slash-and-burn cultivation: cutting down trees and other vegetation in a patch of forest, leaving the cut vegetation on the ground to dry, and then burning it. The ashes that are left add nutrients to the nutrient-poor soils found in most tropical forest areas. Crops are planted between tree stumps. Plots must be abandoned after a few years (typically two to five) because of loss of soil fertility or invasion of vegetation from the surrounding forest.

Solar energy: direct radiant energy from the sun and a number of indirect forms of energy produced by the direct input. Principal indirect forms of solar energy include wind, falling and flowing water (hydropower), and biomass (solar energy converted into chemical energy stored in the chemical bonds of organic compounds in trees and other plants).

Source: a point of origin of material or energy flows used by a system. Coal deposits under the ground are the sources of coal in the short term; in the very long term forests are the sources of coal. Forests are sources of wood in the short term; in the intermediate term soil nutrients, water, and solar energy are the sources of forests.

When you look carefully at sources and sinks, and especially when you look at them over the long term, you see that they are not things, like buckets that can be filled or emptied, but processes. They are buckets that are being continually refilled or emptied by nature at varying rates. Sources and sinks are limits to systems, but they are ultimately limits to the rates at which things can happen, not to the amount that can happen.

Specialist species: species with a narrow ecological niche. They may be able to live in only one type of habitat, tolerate only a narrow range of climatic and other environmental conditions, or use only one or a few types of food.

Strategic petroleum reserves (SPR): a national stockpile of oil that serves as insurance against future oil shortages.

Succession: the process in which a plant-animal community is replaced by another, then another, and so on, as in the change from a bare field to a mature forest. "Climax" is the last and most stable stage in succession.

Sustainable development: the use of components of biological diversity in a way that must not interfere with the functioning of ecological processes and life-support systems. This means that crops must be managed in an ecologically sound way, forests must be protected, and genetic diversity must be preserved for the future.

Sustainable economic development: forms of economic growth and activities that do not deplete or degrade natural resources on which present and future economic growth depend.

Symbiosis: "living together"; a close association between two dissimilar organisms in a relationship that may benefit both, one, or neither. The term now includes commensalism, mutualism, and parasitism.

Thermal inversion: layer of dense, cool air trapped under a layer of less dense warm air. This prevents upward-flowing air currents from developing. In a prolonged inversion, air pollution in the trapped layer may build up to harmful levels.

Toxic waste: a form of hazardous waste that causes death or serious injury (such as burns, respiratory diseases, cancers, or genetic mutations).

Transpiration: a process in which water is absorbed by the root systems of plants, moves up through the plant, passes through pores (stomata) in their leaves or other parts, and then evaporates into the atmosphere as water vapor.

Watershed: a land area that delivers the water, sediment, and dissolved substances via small streams to a major stream (river).

Water table: the upper surface of the zone of saturation in which all available pores in the soil and rock in the Earth's crust are filled with water.

Wilderness: as defined by the Wilderness Act, "an area of undeveloped Federal land retaining its primeval character and influence, without permanent improvements or human habitation, which is protected and managed so as to preserve its natural conditions and which (1) generally appears to have been affected primarily by the forces of nature, with the imprint of man's work substantially unnoticeable; (2) has outstanding opportunities for solitude or a primitive and unconfined type of recreation; (3) has at least five thousand acres of land or is of sufficient size as to make practicable its preservation and use in an unimpaired condition; and (4) may also contain ecological, geological, or other features of scientific, educational, scenic, or historical value."

Zero population growth (ZPG): State in which the birthrate (plus immigration) equals the death rate (plus emigration) so the population of a geographical area is no longer increasing.

NOTES

1. All About What We Eat

The discussion of the primary cycles of the ecosphere is indebted to G. Tyler Miller, in *Living in the Environment*; to Eugene Odum, in "The Land Report #44"; and to *Ecosystems*, an educational aid prepared by the National Audubon Society. I was also aided by an Audubon Nature Bulletin prepared by Dorothy Treat.

The discussion of agricultural practices is based on reports by Maureen Hinkle, Audubon's Director of Agricultural Policy. These include "Pesticide Residues in Food and Water: Toward a Sustainable System" (speech, 7/89) and *Environmental Issues of Biological Control Regulation*, 1991.

The information about parasitic insects in Hawaii is from Richard Stone, *Science*, vol. 255, 2/28/92, p. 1070.

The Rachel Carson quote is from *Silent Spring*, pp. 249–50. Mary Handley's work at the Land Institute was reported in "Land Report #44," p. 10.

The discussion of groundwater pollution is based on Alan Bender in *U.S. Water News*,1989, and Maureen Hinkle's studies.

The discussion of how water is purified by natural systems is based on an exhibit arranged in 1992 by Paul and Julie Mankiewicz, called "A Celebration of Nature's Economy," at the Cathedral of St. John the Divine, New York City

The quote by Donella Meadows is from *The Amicus Journal*, Winter 1992, pp. 27–28. The quote by Wes Jackson is from *New Roots for Agriculture*, p. 154.

Statistics in the Eco Quiz come from Audubon's Population Program video, "What Is the Limit?"

2. Sex, Birth, Death... and the Health Connection

The information about the microorganisms within our bodies is based on Lynn Margulis' work.

Examples of sexual biology come from Chris Catton and James Gray, in *Sex in Nature*.

The article by J. P. Myers appeared in *Natural History*, May 1986, pp. 68–76. Dr. Myers was former Vice-President of Science and Sanctuaries at National Audubon.

The discussion of population issues was aided by Miller in *Living in the Environment* and many reports from Audubon's Population Program, such as "Sustainable Development and Population Growth"; Pat Waak's "Toward a Sustainable Future: Finding the Common Ground Between Developed and Developing Countries," 1987; Nikos Boutis, *Population Newsletter*, Fall 1992. Also helpful was the article, "ECO LOGIC/ Population: Red Hot Realities for a Finite Planet," *The Valley Optimist*, 5/11/92.

The quote by Robert Reinhold appeared in "When Life Is Toxic," *New York Times Magazine*, 9/16/90, p. 51.

The results of the TMI study were reported by Hatch, Beyea, Nieves, and Susser, in "Cancer Near the TMI Nuclear Plant: Radiation Emissions," *American Journal of Epidemiology*, 9/90, p. 397.

The excerpt about radioactive waste dumping is taken from Chris Wille's "Bomb Makers Explode Secret," *Activist*, 1/89, p. 9.

The background on water pollution comes from Philip W. Quigg in the booklet *Water, the Essential Resource*, prepared by the National Audubon Society, 1976.

The material on Nature's Cures was aided by the exhibit on Biodiversity, which appeared at the New York Botanical Garden in 1992, and included a booklet called *Green Treasures*.

The story about the lily plant in the Eco Quiz comes from Donella Meadows, in *The Limits to Growth*, Universe Books, New York, 1972.

3. What We Wear

The material on cotton is taken from J. K.Walker's report, "A Question of Cotton," Texas A&M University; the U.S.D.A.'s "Background for 1985 Farm Legislation and the Cotton Yield Problem"; D. Bottrell and P. Adkisson, "Cotton Insect Pest Management," *Annual Reviews of Entomology*, 1977.

Information about Sally Fox comes from a personal interview, from *Earth Impact*, the newsletter of Seventh Generation catalog company, and from the Pure Podunk catalog.

Helpful on cosmetics was Bruce Anderson, in *Ecologue*.

G. Miller, *Living in the Environment*, was useful on mining issues.

The section on seal-hunting comes from Libby Beaman's *Alaskan Diaries*, 1879.

4. Our Homes

Information about many tree-saving projects may be obtained from the Rainforest Alliance, 65 Bleecker Street, New York, 10012.

David Pearson's *The Natural House Book* was useful for this chapter. Gale Lawrence's *The Indoor Naturalist* was important for examples of how we share our homes with other organisms. The material on Animal Homes was drawn from *Animal Architects*, National Geographic Society, 1987.

The analysis of our energy choices comes from Jan Beyea's chapter "Energy Policy and Global Warming," in *Global Climate Change and Life on Earth*, Chapman & Hall, New York, 1991.

The Carbon Dioxide Reducing Diet comes from *The CO_2 Diet for a Greenhouse Planet*, by J. DeCicco, J. Cook, D. Bolze, and J. Beyea.

5. Appliances and Expensive Toys

Aside from Audubon's files on acid rain, G. Miller's *Living in the Environment* was useful.

The student writing about sources for appliances is Ben Strauss, Ecocentric, *Yale Student Environmental Journal*, Winter 1991.

The information on how termites stay cool comes from the National Geographic Society's *Animal Architects*.

For the discussion on energy efficiency I was aided by publications of The American Council for an Energy Efficient Economy, especially their *Consumer Guide to Home Energy Savings*.

Useful on hydroelectric dams was Philip W. Quigg's *Water: The Essential Resource*, a National Audubon Society publication, and T. Shoemaker's "Two Forks Dam Sparks Platte River Campaign," *Activist*, 3/87.

Useful for the discussion about electric utilities and paying environmental damage costs was "Environmental Concerns Regarding Electric Power Transmission in North America," by John DeCicco, Jan Beyea, and Steve Bernow.

6. Our Yards

The section by Loren Eiseley, "How Flowers Changed the World," comes from *The Star Thrower*. The quotation by Diane Ackerman is from "Insect Love," *The New Yorker*, 8/17/92.

The guide to birding sites appeared in *American Birds*, Spring through Winter 1991.

The discussion of the white-crowned pigeon is drawn from Tom Bancroft's research. See "Rare, Local, Little-known, and Declining Breeders," *Birding*, 2/92. The information about Steve Kress' restoration efforts was based on "From Puffins to Petrels," *Living Bird*, Spring 1992.

Two excellent guides to creating congenial habitats for wildlife are the Backyard Wildlife Habitat Information Kit (available from National Wildlife Federation, 1400 16th Street NW, Washington, D.C. 20036) and Stephen W. Kress, *The Audubon Society Guide to Attracting Birds*.

Karen Blumer's *Long Island Native Plants for Landscaping: A Sourcebook* was helpful for information about native plantings.

7. Our Garbage

The analysis of our waste stream is primarily based on Audubon's *Solid Waste Guidebook* and a related "Audubon Adventures" curriculum guide on trash.

The discussion of deforestation was taken from the work of Brock Evans and John C. Ryan's "Conserving Biological Diversity," in *American Forests*, 3/92.

The material on paper and plastic comes from Jan Beyea's speech "Plastics and the Environment."

Natalie Angier's excerpt is from "In Recycling Waste, the Noble Scarab Is Peerless," *New York Times*, 12/10/91.

The composting section was aided by Robert Rynk's *On Farm Composting Handbook* (New York: Northeast Regional Agricultural Engineering Service, 1992).

The Eco Quiz is taken from Jan Beyea's "Bagging: the Paper vs. Plastic Debate," *Activist*, 3/89.

8. Our Office Buildings

The following sources have been useful for this discussion on workers and sick and healthy workplaces: Dan Fagin, "Sick Buildings, Sick People," *Newsday*, 11/15/92; Debra Lynn Dadd, *The Nontoxic Home and Office*; The Worldwatch Institute, *State of the World*.

Information about Audubon House comes from Audubon's *Building for an Environmental Future* booklet and the *New Building Guidebook*.

Brendan Gill's article "Endangered Species" appeared in *The New Yorker*, 10/12/92.

The fact about the energy use in buildings for 1986 and potential savings if compact fluorescents were used is from *Monthly Energy Review*, Department of Energy, Energy Information Administration, Washington, D.C., 12/92, p. 2.

The discussion on solar energy was aided by Christopher Flavin's article "Solar Cell Production Expanding," *Vital Signs 1992*, Worldwatch Institute, 1992.

The discussion of ants is adapted from *Audubon Nature Encyclopedia*. Urban Wildlife is adapted from *Audubon Adventures* Leader's Guide, "Urban Wildlife" issue.

9. Transportation

Statistics on air pollution's threat to health come from Audubon's Conservation Information sheet "Clean Air."

Lesley Hazleton's "Confessions of a Fast Woman" appeared in the *Fairfield County Advocate*, 2/4/93.

The Office of Technology's piece on the deconstruction of a car is from "Strategies for Green Design," *Green Products by Design*, 9/92. The chart showing costs of road transportation is from Konheim & Ketcham, Inc., 5/92. The list showing total energy demand in BTUs of modes of transportation is from the *Transportation Energy Data Book*, Oak Ridge National Laboratory, edition 12, 3/92.

Material on the carbon dioxide emissions of automobiles and strategies for reducing them comes from John DeCicco, James Cook, Dorene Bolze, and Jan Beyea, *CO_2 Diet for a Greenhouse Planet*, Audubon Policy Report, 1990; also "Transportation on a Greenhouse Planet: A Least-Cost Transition Scenario for the U.S.," by DeCicco, Bernow, Gordon, Goldstein, Holtzclaw, Ledbetter, Miller, and Sachs, 6/92.

"The Green Car," by John DeCicco and Deborah Gordon, is from the *EPA Journal*, 12/92.

The discussion of biofuels comes from "Potential Impacts of Biomass Production in the U.S. on Biological Diversity," by Jan Beyea, Jim Cook, and Kathleen Keeler, *Annual Review of Energy*, 1991.

The section on Super Speed is inspired by Jan Beyea's "Superconductivity and Stray Magnetic Fields."

The material on global warming is based on Rob Lester and J. P. Myers, "Global Warming, Climate Disruption, and Biological Diversity," *Audubon Wildlife Report*, 1989–1990, and testimony on global warming by Eric Fischer and Jan Beyea in Wisconsin, 1/13/92.

The section on bicycling was aided by Marcia D. Lowe, "The Bicycle: Vehicle for a Small Planet," Worldwatch Institute Paper 90, 9/89.

The quotes by Thoreau are from "Walking," *The Writings of Thoreau*, and "Journal," 1/7/57, *The Winged Life*, edited by Robert Bly.

10. Recreation and Tourism

The ecotourism material is based on the first chapter in *Nature Tourism: Managing for the Environment*, edited by Tensie Whelan (Washington, D.C.: Island Press, 1991). Sources for this section include World Tourism Organizaton, *Policy and Activities for Tourism and the Environment* (Madrid, 1989); C. Kallen, "Ecotourism: The Light at the End of the Terminal," *E* Magazine, 7/90; D. Edgell, *International Tourism Prospects 1987–2000* (Washington, D.C., Department of Commerce, 1987); C. D. Ingram and P. B. Durst, *Nature-Oriented Travel to Developing Countries*, FPEI working paper 28, Forestry Private Enterprise Initiative, North Carolina, 1987; L. Kruckenberg, "Tourism as a Conservation Tool," World Resources Institute, Washington, D.C., 1988; P. J. Puntenney, "Defining Solutions: The Annapurna Experience," *Cultural Survival Monthly* 14(2), Cambridge, MA, 1990; P. English, *The Great Escape? An Examination of North-South Tourism: The Case of Tropical Forest Tourism* (Ottawa: The North-South Institute, 1986); K. Lindberg, "Tourism as a Conservation Tool," World Resources Institute, 1990.

The Laurens Van Der Post excerpt appeared in *A Testament to the Wilderness* (San Francisco: The Lapis Press, 1985).

The section by Gary Zahm on photographing at wildlife refuges appeared in Audubon's *Guide to Nature Travel and Photography*, 5/89.

11. Our Public Lands

Quotes in this chapter are from Aldo Leopold, *A Sand County Almanac*, p. 221; Henry D. Thoreau, *Journal XII*, p. 387; Roderick Nash, in Part II, "National Case Histories," *For the Conservation of Earth*, Vance Martin, ed., p. 201; Anne LaBastille, *Beyond Black Bear Lake*, pp. 232, 244; Charles H. Callison, *Overlooked in America*; William B. Robertson, *Everglades—The Park Story*.

12. Our Wildlife

Quotes in this chapter are from E. O. Wilson, *The Diversity of Life*, p. 357; Jared Diamond, on the book jacket of *Biodiversity*, ed. by E. O. Wilson; Aldo Leopold, *A Sand County Almanac;* Howard Ensign Evans, *Wasp Farm*; Henry Beston, in *The Outermost House*; Vincent Dethier, *Man's Plague: Insects and Agriculture*, p. 223; Rachel Carson, *Silent Spring*.

The wetlands material includes figures from Katherine Barton, "Wetlands Preservation," *Audubon Wildlife Report*, 1985, p. 213; and *Saving the Wetlands*, NAS Education Division, 1990. The quote is from James Gorman, "Wetlands," *Audubon* Magazine, 5/92.

In the "Thinking Like an Ecosystem" feature, the example about moths and bats is from Paul R. Ehrlich, *The Machinery of Nature*, p. 139.

Some of the material on Parasites and Predators is from Frank Graham, Jr., *The Dragon Hunters*, Alfred A. Knopf, New York, 1984.

13. Our Oceans and Marine Life

This chapter is primarily based on Dr. Carl Safina's work, particularly in "Conserving Marine Resources." Also useful in the opening pages was the *Oceans in Peril* curriculum guide, prepared by *Audubon Adventures*.

The fact about mollusks comes from Harald A. Rehder, *Familiar Seashells*, p. 8.

The "Thinking Like an Ecosystem" feature was helped by G. Miller,

Living in the Environment.

The section on "Inhabitants from the Deep" is taken from *Audubon Adventure* guides.

In "Fish Alliances" the quote by Ann McGovern and Eugenie Clark is from *The Desert Beneath the Sea* (New York: Scholastic, 1991), p. 27.

The statistic on oil spills is from Kristi Streiffert, "The Threat of Rain," *Texas Parks & Wildlife*, 7/92, p. 35.

The material on conservation "rules of thumb" is based on *Stemming the Tide, Conservation of Coastal Fish Habitat in the U.S.*, compiled by Carl Safina, 1991.

Articles by Safina, "A Mako for the Sport of It" (10/19/89) and "Rigging for Release" (1/3/91) appeared in *The Fisherman*.

14. Politics and Media

The Household Ecoteam Workbook, by David Gershon and Robert Gilman, is available from Global Action Plan for the Earth, 84 Yerry Hill Road, Woodstock, NY 12498. Agenda 21 is available from the United Nations Department of Public Information, room S-845, the United Nations, New York 10017.

The quote by Brock Evans on compromising is from an interview, "The Sierra Club as Lobbyist," published in the *Journal of Forest History*, Duke University Press, vol. 30/4, 10/86. His comments on the long-fought battle to protect the North Cascades is from "In Celebration of Wilderness," a lecture for the University of Idaho Wilderness Research Center, 1984.

The quote by Jan Beyea on the politics of blame is from an interview by Valerie Harms, published in *Annals of Earth*, vol. IX (2), 1991. The quote by Bill McKibben on television fare is from "What's On?" *The New Yorker*, 3/9/92.

Material on the Wise Use platform is drawn from a National Audubon Society Fact Sheet produced by the Government Relations Department. *The Wise Use Agenda* was published by the Free Enterprise Press, Bellevue, WA, 1989.

15. Business and Money

The example of the regeneration of the island in the East Indies is drawn from Peter Farb, *Ecology*, (New York: Time-Life, 1963).

The quotes by Anita Roddick are drawn from *Body and Soul*. Paul Hawken's article on "The Ecology of Commerce" appeared in *Inc.* magazine, 4/92. The example about wine stock by Claudine Schneider came from Seeds of Change's 1993 catalog. The quote by Wendell Berry is from *Sex, Economy, Freedom and Community* (New York: Pantheon Books, 1993).

Information about Peter Barnes was found in "Gaialogues, Economics Meets Ecology," *The Elmwood Newsletter*, Summer 1989.

The report on the returns of funds that invest in environmentally responsible companies appeared in the *New York Times*, 4/26/92.

The Council on Economic Priorities is the author of *The Better World Investment Guide* (New York: Prentice-Hall, 1991). Reports on companies are available from CEC, 30 Irving Place, New York, NY 10003.

In the Eco Quiz section, the *Business* magazine's Top Ten list appeared 11/92.

16. Our Global Family

Quotes are from E. O. Wilson, *The Diversity of Life*; Bryan Norton, "Commodity, Amenity, and Morality," in *Biodiversity*, E. O. Wilson, ed.; Archie Carr, *The Windward Road*; Alexander Skutch, *A Bird Watcher's Adventures in Tropical America*; Margaret E. Murie, *Island Between*; Garrett Hardin, "The Tragedy of the Commons," *Science*, vol. 162, 12/13/68; William Vogt, *Road to Survival*.

17. Art and Spirit

A detailed explanation of the evolution of the speech attributed to Chief Seattle is found in "Chief Seattle's Speech(es)," by Rudolf Kaiser, *Recovering the Word*, edited by Brian Swann and Arnold Krupat (Berkeley: The University of California Press, 1987).

Starhawk's quote is found in *Reweaving the World*, edited by Irene Diamond and Gloria Orenstein, p. 74. The quote by Gary Snyder is found in Bill Devall and George Sessions, *Deep Ecology*, p. 13. The quote by Helen M. Luke is from "Kaleidoscope: The Way of Woman and other Essays," in the journal *Parabola*. The quote by Black Elk is from *Black Elk Speaks* (John Neihardt), pp. 199–200. The quote by Annie Dillard is from *Pilgrim at Tinker Creek*, p. 129. Edward Weston's quote is from *Daybooks* p. 221. Linda Hogan's quote is from *Ariadne's Thread*, edited by Lyn Lifshin, p. 291. Gretel Ehrlich's quote is from *Antaeus*, The Ecco Press, Autumn 1988, pp. 124–25. Thoreau's quote is from *The Winged Life*, edited by R. Bly, p. 59.

The Center for Safety in the Arts is located at 5 Beekman St., New York, NY 10038.

REFERENCES

Audubon Nature Encyclopedia, Curtis Publishing Company, New York, 1965.

Audubon Wildlife Report (1988), Academic Press, San Diego, 1988.

Beaman, Libby, *Alaskan Diaries* (1879), Houghton Mifflin, Boston, 1989.

Beston, Henry, *The Outermost House*, Holt, Rinehart and Winston, New York, 1928.

Berry, Thomas, *The Dream of the Earth*, Sierra Club, San Francisco, 1988.

Blumer, Karen, *Long Island Native Plants for Landscaping: A Sourcebook*, Growing Wild Publications, New York, 1990.

Borland, Hal, *Book of Days*, Knopf, New York, 1976.

Brown, Lester R., et al., *State of the World: A Worldwatch Institute Report on Progress Toward a Sustainable Society (*annual*)*, W. W. Norton, New York.

Callison, Charles H., *Overlooked in America*, Aperture, New York, 1991.

Carr, Archie, *The Windward Road*, University of Florida Press, Gainesville, Florida, 1956.

Carson, Rachel, *Silent Spring*, Houghton Mifflin, Boston, 1962.

Catton, Chris, and James Gray, *Sex in Nature*, Facts on File, New York, 1985.

Council on Economic Priorities, *The Better World Investment Guide*, Prentice-Hall, New York, 1991.

Cousteau, Jacques-Yves, and the Cousteau Society staff, *The Cousteau Almanac*, Doubleday, New York, 1981.

Dadd, Debra Lynn, *The Nontoxic Home and Office*, Jeremy P. Tarcher, Los Angeles, 1992.

Dethier, Vincent, *Man's Plague: Insects and Agriculture*, Darwin Press, Princeton, 1976.

Devall, Bill, and George Sessions, *Deep Ecology*, Peregrine Smith Books, Salt Lake City, 1985.

Diamond, Irene and Gloria Orenstein, eds., *Reweaving the World*, Sierra Club Books, San Francisco, 1990.

Dillard, Annie, *Pilgrim at Tinker Creek*, Bantam, New York, 1975.

DiSilvestro, Roger, *Rebirth of Nature*, John Wiley, New York, 1992.

Durrell, Lee, *State of the Ark*, Doubleday, New York, 1986.

Ehrlich, Paul R., *The Machinery of Nature*, Simon & Schuster, New York, 1986.

Ehrlich, Paul R., and Anne H., *The Population Explosion*, Simon & Schuster, 1990; *Healing the Planet*, Reading, ME, Addison-Wesley, 1991.

Eiseley, Loren, *The Star Thrower*, Harcourt Brace Jovanovich, New York, 1978.

Evans, Howard Ensign, *Wasp Farm*, Cornell University Press, New York, 1963.

Farb, Peter, *Ecology*, Time-Life Books, New York, 1963, 1970.

Forman, R. T. T., and M. Godron, *Landscape Ecology*, John Wiley, New York, 1986.

Gore, Al, *Earth in the Balance*, Houghton Mifflin, Boston, 1992.

Hanney, Peter, *Rodents*, Taplinger, New York, 1975.

Hollender, Jeffrey, *How to Make the World a Better Place*, William Morrow, New York, 1990.

Hubbell, Sue, *A Country Year*, Random House, New York, 1986.

Jackson, Wes, *New Roots for Agriculture*, University of Nebraska Press, Lincoln, 1985.

Kohn, Alfie, *No Contest*, Houghton Mifflin, Boston, 1986.

Kress, Stephen W., *The Audubon Society Guide to Attracting Birds*, Scribner's, New York, 1985.

LaBastille, Anne, *Beyond Black Bear Lake*, W. W. Norton, New York, 1987.

Lawrence, Gale, *The Indoor Naturalist*, Prentice-Hall, New York, 1986.

Leopold, Aldo, *A Sand County Almanac*, Oxford University Press, New York, 1949.

Lovelock, James, *The Ages of Gaia*, New York, W. W. Norton, 1988.

Lovins, Amory, and L. Hunter, *Brittle Power*, Brick House, Andover, MA, 1982.
Margulis, Lynn, and Dorion Sagan, *Microcosmos: Four Billion Years of Microbial Evolution*, New York, Summit Books, 1986.
McHarg, Ian, *Design with Nature*, Doubleday, Garden City, New York, 1971.
Meadows, Donella, et al., *Beyond the Limits*, Chelsea Green Publishing, Chelsea, Vermont, 1992.
Miller, G. Tyler, *Living in the Environment* (7th edition), Wadsworth Publishing, Belmont, California, 1992.
Mills, Stephanie, *Whatever Happened to Ecology?* Sierra Club Books, San Francisco, 1989.
Murie, Margaret E., *Island Between*, Fairbanks, University of Alaska Press, 1977.
Naar, Jon, *Design for a Livable Planet*, Harper & Row, New York, 1990.
Nabhan, Gary Paul, *Enduring Seeds*, North Point Press, San Francisco, 1989.
Nash, Roderick, *Wilderness and the American Mind*, Yale University Press, New Haven, 1973; *The Rights of Nature*, University of Wisconsin, Madison, 1989.
Neihardt, John G., *Black Elk Speaks*, Washington Square Press, New York, 1972.
Nelson, Richard, *The Island Within*, Random House, New York, 1989.
Oak Ridge National Laboratory, Transportation Energy Data Book, edition 12, March 1992.
Orr, David, *Ecological Literacy*, State University of New York Press, New York, 1992.
Pearson, David, *The Natural House Book*, Simon & Schuster, New York, 1989.
Pettit, Ted, *The Web of Nature*, Doubleday, New York, 1960.
Pringle, Laurence, *Ecology*, Macmillan, New York, 1971.
Rehder, Harold A., *Familiar Seashells, The Audubon Society Pocket Guides*, Alfred A. Knopf, New York.
Ricklefs, Robert, *Ecology*, W. H. Freeman, New York, 1990.
Rifkin, Jeremy, *The Green Lifestyle Handbook*, Henry Holt & Co., New York, 1990.
Robertson, William B., *Everglades—The Park Story*, University of Miami Press, Gainesville, Florida, 1959, 1963.
Roddick, Anita, *Body and Soul*, Crown Publishers, New York, 1991.
Roszak, Theodore, *The Voice of the Earth*, Simon & Schuster, New York, 1992.
Schoenfeld, Clay, *Everybody's Ecology*, A. S. Barnes & Co., Cranbury, New Jersey, 1971.
Scientific American, The Biosphere, W. H. Freeman & Co., San Francisco, 1970.
Simon, Anne W., *Neptune's Revenge*, Franklin Watts, New York, 1984.
Skutch, Alexander, *A Bird Watcher's Adventures in Tropical America*, University of Texas Press, Austin, 1977.
Swann, Brian, and Arnold Krupat, *Recovering the Word: Essays on Native American Literature*, University of California Press, Berkeley, 1987.
Thomas, Lewis, *Lives of a Cell*, Bantam, New York, 1975.
Thompson, John, *The Environmental Entrepreneur*, Longstreet Press, Atlanta, Georgia, 1992.
Thoreau, H. D., *The Writings of Thoreau*, Random House, 1965, and *The Winged Life*, edited by Robert Bly, Sierra Club Books, 1986.
Todd, John, and Nancy Jack, *Bioshelters, Ocean Arks, City Farming*, Sierra Club Books, San Francisco, 1984.
Vogt, William, *Road to Survival*, William Sloane Associates, New York, 1948.
Wellesley-Miller, S., *Earth's Answer*, Harper & Row, New York, 1977.
Weston, Edward, *Daybooks*, vol.I, Horizon Press, New York, 1961.
White, Gilbert, *The Natural History of Selborne*, Penguin, New York, 1977.
Williams, Terry Tempest, *Refuge: An Unnatural History of Family and Place*, Vintage Books, New York, 1991.
Wilson, E. O., *The Diversity of Life*, Harvard University Press, Cambridge, 1992; *Biodiversity*, editor, National Academy Press, Washington, D.C., 1988.
Worldwatch Institute, *State of the World 1992*, W. W. Norton & Co., New York, 1992.
Wyman, Richard L., *Global Climate Change and Life on Earth*, Chapman & Hall, New York, 1991.

INDEX